Forty Types of PROOFS of

ACTUAL TOTALITY

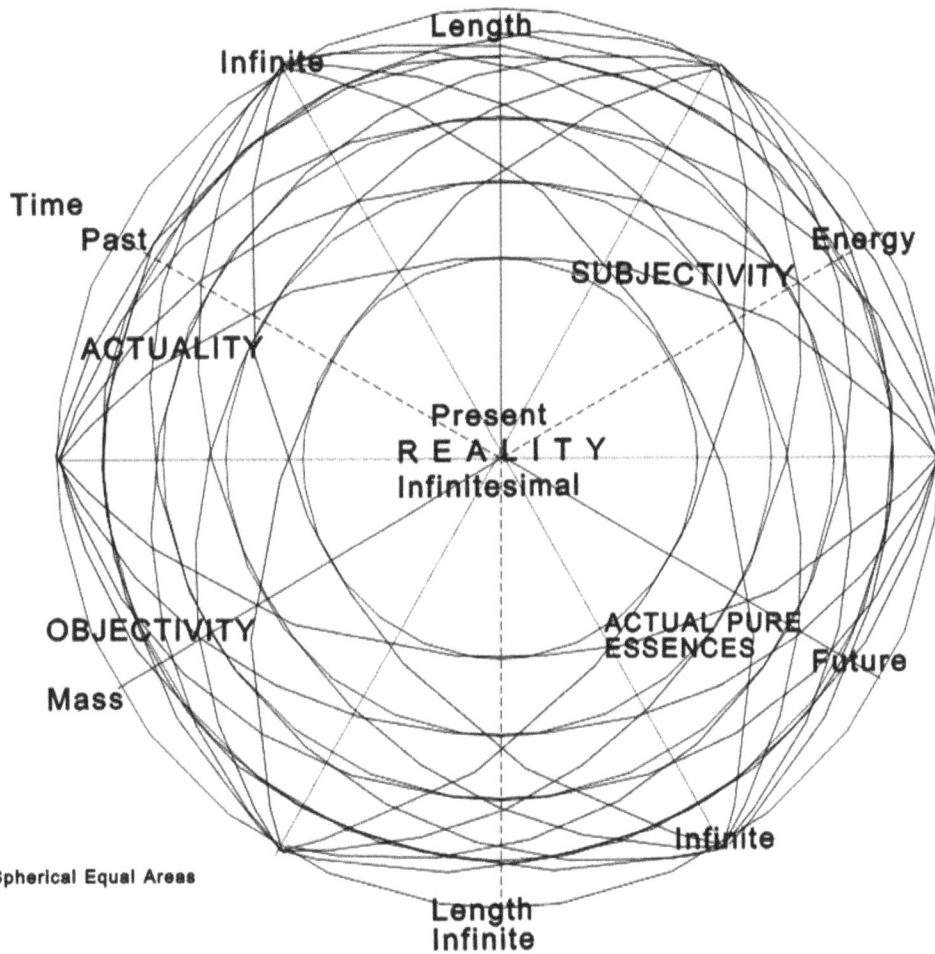

SPEST = Spherical Equal Areas

Forty Types of Proofs of Actual Totality
Copyright ©2014 Marcus Bergh, M. D.

ISBN 978-1506-912-51-6 PRINT
ISBN 978-1622-874-73-6 EBOOK

LCCN 2014931047

April 2014

Published and Distributed by
First Edition Design Publishing, Inc.
P.O. Box 20217, Sarasota, FL 34276-3217
www.firsteditiondesignpublishing.com

Investigation, analysis, demonstration, proofs and laws of Actual Totality, TO, its five Modes, Continua, Kinds, Varieties and Transformations, from non-existence to total combined proofs, with reconciliation and applications.

DEDICATION

This book is especially dedicated to those bright young minds, who can recognize, learn, know, appreciate, and endeavor to save their souls, as these are, and enable us to achieve, identity and presence in actual totality. It is also dedicated to the leaders of the world, its executives, ministers, doctors, politicians, lawyers, teachers, managers and all persons in lives and positions of responsibility. It is to these that guide and spearhead a higher age for humanity that will prevent us from continually becoming victims, and be the necessary victors. It is for the expanding union of the world, universe, God, systems and totality that are included in the absolute-relative varieties of best human life in actual totality, TO.

More personally the book is dedicated to all my children and progeny. May they realize, or be "born again" by their actual pure essences, as it is their part and role in actual totality? Most personally this book is dedicated to my wife Barbro, who has helped to support work on actual totality, and without whose companionship, love, and sacrifice we could not have endured all that has led to, and brought actual totality to fruition.

PREVIEW

This book is a culmination of works on the organization of knowledge, the relative universe and their adaptation to totality. It arose from a deep interest in revealing ideas and concepts that produced the most effective and proportional order of all. The two books, and web sites on The Relative Universe did not fulfill the requirements of actual totality. This requires a correct and comprehensive approach from the top, unity in totality, with vertical integration. To this end Actual Totality, its Forty Types of Proofs and Five Critical Modes are produced. It provides the needed complete, valid, actual and proportional presentation of all.

This book's method was to discover and prove that there is an ultimate unity in totality that has been unknown or ignored even with all current knowledge. Actual totality has been an accumulation of approaches through many years. It began with investigations in the organization of knowledge. It was accelerated by recognizing the need for unity in totality. This was aided by target theory, convergence of spatial and temporal dimensions, and their representation in human life. It was stimulated by the need to resolve the polarization of science and philosophy. How the absolute and relative do not exist only separately but are two poles of the same thing, This was confirmed by the collection of all proofs of this totality. It was reinforced by studies on absolute totality, and its contrast with the physical universe, or relative totality. The application of ideas and representation of the relative universe to actual totality has added, and been adapted to confirm actual totality. Problems in representation, positioning of essences and universals led to methods of superimposition and convergence in totality. This has been found to be similar to Whitehead's method of extensive abstraction. Many typical methods enabled by the form and function of actual totality include each of the book's chapters.

All is infinite to infinitesimal, as is all selectively qualified by actual totality. Such absence of limits cannot be restored in the differential extremes of a separate infinitesimal, finite or infinite. It is found by their unity in totality, convergent vertical integration that exists in the form and function of actual totality. This is a vertical integration that is reversible, VIR. Reversible from bottom to top and from top to bottom, ultimately centered on the unity at the top of actual totality. It is the typical selective qualification of the categories, universals, essences, modes, kinds, varieties and all in actual totality that offer the form needed for optimal order. Not only does this exist in different ways, it can be represented in more ways. Actual totality is not static, but dynamic as a continuum from one extreme to the other, in various ways and regions, and overall transition in time. It is by bringing these together in the most valid and proper way that the method of actual totality is fulfilled.

Einstein spent most of his life searching for a grand unification theory, but without success. He didn't realize that he had it in the beginning with the special theory of relativity, the bond that exists between an observer and observed. Without this bond the physical universe is believed to be absolute totality, separate from relative and actual totality. Other totalities, worlds, universes and systems in a similar extreme way believed to be, exclusive totalities. They are only parts of actual totality. This book features the union of the physical and relative universes, and gives life to absolute-relative totalities as all exists in and by actual totality. We must rationally and logically analyze and determine all that the reality of objectivity and subjectivity produce. What all relativity converges on and includes, is to discover the necessary existence of actual totality. By their pragmatic actual pure essences all are typified by the valid, proportional and total contribution in the whole that make actual totality. Each contributor, in their own way, builds on and stimulates others, adding significant knowledge to the correct unity in totality. By removing the wrong pieces and adding all that is right the jigsaw puzzle finally fits together as a whole.

The progression of human life and civilization has, is and will not succeed without an adequate quality and quantity of knowledge. Ever greater wisdom is ready for use by each and all persons as a guide to correctly fulfill their parts and roles in life. This total knowledge must be identical with and in order by the absolute-relative and general-special totality of all that necessarily exists. All can be known and solved by a properly applied, and reversible vertical integration. Actual totality, its knowledge and the general to specific needs and problems of life are interdependent, interactively and simultaneously filled and solved. It is the most direct and best way cultures have of achieving the confident and optimistic perspective that yields the greatest wellbeing and permanence. All persons more or less lack a serious beneficial attitude that adapts life to totality. This is caused by the limits of life, experience, a fixation on self gain and partial goals. This can be corrected when properly approached from the top, with vertical integration. Actual totality provides a more complete, valid, actual and proportional description, representation and presentation

The interpretation, analysis, proofs and solutions to actual totality, its potential and the relative universe of best human life, when supplemented by summary, interpretations and conclusions make all certain and complete, with the order knowledge must take. With such powers at our disposal the climactic struggle of humans to succeed, find wellbeing, immortality and continuity in the generic to specific continua of life, can be permanently fulfilled.

Actual totality is neither perfect, complete, nor static. In general its necessary existence, categories, features, modes, kinds and varieties are most certain. There will be improvement and much work toward completion and adaptation to transition in time. It should not be used as final, nor dismissed for lack of perfection.

HOW to USE this BOOK

The world, universe, God, systems and totalities are very different, poorly known and lack unity. Yet they are most important, need to be reconciled and comprehended together. Thereby, it is vital that each and all persons discover, learn, know, apply and live by the ultimate unity of actual totality and its change in time.

The best shortcut to a clear understanding of actual totality is to learn its unity, vertical integration, proportions, figures and most important key ideas and proofs. Many are given through the text and in the addenda. The following are some of the ideas, concepts and proofs that are most critical to actual totality. Those ideas and concepts that are important evidence of actual totality become its proofs. Proofs that achieve a level of certainty are axioms. Axioms that are more certain, directive and compelling guides for life in actual totality are its laws. Many proofs of actual totality are not signified, but may be found in its laws. The quickest way to acquire an in-depth perspective of TO, is to learn and coordinate its greatest laws, cf., Addenda.

To be interesting and easily read the book needs to directly and most efficiently give actual totality's full meaning. Yet it must capture the attention and usefulness to the reader. Such attentions exist by the greatness and need of actual totality, and a person's identity with it. It must have adequate detail, order and avoid most all that is negative, null or of lesser worth. It must emphasize the combined forms and interactions of its greatest categories and themes. It must have a balanced approach to unity and diversity, genera and specie, subject and object, absolute and relative and its other major continua. This is what characterizes Actual Totality, TO. It must resolve the greatest paradoxes and enigmas of the most basic aspects and modes of totality, such as existence, essence, actuality, relation between subjectivity and objectivity and their accelerative expanding combination of reality.

Identify and find interest in universal knowledge, TO. At first learn the basics, orient yourself, avoid all less. Purview the supplemental contents, and addenda. See if you can bring the major figures into the unity of actual totality. To be conclusive all pieces of the jigsaw puzzle must fit into a final complete picture. Leaf through the book, noticing sufficient detail. Read supplemental contents and a few paragraphs of each chapter. Read enough to get the gist of all chapters. Continually develop a solid perspective and understanding of all TO. Seek and find how it applies to and guides your life. Search for answers that modify and correct your beliefs of universes, deities and world. Be open to the best, greatest and most stimulating ideas that form the basis for actual totality. Learn the different approaches to actual totality, its truth, concepts, principles, laws and proofs. Develop an early understanding of its most basic concepts, laws and the place and power of its constituents. All is related, comes together, and can be applied.

Actual totality like any great science, philosophy and field of knowledge are complex and massive. To master them is more appreciated when you have experienced their absence, become interested in their contributions, and have found consolation in their achievements. So it is of actual totality, although much more so. For actual totality is a sum of the valid and proportional essences of all sciences, philosophies and fields of knowledge uniquely and typically formed and functioning in the whole.

Truth and the proofs of actual totality like its knowledge are not simple. All in actual totality is highly interactive and interdependent. Truth and proofs are less singular or based on appearance than how basic premises and categories fit as one. This is done by focus, converging our ideas and beliefs on actual totality by pure reason, logical thinking and its collective ordered unity. The dynamic of actual totality is continually interactive with selective dependence on much that is within or without, other, less or not, of itself. The greatest idea that is the basis of actual totality is, its massive and exclusive extended presence by functional value, practical and necessary existence or actuality. This is supplied by the powers of fusion of the relative universe and totality with the physical universe and absolute totality. By broad and selective inclusiveness, much we currently know is great, is in fact actual totality, only in it's typical, and modified, way.

It is by the greatest ideas, concepts, laws, key dynamic mechanisms and proofs of the approaches, forms, is, saves and guides actual totality that it can be well understood.

Terminology of Actual Totality

All great fields and sciences have their own language, terms whose acronyms or abbreviations enable a quicker and easier understanding and use. For a first time reader of actual totality this may be an obstacle, so that for this book each first entry of a new term is associated with its full name. The uses of new terms are kept at a minimum, unless for reasons of reference or inclusion in figures they are needed.

The most basic and frequently used terms include,

TO = Actual Totality,

TOT = Potential Actual Totality

ABST0 = Absolute Totality,

RELTO = Relative Totality

ARTO = ABSTO-RELTO united in TO

AU = Absolute and physical universes

RU, TRU = Relative universes of best human life

NEAC = Necessary existence equals actuality

APE = Actual Pure Essences, APET = APE of TO

S-O = The interaction and combined subjective objective

S-R-O = Combined subject object producing reality

MIDST = Mass in TO, as in all, is that most predominant.

A SYNOPSIS of ACTUAL TOTALITY

Is there one world? Or are there many worlds? How do we distinguish between one and many worlds? If there is only one world, what is it, how is it derived, how does it exist, of what does it consist? And how do we live by it?

There are currently two major contestants for one world, an absolute focus like the physical universe, and a relative focus like God or a deity that identifies with human life and mind. A number of other approaches to one world, depend on the given observer or observed, subject or object. In this way, the contestants and our philosophical approach to them, are key to answering the question, is there one world? This is the difference between the objective and subjective. Which one is right, or is it some mixture of the two? The objective and subjective are equivalent to physical matter and mental mind. Matter may be the subject, and mind may be the object, depending on how they relate to the observer. The dependence on an observer, reinforces the importance of mind in relation to matter. It is the basis of relativity in the physical universe, just as an observed, reinforces the importance of matter in relation to mind. This is the basis for the absolutivity that exists in the eyes of the observer, or best human life.

Neither the physical universe, nor the world of best human life and mind are perfectly separate. The bond of an observer and observed, brings the physical universe and world of best human life and mind together. To be perfectly separate, the absolute and relative, must be a polarity, of absolute totality and relative totality. In this way absolute and relative totality, by the relativity of the physical universe, and the absolutivity of the world of best human life and mind, must be a continuum, a duality, and polarized in that way that most accurately reflects an equilibrium descriptive of its features, aspects, form, function, states, conditions, relations and limits.

The question is raised, "are, and how does this equilibrium and union, separate the indefinite from the definite absolutes and relatives?" The answer is readily apparent in the nature and normalcy of the physical universe and the world of best human life and mind, how absolute totality focuses on the physical universe by relativity, and how relative totality focuses on the world of best human life and mind by absolutivity. The focus of the world of best human life on absolutivity is readily apparent from its necessary existence, and its actual pure essences whose powers are indispensable.

Upon these facts and essentials, mass energy not only characterizes the physical universe, it also exists in typical ways in the relative universe, or world of best human life and mind. This continuum of absolute to relative, becomes absolute totality and relative totality in TO, ABSTO and RELTO. As this continuum, ABSTO-RELTO in TO fuses, their completed union in actual totality, is signified by ARTO. It becomes the primary parameter and dimension of mass energy in actual totality, TO. This dimension, along with the typical form and function of space and time, in actual totality produces the form and frame by which we can accurately represent and present actual totality. Mass energy as it exists in actual totality partakes of the typical continuum from physical to mental, object to subject. The objectivity and subjectivity are those modes of actual totality, which when combined, provide the reality of our needed existence. The necessary existence, and actual pure essences of RELTO and the relative universe, produce actuality, which with reality, are basic modes to actual totality. They determine many of the laws, and provide the greatest proofs of actual totality, as it is this one world.

Non existence, existence and necessary existence, lead to actuality. Actual pure essences provide practical life to things which when actualized is the key to totalities and actual totality. These categories when joined by its modes, objectivity, subjectivity and their higher levels of reality, unite in ever higher states that produce actual totality. It is the vitality of reality that describes and proves the fusion of absolute and relative totality. It is the certainty of absolutivity, of absolute necessary existence, or actuality that proves and is signified by actual totality.

There is neither an absolutely relative nor absolutely absolute totality, only their form as a continuum, duality and spectrum determine actual totality. Such a spectrum of absolute and relative gives the generic to specific varieties that make up actual totality. The spectrum defines the region of physical and relative universes. Actual totality is also made up of other features, as its mass, length and time dimensions, whose higher states of mass energy, volume and variation produce the actuation, extension and intension typical of actual totality. Mass energy exists as a spectrum from most simply physical to most complex by actual pure essences whose reality of subject and object typifies actual totality. Together with the spatial and temporal this unique form of mass energy in subject and object typifies, enables us to represent and prove actual totality.

Actual totality is further confirmed by the added proofs of its dynamic and convergent relative states, centered on here and now. When we show the role of language, logic, mathematics, proportions and geometry into these features, form and function of actual totality, we add more convincing knowledge and proof to its certainty. They reveal the great reality of language in life and the value of mathematics based on logic of the proportionality and hierarchy of actual totality. Geometry completes the formative role of shapes, designs and their role in inorganic and organic chemical structure. With many other propelling, and increasingly compelling formative forces, life was created and evolved. As animal and human life evolved it became more relative, both generically and specifically to produce a greater exclusive and inclusive, relative totality. It was a relative totality that lacked the essentials of current actual totality. The combined fusion of the generic to specific varieties is what we are currently in and constitutes actual totality. Without convergence and fusion the common human world, physical universe, deities and philosophical systems are more or less partial, lack direction and left in opposition. The correct convergence and balance of ABSTO and RELTO combined as ARTO is what make actual totality. This is why we must discover, learn, know, be, live by and work for actual totality. It cannot exist with what is less than itself, and humans cannot exist without it.

Also actual totality is more or less in opposition and convergence, by its own typical relation to time and change. This polarity is largely unavoidable and reveals the closed and open set forms of actual totality. Actual totality is always more or less permeable, interactive and admits what is positive, and expels what is negative. It is in, a continuous interchange of potential and actual. Actual totality is a highly closed set in the extended and practical present, less so in the long term. The dependence on time is internal and external of TO. They are associated with transformations. The transitions, of TO and TOT that occur with time in the long term, reveal the potential to actual, continuum that largely governs actual totality. Potential actual totality exists only when it becomes actual totality. They, in the long term, are different aspects of the same thing. This is an adaptation of actual totality and alteration of potential actual totality. How well this adaptation and alteration occur is their success or failure. As long as relative totality is predominant and on the ascendance, its end product is a relative - relative, or absolute relative totality. Since this is not possible absolutely, there will always be some indefinite absolute totality, or physical universe. Reversely there will always be some, or some kind of indefinite to definite relative totality. Their proportions may vary from one extreme to the other. It is the equilibrium and success of optimal actual totality with the potential in the long term that is necessary for its existence and permanence.

It is between these worlds of common, lesser existence, and the higher existence of actual totality that each person lives, and doesn't, or poorly, knows it. It has been vaguely held by religions, God and soul. It has given them great focus and devotion. So, also, has it been vaguely held as the physical universe in science. Yet why, how and what it is as actual totality, is more or less, largely unknown. By learning and bringing together all powers producing the identity of each and all persons with actual totality that knowledge must be directed and succeeds. Knowledge is extremely vast, from the oneness unity of actual totality through all its detail, and all that has less or unnecessary existence. There are many kinds of knowledge, from universal, partial, special, essential, and others. Universal knowledge is the quality and quantity each person knows, has interests, attitudes and beliefs, guides, lives and works for are their part and contribution to actual totality. Like the best of education it is a need we should gladly provide. Pure actual totality should be available to all. Knowledge grows on itself, gains confidence, inspiration and certainty in its powers to overcome all. Such is the spark that is growing in humanity, and can be ignited to assure permanence. This is the good work that yields survival/extinction, eternal life in actual totality.

There are many kinds of proofs, including, generic to specific, general to partial, direct and indirect, disproof, qualified, reciprocal, multiple, enigmatic and total combined proofs. It is how we bring these to actual totality in the most ready, ordered, proportional and effective way that gives them their power. Powers that are essential if we are to learn, understand and live by actual totality. By representation and application actual totality can be given better perspective, presentation and fulfill its major purpose. This purpose is the survival, improvement, success, perfection and expansion of actual totality and the quality and wellbeing of our lives in it.

Each and every person, more or less, fails to come to terms with, produce, and get the best out of life. All who suffer from the lack of, and hunger for, the right and proper order from the top. This includes most all who understand neither themselves nor the rest of the world. Our life is much different from what we are led to believe. We must discover and best apply the proper world with the right balance in universal moderation if we are to return what has been given to us, survive and flourish by actual totality.

To recognize that there is only one necessary and definite world, to learn a few of its proofs, or to picture a representation of it, is a modest beginning. Yet to have an adequate understanding of actual totality, and to have the knowledge, endeavor, accept, identify with, and enhance actual totality is the great spectrum whose primary purpose is our salvation. This spectrum of universal knowledge is available and most needed by each and all humans on earth. The spectrum of universal knowledge, technology and all that make up civilization exists at many levels. It is a duty of everyone and all, to do their best from infancy to the end of life, to achieve the highest level equal to their ability and proficiency. Life is, neither simple, easy, nor anything less, lacking or negative. The joy and rewards of life are much more readily gained by fulfilling our part in the whole, when it is actual totality.

INTRODUCTION and OVERVIEW

The premise of this book is to prove the existence of actual totality. If the human mind is so certain of itself and if the environment about us is as certain as it appears then what is the whole? What is the totality of all? How is it formed and how does it function? To prove the premise of actual totality, we need to prove many of its aspects, features and constituents. We also need to prove or show how other universes, systems, deities and worlds are less or other than actual totality. To what extent they are the same or a part, and why are actual totality preferable, and a preferable name?

Actual Totality, TO, consists of several aspects and categories including, modes, continua, kinds, forms and varieties. Its modes are actuality, actual pure essences, objectivity, subjectivity and reality. Its kinds include many continua and dualities whose polarities, spectra and gradients make up much of TO. Among these, the most fundamental is, between absolute and relative, especially absolute totality to relative totality continua whose convergence most clearly demonstrates the unity and proof of actual totality. This continuum exists in actual totality as a spectrum. The gradients of the spectrum absolute to relative, form a sequence from the most generic to most specific. Generic to specific gradients, is the varieties of actual totality. They can be of various degrees of detail. When limited to ten gradients, they provide a formal way of showing actual totality - from extreme absolute totality to extreme relative totality. When only within actual totality, TO, by their actual pure essences, the varieties exist as ABSTO-RELTO that forms their combined whole, ARTO in TO. Varieties are formed and function with the actuality of the modes of TO whose objectivity and subjectivity combine, and exist at ever higher levels of reality. All together they exist in that sum typical of TO, whose actual pure essences characterize and reveal the components, properties and dynamics of the whole.

Only when absolute totality or the physical universe and relative totality or the relative universe based on life and human mentality converge and unite in TO, do they fully achieve essential existence. This is the fusion of senses and things, subjects and objects in the reality of TO. Nothing has actual existence separate from TO, and to this degree they are not TO. To assume a physical, scientific, philosophical or theological universe different from their part and role in TO, is an error whose disproof must be known, shown and remembered. This is why things and senses and so much we take for granted is correct only when qualified by how it exists in actual totality. How all exist by its actual pure essences, and typical unique form and function in actual totality. To the degree they are their parts and roles in TO, the physical, scientific, philosophical, and theological universes, like absolute and relative totality must be clearly and fully developed, known, shown and applied. When proven, recognized and accepted with all the other modes, kinds, varieties, features and constituents the whole, they reveal a united, proportioned, balanced and well functioning actual totality.

The relative components of the absolute-relative continua in TO, like the absolute or physical, must be limited to its part and role in TO. Like the physical universe in ABSTO so is the relative universe of best human life in RELTO a large feature and component. By its inclusion in RELTO and TO, TRU or The Relative Universe and its actual pure essences are large contributors to the form, function, parts and processes of TO. TRU is a special case or category of RELTO and also contains aspects of the physical universe. So does the physical universe, although a special case or category of ABSTO, also contains aspects of the relative universe.

Actual totality is typified by its different forms in each of its varieties, whose center is here and now in space time. The greatness of the human person, mentality, knowledge, civilization in RELTO and RELTO in TO is evident by the focus of life in here and now. This is most characteristic of RELTO in TO and the part, roles and effects of human life on all TO. This is confirmed by the progression of mass energy, length and time, mlt, to ALT, and ALT in SPALT in ARTO and TO. Convergence on the center produces the unity that characterizes TO. When all constituents, including persons, conform well with all other constituents, their actual pure essences of their dynamics and homeostasis of become actual totality. This does not deny the greatness of an individuality, nor the human mind, only corrects its actual essential existence and form in TO. Like religious teaching that demotes humans without souls, so does TO selectively exclude all that does not properly become its part identify with and fulfill its role. This includes the world about us, our thoughts, beliefs, decisions, actions, behavior and the environment or the entire objective world we interact with. This includes the role of humans, persons, consciousness, thought, instantaneous attention and awareness. Much of the humanity, each and all persons, is not, or not fully TO. They are of TO only to the extent their actual pure essences are the same as that in their part and role in TO.

The following proofs of the book add to support and uphold many of the different and more specific proofs that are necessary to complete an ordered sum of proofs of actual totality. There are many, and many kinds, of proofs of actual totality, TO. They may be singular, multiple, combined, complex and total. They may be direct, indirect, closely, or less closely related. They may arise from disproofs of TO or that lacking equivalence with TO. Proofs uphold TO, and TO clarifies, adds to and shows proofs. They are derived from the great accumulation of knowledge through the ages, with emphasis on all that converges on ultimate unified totality. Much of this knowledge was incremental in smaller or larger steps, as fractals, discoveries that were created by the collective experiences, thought and trends of ideas, not only by a single person, study or experiment. Single persons, studies and experiments often brought them out, by the depth and intensity of their concentration, and resulting publicity. It is this increasing growth of knowledge of TO and the power and predominance of its autonomous existence that has spearheaded the relative and fused with the absolute to provide the basic form and function of TO, its modes and kinds. The direction and course of this growth of knowledge, with TO, have accelerated with higher powers of learning, reasoning, logic, new concepts, and the benefits of technology, especially electronics, internet and modular operations of the whole of TO. It is with such knowledge of TO and its assurance of existence that we know and know that we know of our actual existence that TO arises to be the kind of essential existence it is. Such certitude is the core of the relative absolute, what make, save, is, shows and enables us to be TO, influenced by the growing greatness of the relative universe bonding with absolute totality and the physical universe to be actual totality.

Kinds of distributions of proofs - Problems in differentiation and some of the most common features that permeate much of the actual totality, exist in more than one of the proofs of TO. Aside from the fact that proofs are often multiple is the fact that it is highly interdependent and interactive with others. There is a presence of proofs in many different kinds of proofs, e.g., bonds and love. Proofs reveal, determine, limit, improve, produce, and make right, the problems, errors and negatives of life. Although some positive and negative concepts and problems have fairly definite positions in kinds of proofs, some have widespread distributions or exist in the origin and ends of the kinds of proofs. The distribution depends on the way they exist in, are formed, function in, and the level to which they determine TO. In the earlier proofs it is the way they contribute to the development and determination of TOT-TO, In the middle portion of proofs, it is how they exist in TO its modes, continua, kinds, varieties, physical and mental attributes, including relation, language, math, proportions, geometry and the creation and evolution of life in TO. In the latter portion of proofs all that helps to show and propel TO in the future includes, transitions, humans, persons, knowledge, higher stages, and uses propel the progression of proofs to totality. All of these varieties of kinds of proofs confirm the importance of a combined and total proof. They also explain why ultimate unified totality, as TO, has never been adequately developed, as well as why we must now take advantage of all that enables a well developed, recognized, accepted and extremely vulnerable actual totality.

We signify TO, actual totality, as the unified ultimate totality, which for the present and adapted in the future, it is. It is in this context a closed set with a high degree of exclusiveness. Yet we must always remember, ECRO, that TO exists in two or more forms. The first is its preeminence in present time that extends through the near term, and thence into the mid term and far term. The second is the existence of TO in time, adapted and altered in relation and the necessity of coexisting with potential and interacting totalities. When TO exist by transitions in time, it is signified by TO-TOT, TO-TOP or TO-TOF, interacting generally, in the past or future. The separation of TO and TO-TOT is variable, both in its existence, and degree of interpretation. This is accounted for by the quality of TO and degree TO is undergoing transition in time. When these are accounted for and applied, further proof of TO is achieved.

The sequence of proofs of TO, must converge on an ideal flow to be well ordered. The distribution of each type of proofs flows, one following upon another in the transitions and development of TO. A flow in which each chapter is a preview of the next, and all chapters preview TO, from its origins to its completion. In this way each type of proof begins and ends tied in with the prior and following and all

TO, beginning from non existence, through many levels, to all that proves actual totality TO, and its applications in life. This fortifies and gives added support to proofs of the whole, and an excellent guide that makes more reasonable all proofs from chapters one to forty. From nothing, to the barely existent, any, the potential and to its actual existence in TO, the flows of kinds of proofs emphasize their bond and bridge between each other and all TO. This is shown by the table of contents. Each type of proof emphasizes its topic, ties in with prior topics, and its connection with its following topics and proofs.

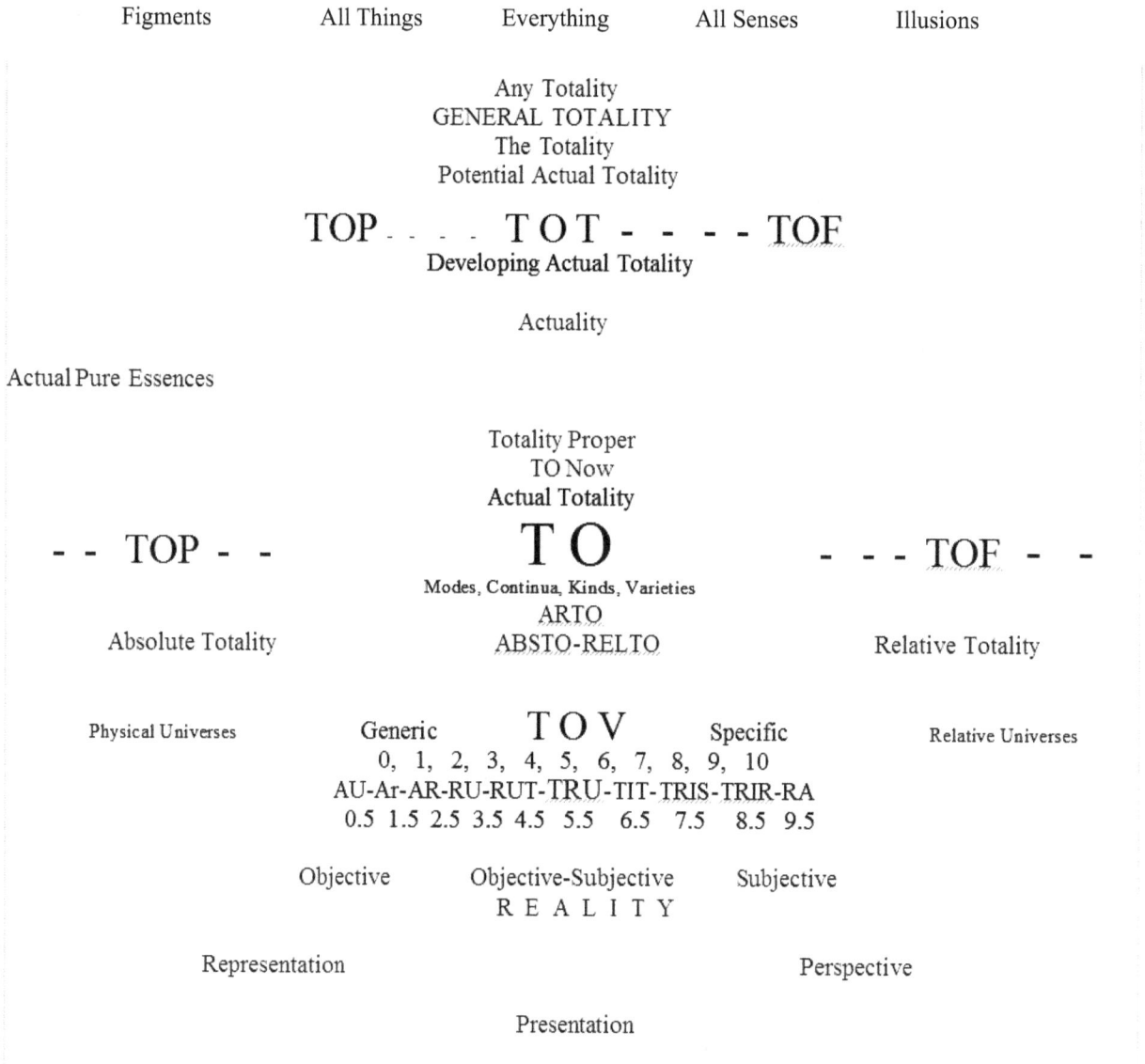

Figments   All Things   Everything   All Senses   Illusions

Any Totality
GENERAL TOTALITY
The Totality
Potential Actual Totality

TOP - - - - T O T - - - - TOF
Developing Actual Totality

Actuality

Actual Pure Essences

Totality Proper
TO Now
Actual Totality

- - TOP - -   T O   - - - TOF - -

Modes, Continua, Kinds, Varieties
ARTO
Absolute Totality   ABSTO-RELTO   Relative Totality

Physical Universes   Generic   T O V   Specific   Relative Universes
0, 1, 2, 3, 4, 5, 6, 7, 8, 9, 10
AU-Ar-AR-RU-RUT-TRU-TIT-TRIS-TRIR-RA
0.5 1.5 2.5 3.5 4.5 5.5 6.5 7.5 8.5 9.5

Objective   Objective-Subjective   Subjective
R E A L I T Y

Representation   Perspective

Presentation

Fig. SC-1 Overview of Actual Totality and all associated, internal to external.

This figure features the best possible approach to actual totality, its categories, themes, topics and constituents of TO plus what is interactive and potential externally. It also gives an indication of the past and future development. It can be done in many ways, and fused or correlated with many other figures to show the most unified to diverse form and function of TO. When reading this book, or thinking about TO, reference to it should help to orient us by an overall picture of TO and its surroundings. It provides some of the detail of actual totality to emphasize union of absolute and relative and spectrum of varieties. No single figure, especially those in two dimensions, can adequately simulate actual totality. Three dimensions do better, especially when large enough, well detailed, proportional and inclusive of surroundings. To achieve a good, and efficient understanding, the above figure is the basis and primary means of attaining the proper and optimal overview.

Although TO is essentially a unit, it can only be adequately understood by increasing knowledge, meanings and proofs of all its aspects, form, function, mechanisms, parts, processes, concepts, principles and laws in an ordered, and integrated to differentiated whole. It cannot be generally understood by what is less, null or opposing, except in disproof or ways that support the whole.

Forty Types of PROOFS of ACTUAL TOTALITY

Marcus Bergh, M.D.

Table of Contents

FIGURES in TEXT

None of these figures provides a complete replica or representation of actual totality. Each adds to the overall perspective or adds to some salient aspect of actual totality or its presentation. The figures shadowed are primary overviews of actual totality. To gain a good image of each primary figure, combine and develop them into their unified totality is the aim or this book and all.

I Preliminary Proofs of Precursors to, and Developing Actual Totality

CHAPTER 1 - NON EXISTENCE to Existence

What is not, what is lacking, void, negative, figment, illusions, disproportions, to levels of existence in TO

To exist or not may seem simple initially, yet, like to be or not to be, is a question, problem and enigma that are most complex, subtle and profound. It is largely resolved by distinguishing its main features. Is it completely separate existence, existence as a part, or some combination? Is it physical or objective existence, subjective or mental, thing or sense, or are they combined? Is it necessary, essential, or actual pure essence's existence, or something less or other? Is it a part of anything, any totality or actual totality, TO?

These are questions that initiate and guide the great search for the proofs of all. They cannot be well answered without subsequent proofs. They are largely determined by actual totality, by neither what they are physically, our awareness nor their combined reality. Since actual totality, TO, is only practically and essentially autonomous or absolute, even what does or does not exist depends on time, and the interactions of TO with TOT.

Each and all persons see themselves as existent, even to the extreme of separation from all else. They see what exists or doesn't before them, as life and death. They do not take advantage of all the knowledge available, reason with it logically and fully, nor seek its resolution in actual totality. Herein is the greatest meaning of being, how we exist, yet only to a degree, and in certain ways. And how much we and the world suffer from our lack of understanding and appreciation of our partial and contributory role in TO. Without this we lack existence, are more or less dead, and don't know it.

What is not, is largely what lack's existence. What lack's existence, may lack necessary existence or actuality, and this may lack the pure essences of totality. Only those that consists of the same actual pure essences that are in TO have complete existence. Being in any totality, actuality, necessary existence, any existence, or non existence is what simply doesn't have full and proper existence. Of those lacking in existence are figments, illusions, voids, negatives, the chaotic and lacking limits, relations, focus, centering and proportion.

Questions, errors and corrections in knowledge are often found in what they leave out. By learning what is not, and what is lost in transition, as in time and transformations, as of form and function, is how we can show an object, totality, and prove TO. That lacking existence is not all the same, neither homogeneous nor heterogeneous. It depends on how it relates to the stages from existence to TÓ. What is non existent can be what is lacking in, existence, necessary existence, actual pure essences, any totality, totality potential to TÓ, and what is not, or lacking, of TO. In this way what is non existent has existence in those stages below the one specified in the spectrum TÓ to existence. This is the same and also applies to many other aspects and features of the spectrum TÓ to existence.

Non existence may be interpreted in different ways, from zero to one to all. If it is from zero, it has nothing to add, thus of zero resultant. If it is of many or all and remains at many to all, it is of an indefinite, to an infinite product. It is yet neither complete, nor existent in actual totality. A non existent is separated by becoming one existent by the fusion of parts in such a way or at a level of energy the one existent comes into being. This is the most basic mechanism of actual totality. It is also the mechanism by which relative totality creates life with absolute totality in actual totality, (Lunen)

What is less than TO often includes, aspects, features and parts of what is not TO that, are similar to what, is in TO? Actual pure essences of anything must equal their actual pure essences in TO be in TO, how they are adapted and altered to become a part of TO. By comparison with TO what all lack that is present in TO is much of their proof. What is other than TO includes all that is external and what seems to be a part or constituent that is not, as in failure of the potential totality, TOT, to become actual in TO.

What can't be TO includes, nothing, nothingness, no unity nor totality, internal singularity, undifferentiated or disintegrated, zero and absolute, unless qualified in a special way by TO? By actual pure essences exceptions in extremes are attained when adapted to existence in TO. To the degree any entity has separate existence it is not of actual totality, TO. What makes the most crucial difference in the world, is whether any entity, sense or thing is separate and independent or whether it is its part a contributor in TO.

Nothing and nothingness are not the same for actual totality, as their common meaning, what they are in the physical universe, the relative universe, or for God. For TO nothing is a vanishing point between its ultimate unified totality and their absolute negative. All that exists in TO does so by degrees of presence and proportionality. It is this absence of perfect existence, and approach to zero proportionality whose disproof in TO is most characteristic. It is the measure and test by which what is not TO can be judged on the one hand, and what exists, is necessary and must be fulfilled by TO. There is meaning in everything, in and by TO, if only partially, fleetingly or negatively. Such a vanishing point may be compared to death, or to the origin of the big bang, or end of its expansion.

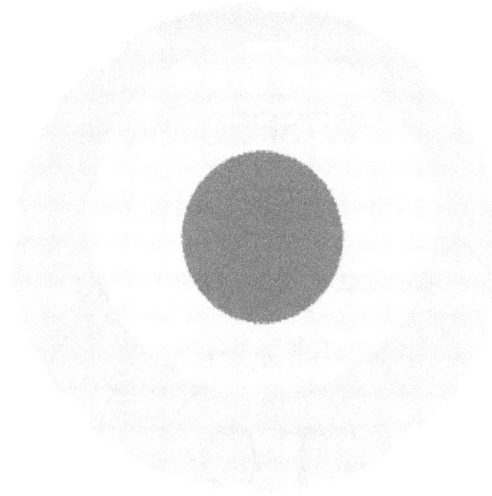

Fig I-1 Four categories or regions of Existence and Totalities, non existence, existence - essence, totalities and potential TO, TOT, and actual totality, TO.

In this figure the white external region is non existence, all that lacks existence. The second in light blue signifies increasing existence and essence. The third in yellow is that potential and interacting with TO. The fourth or red center signifies actual totality and all its categories, modes, kinds, varieties, forms, states and transitions. It is a closed set when of the most essential, actual and real existence in the extended here and now by relative totality and the autonomous at once being of best human life TRU. The separation of these categories is permeable, each is interactive with the others, shown hatched to represent their interactions. Each category of totality consists of major levels of the spectrum from non to absolute existence. Each of these categories has external and internal limits, which become more set and distinct as we progress from non to absolute existence. The internal limit of the less existent is the same as the external limit of the more existent, and each category in turn. So that for actual totality, TO, it is the beginning of unity and convergence on the set of absolute-relative and generic specific form and functions. For TÓ, TRU to TÓ is the focus and power by extreme specificity in here and now, balanced by that of extreme generality of the highest levels each in TO.

Whether the big bang of the physical universe began from nothing, something, single-multiple universes conjectured by physical theory, does not alter the role of absolute and relative totality in actual totality, TO. Transitions in time, and transformations of mass, energy, volume, pressure, temperature and other features of mechanics are not the same for different totalities and are adapted, altered and accommodated by TO. This occurs by the form, function, change and reversibility of TO with TOT. The role of actual pure essence, and the powers of intensification by relative totality in TO, are factors and powers that sustain TO-TRIR-TOF in the far term indefinitely by TO. Hereby the qualities of being absolute and nothing which seems to be impossible is corrected or altered, consistent with the unity of TO. It is what provides the basis for and proof of its actual, pure, essential and perfected existence. This is the power provided by relative totality, by ARTO in TO.

3

As we progress from nothing, non existence to existence, to existence potential TO, TOT, and TO, providing an analysis and interpretation of what TO is not and is, they become much clearer. Likewise as we proceed with these proofs it will become more certain that the typical form and function of TO, with its actual pure essence produces an interaction and affect with what is external or not TO for exception and qualification. These will be noted as we go along, and as TO becomes perfected and completed. An example of this is seen in the development of continua and especially absolute - relative. This is expressed in LANERA, law of never absolutely relative nor absolutely absolute. Likewise in LAMOL or the Law of More or Less, wherein TO exists, not as a static unit nor has parts that are static function, but are more or less variable. This is a product of time, and by transitions with time. TO is much more than the static things and objects we observer, or senses and subjects we know.

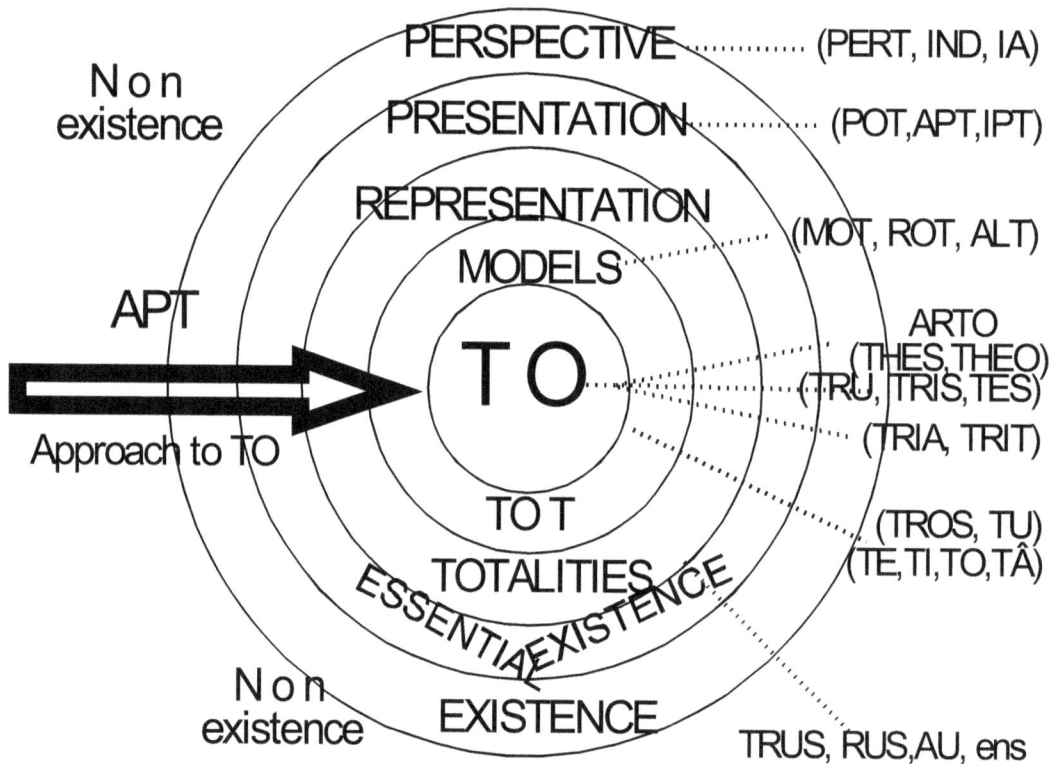

Fig 1-2 Facades of an Approach to Actual Totality

There are many kinds of approach, direct or indirect, frontal or rear, lateral, transverse, or sagittal, differentiated or integrated, systematic, standardized or natural. It is not possible to give an immediate and perfect showing or viewing from one simple approach or perspective. We must remember that such limits are mostly of the presentation and representation TO. In this figure alternating shells from the central core of TO and TOT, as with all topics, may be shown from the outside as in APT or from any direction selected. This is what happens when we take the world in our hands and look at it. As can be seen from this diagram and the text, it duplicates these alternating shells to provide the best possible positive objective approach, POA, so that we come nearer to TÓ - as to God - by its S-O or A dimensional mass energy predominant form. This is signified by, COPÁ, Convergence of Objects, applicable through presentation, and SO that is Á, the Predominant form and mass energy dimension in TO.

Although perspective tends to view non existence, existence, necessary existence, potential and existence as separate objects, their relations to, form and function in actual totality are highly interdependent and interactive. The interface between each is highly permeable and dynamic. This is a major characteristic and proof of TO.

Non existence and existence are not the same for TO, TOT, other totalities, and what we commonly believe. It depends to a large extent on their actual pure essences, development, present and future existence. This explains many of the seeming differences and paradoxes between fundamental aspects of totalities, universes and God. An example is of a person, I, which commonly is of certain existence, largely separate, independent and seemingly most important. Yet, I am mostly only my actual pure essences or soul, as this is increasingly altered by the actual pure essences of TO and with TO. For universes, I am of little significance. For God and philosophical systems I am of limited importance. Only for myself and my certainty, and actual totality, do I become important, by LOPAN, the Law of Partial Necessity. How much I as my counterpart into must exist, if TO is to be. In many other ways and instances all exists or lacks existence depending on their altered position and relation to increasing levels and stages of TO and its transitions and transformations in time.

If actual totality is exclusive of all not of itself, then non existence is all not of actual totality, and the opposite is its non existence. This is the simple and obvious confirmation of the fundamental fact, form and function of actual totality.

It is from non existence and incipiency to its final absolute in the absolute relative, TAR that both explain the enigma of TO, and its positive imperative. Powers providing the existence of progressive transitions, are what has made and saves, reveals, shows, and how to live by TO. That which does not conform and all who do not follow these imperative powers of TO, driven by relative totality, TRU and TÂR, are to that extent relegated to a decaying state or life of non existence. What is negative to TO-TRIR-TOF will, sooner or later, be or cause selective rejection or be eliminated, by natural or more rational selection. They, by the lack of actual pure essences necessary for existence, TO, are already irrelevant, of decreased proportion or worthless and thereby extinct. They become negated by actual totality. Why this enigma exists is explainable by the form, function, and inclusions of much that seems to refute TO-TRU-TIT-TRIR-TOF. All that proves TO and its constituents help to prove the enigma.

The absolute is not all, by itself it is less to a non existent. It is the absolute-relative and relative-absolute that transcend all, all in TO. Although our attention, self interest, and much that are partial are great aspects of our being, they are not all. They often result in a lack of what we must become. For all is incomplete, and leads to less unless it is finalized in the totality of TO. The approach to TÓ, ÀPT, is by the processes of perspective, presentation, representation and models whose in and out manipulation, OITIO, reversibly reveals and proves TO.

What anything is not, is all that is other than itself, except the effect all else has, and is a part of it. Thereby for non existence it is all that is of itself only. It is what is less than, lost by or absent in all else. By the hierarchy of actual totality, Chapter 15, non existence is largely the failure to gain, and all that is lost by actual totality, including all its categories, features, constituents, etc. In this way non existence is not constant, it will vary with external effects and changes, most prominently changes in, and interactions of the constituents, of actual totality. This is the Law of Non Existence, LANEX, and its being external to actual totality. This is fact that non existence and actual totality lack any and all interdependence.

A part can't exist unless it is its part in the whole. If this whole is actual totality, the part exists only within it. Thereby, all that is a part, is non existent, when not of actual totality. This includes what I am, and my soul or counterpart in TO. I and the part are not all or none, except by inclusion in TO. It is a spectrum from least to most potential, to becoming actual. It also may be sufficiently actual in various ways depending on necessary existence of the part and the whole, LOPAN, and the Law of a Part's Existence in the whole, (LOPEXT). These, and the spectra they assume, are most important concepts that help to reveal and show TO. An entity, sense, thing, subject, object, etc., may exist separately, in TO, both, or neither. All must be correct by their existence, and its form and state, in TO. There are spectra of the variants of existence and non existence, and spectra of the variants of TO and a non TO.

We should never forget, ECRO, that the non existent is most of the world around, and of, us. There is a continuum from completely non existent to precisely existent in actual totality, in which all less than their part in TO does not exist, LACNE. This continuum is a subclass of the negative to positive continuum, in relation to benefit or detriment to, TÓ. Indeed, this will become much more evident through the following chapters by proofs of TO.

CHAPTER 2 - EXISTENCE, to Necessary Existence

Kinds and Forms of Existence, Potential to Actual

What exists? What potentially or actually exists? What are the gradients, interactivity and transformations between what does not exist, exists and actually exists? Is there such a thing as an ultimate unified totality? Is this the whole of what actually exists? Is this Actual Totality?

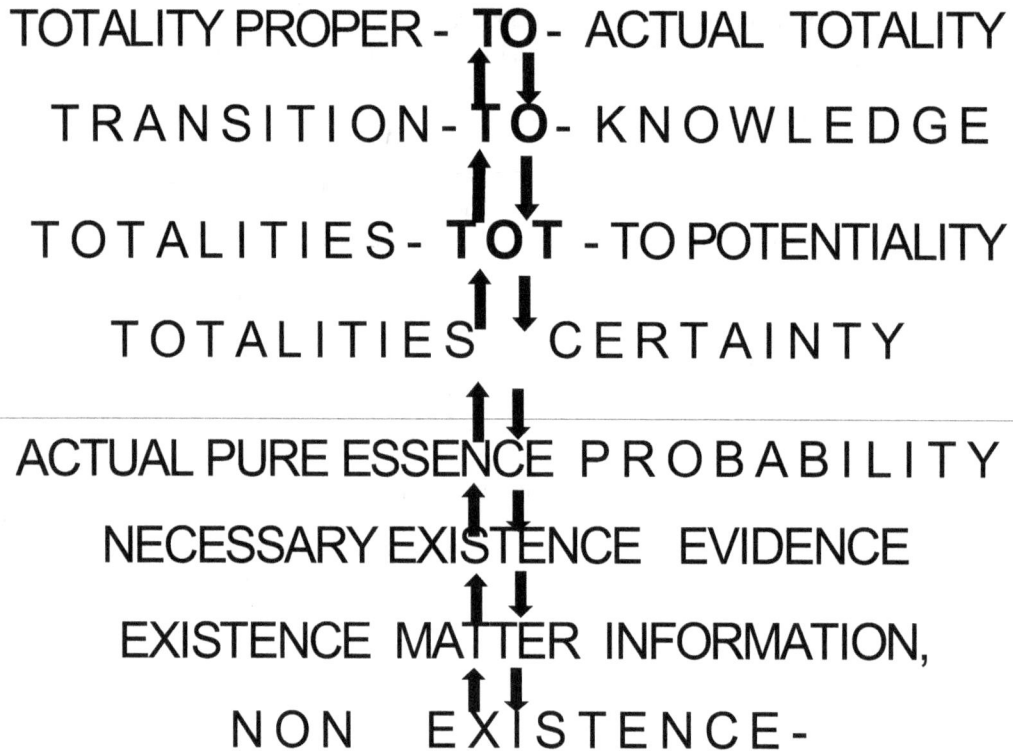

TOTALITY PROPER - **TO** - ACTUAL TOTALITY

TRANSITION - **TO** - KNOWLEDGE

TOTALITIES - **TOT** - TO POTENTIALITY

TOTALITIES   CERTAINTY

ACTUAL PURE ESSENCE   PROBABILITY

NECESSARY EXISTENCE   EVIDENCE

EXISTENCE MATTER INFORMATION,

NON   EXISTENCE -

Fig 2-1 Means to Existence in Proof of TO, (PRUT)

How we can collect, organize, order, coordinate, proportion, posit, and use all knowledge, is almost as great a way as in discovering and gaining knowledge initially. It is the key to permanence. Knowledge that is irrelevant, unconditioned, without the states, conditions, relations and limits of its object, results in confusion and failure. It is little better than having no knowledge. It is in developing knowledge of all things comprehensively in totality and actual totality that knowledge bursts forth in all its glory, to our absolute benefit. How existence, necessary existence, actual pure essences and totalities become actual totality is shown by many figures, including its model SPALT. The text develops and seeks to gather information, evidence and proofs that show how discovery can have the subjectivity necessary to best respond, answer, and solve all that needs the right response. It is the natural order of this text to properly integrate, differentiate and demonstrate the relations among the first person and TO, and the subject and object as they are out of and in TO, OITIO. It is the attainment of an overall and masterful view, and a total comprehension, management and control, that eluded mankind, yet has been the purpose, and to a great extent the end, of philosophy, knowledge and life. In order to test any theory, it is the purpose of all fields and sciences to enhance objectivity, subjectivity and their combined reality. This is what develops theses and successfully exercises the predominant powers of best human life in actual totality.

Actual totality is proven by the existent over non existence, what exists rather than does not exist, It is also proven by the actual rather than potential, where the potential exists externally, and is yet to change TO. Only when the non or lesser existent potential has an effect on TO or a part and role in TO, do they exist by presence in TO. Existence with an observer is presence. Presence is TO with actual pure essences. This is the Law of Non Existence, Existence and Coexistence, or absence and presence. The gist or deepest meaning and greatness of existence, coexistence and non existence approaching the absolute, is the primacy of TO. It is the difference between the extreme polarity of perfect presence and non presence or absence. This is much more than simple totality. It is the complex and most compound TO that consists of all with actual existence in actual totality. By inclusion, this consistency it is made, to a large extent, by the presence of potentials that become actual, and have actual pure essence. It is in this way we have the accuracy of absolute existence and coexistence of relative existence, and proof of TO.

It is the presence, completion and autonomy of the actually existent that gives actual totality its aceity, or state, condition and form of being selfly originated or self derived. This is a result of the separation of TO from TOT and all else, by its actual pure essences and other forces. These are the propelling and compelling forces directly that occurs along with the stability of the form of actual totality to give TO its power and continuity.

It is by the contrast between potential and actual, absolute and relative, and many of the dualities whose continua make up general, potential, actual totality, TO, and all that enables its more specific varieties, as best human life, TRU that all can be most clearly recognized, shown and proven. Potential and actual are not completely separate. They are two portions of the same continuum, activity and process. This continuum exists in more than one way, by the degree potential too actual, by the degree of potential and degree of actual, and by their relations to the whole totality, TO or categories of the whole, e.g., TRU. Throughout these proofs of TO the questions, answers, purposes and intentions of relative themes of life reveal what, why and how all positively over negatively fit together in this totality, TO.

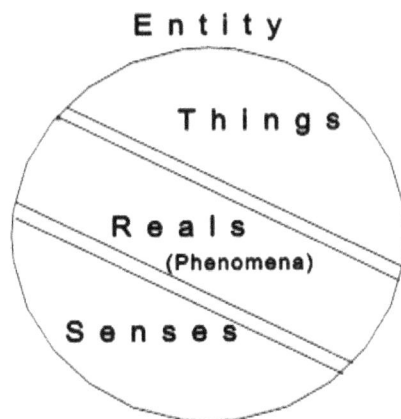

Fig 2-2 Entities, Senses, Things, Reals, Classes and forms of what exist.

Numbers, letters, words, objects and all with separate existence are entities which are represented and used both as senses and things. Those entities that consist of both senses and things are called reals or phenomena. This is the inescapable bond that exists when senses and things interact. This is not the

7

only meaning of the term phenomena. It also refers to some occurrence in nature that can be subjected to analysis and determination. Like the real, phenomena join senses and things, subjects and objects. This is the beginning of reality, a primary proof of TO.

Senses and things, like subjectivity and objectivity and physical and mental when combined form the fundamental basis of existence, life and actual totality. They bring absolute totality, as the physical universe, to bond with relative totality as cognitive powers in the relative universe of best human life. Subjectivity - objectivity, S-O, and its many manifestations originated early in the creation of life and have evolved ever since to become a salient feature of TO, and its progression in the future. To trace this approach in existence and its progression is key to understanding and proving TO. Subjectivity and objectivity, as subject and object, derive from senses and things. Things and senses are forms of matter and antimatter, which when subject and object, form the basic dimensions of all that culminates in the whole, varieties and highest altered states of TO. This is the distinction, and proof of TO in comparison with other totalities, entities, things and senses.

Nothing is perfectly a thing in itself. They exist largely as they are a part of, contribute to and enable actual totality, TO, its varieties, TOV, and the world of Best Human Life, TRU. This is the Law of the Actual Pure Essence over a thing in itself, LACET. As a thing approaches perfection, like a person and their soul, it loses its thing in itself status and nears an absolute and actual pure essences form in TO. Hereby is the message of the new testament, how Christ was transformed from a thing in itself into being a part and contributor to God. Only all entities, senses and things - as well as all people - can and must be transformed into their part and role in TO. This holds for Christ and God to the extent they are Equivalent, TREQ, to their part and role in TO. The transformation from non existence to existence as things in themselves, and potential to actual is the basis for and proof of TO. How the greatest in all things, as human life and its concepts and words, exist as actual pure essences potentially to actually in TO. They provide many of the proofs that we increasingly pursue through the text, and on into all our lives in TO.

Many of the most basic doubts problems, paradoxes and enigmas in knowledge arise from differences in the use and interpretation of the basic features and categories of all. Are they to be used and interpreted directly, as in simple experience or more indirectly, logically, theoretically or relatively? This is at the core of actual totality, the differences and sameness of absolute and relative totality, as they exist separately, or together in actual totality. The solution is to be found in actual totality, the effect of relative totality on absolute totality and the physical universe. Non existence and existence thereby depend on what they are a part of, what they are in, and their external relations. By the limits imposed by relation and actual totality, all retains much of its direct and physical use and interpretation. Yet it must be modified by the limits and effects of relative and actual totality, whether and to what degree there is, a unified totality, no unified totality, or some intermediary. Since actual totality is intermediary by transition in time, it is practically unified. By this fact and to this degree all must be ultimately determined and proven by how they exist in the absolute-relative unity of actual totality, TO. Since this unity is not absolute, nor thereby absolutely stable, its variations render all more or less, LAMOL. However this is not as bad, nor as hopeless as it may seem. For relative totality brings to actual totality powerful and large capacities and abilities, the practical form and function of actuality, actual pure essence, reality and many others that converge on, and are highly unifying to actual totality, and its many varieties including all that enables best human life. In this way non existence, nothing and even the negative have some existence depending on how they relate to various aspects or regions of, and are used and interpreted by actual totality.

Since humans are largely the given and the focus from whom objectivity is being sensed and perceived, we adopt, POA, and positive objective and PSA, subjective approach. These approaches can be combined producing the positive approach of reality, PRA, which when formalized and perfected describes, proves, shows and helps to present TO. We make it by the form and function of actual totality, ordered and focused from the first person. In this way we have figure 2-3. It acts as a guide to show the form, features, shape, structure, order, sequence and contents of TO. It may be compared with other stages in the development of mass energy in actual totality, and needs to be viewed in conjunction, with other figures presented in this book. How senses, things and reals are needed, practiced, and used is largely the determinant of their existence, form and value in progression with TO.

The Ontological Argument arose from the philosophy of St. Anselm (1033-1100) who proposed an argument for the existence and proof of God. Since God is being, than who no greater can be conceived,

God must exist. For it is better to exist than not to exist. It is better to be actual than potential. And it is better to exist necessarily than to exist contingently. The unified totality of all must be all, and by definition must exist necessarily. For the definition of actuality is to exist necessarily. What is common to actual totality are RELTO and TRU in union with ABSTO and the physical universe. Yet what is common also differs by the relevant, functional, proportional and other qualities typical of the sum of actual totality.

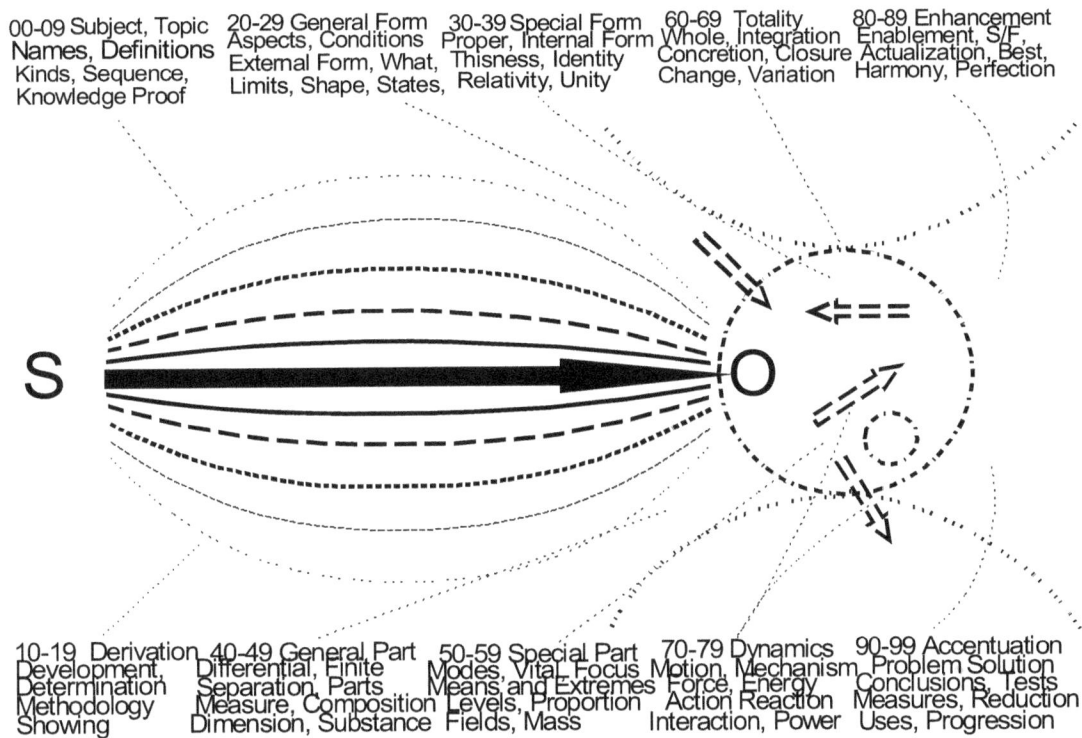

00-09 Subject, Topic Names, Definitions Kinds, Sequence, Knowledge Proof

20-29 General Form Aspects, Conditions External Form, What, Limits, Shape, States,

30-39 Special Form Proper, Internal Form Thisness, Identity Relativity, Unity

60-69 Totality Whole, Integration Concretion, Closure Change, Variation

80-89 Enhancement Enablement, S/F, Actualization, Best, Harmony, Perfection

S   O

10-19 Derivation Development, Determination Methodology Showing

40-49 General Part Differential, Finite Separation, Parts Measure, Composition Dimension, Substance

50-59 Special Part Modes, Vital Focus Means; and Extremes Levels, Proportion Fields, Mass

70-79 Dynamics Motion, Mechanism Force, Energy Action Reaction Interaction, Power

90-99 Accentuation Problem Solution Conclusions, Tests Measures, Reduction Uses, Progression

Fig 2-3 Positive Subjective- Objective Approach, Development and Determination of Existence

This is an existence that is by the necessity and purpose of life, a coordinated form amenable to demonstration. The inclination to a unified totality of all, must be all, and by definition must exist necessarily. This is largely by the actual pure essences of actual totality. What determines importance, and makes the most crucial difference in the world, is the transformation to actual from potential, transformation to actuality in the specificity of RELTO in TO and TOV, and transformations with what are actual in TO. By the actual pure essences of entities, senses and things which are often repeatedly transformed until they become their actual pure essence's role in TO, TOV and TRU. From non existence to existence, relevance and potential to actual in TO, TOV and most specific varieties, all becomes existent by their actual pure essences in TO. Whether it is any entity, sense or thing or whether it is its part and contribution to the actual totality and its varieties, including, best human life, whether it is non existent, irrelevant, potential or actual in TO. Herein is the test of all, proof of TO and reduction of the physical universe, materialism, sciences, philosophies, realism, idealism, and religions to their part and contribution by this ultimate union in actual totality.

Existence and necessary existence are different in very important ways. Absolute totality and the physical universe, for the most part, exist but exist only by lower level of necessity, e.g., gravity. To exist necessarily is actuality, the qualified existence that has an actual/potential and real, subject-object cause or reason for its existence. This is how relative totality, and relativity, are initiated by relation that causes and is a reason for necessary existence. Herein are the initial and one of the greatest steps, in the transformation from existence to, and proof of, actual totality, TO.

CHAPTER 3 - NECESSARY EXISTENCE to Actual Pure Essences

Indefinite to the Definite, Actuality, Unessential to Essential

Proof of actual totality is by definite/indefinite, necessary/unnecessary and essential/non essential. Necessity is key to actuality and essences, as these are to TÓ. Necessity is largely what is essential for existence, as actuality is what is necessary for existence. What actually exists does so necessarily. To exist necessarily requires a distinction between the definite and the indefinite. To exist necessarily also implies dependence, design, degrees of purpose and its own typical essence. Such distinctions require more than any object. They require a relation, a relation between object and subject, a doer, subjectivity, thus degrees of reality. It is the relation, or relativity the physical universe has, that distinguishes it from absolute totality. Yet to exist necessarily requires more than physical relativity, the degrees of purpose, dependence, design and its own typical essences are what typifies actual totality by relative totality and the relative universe.

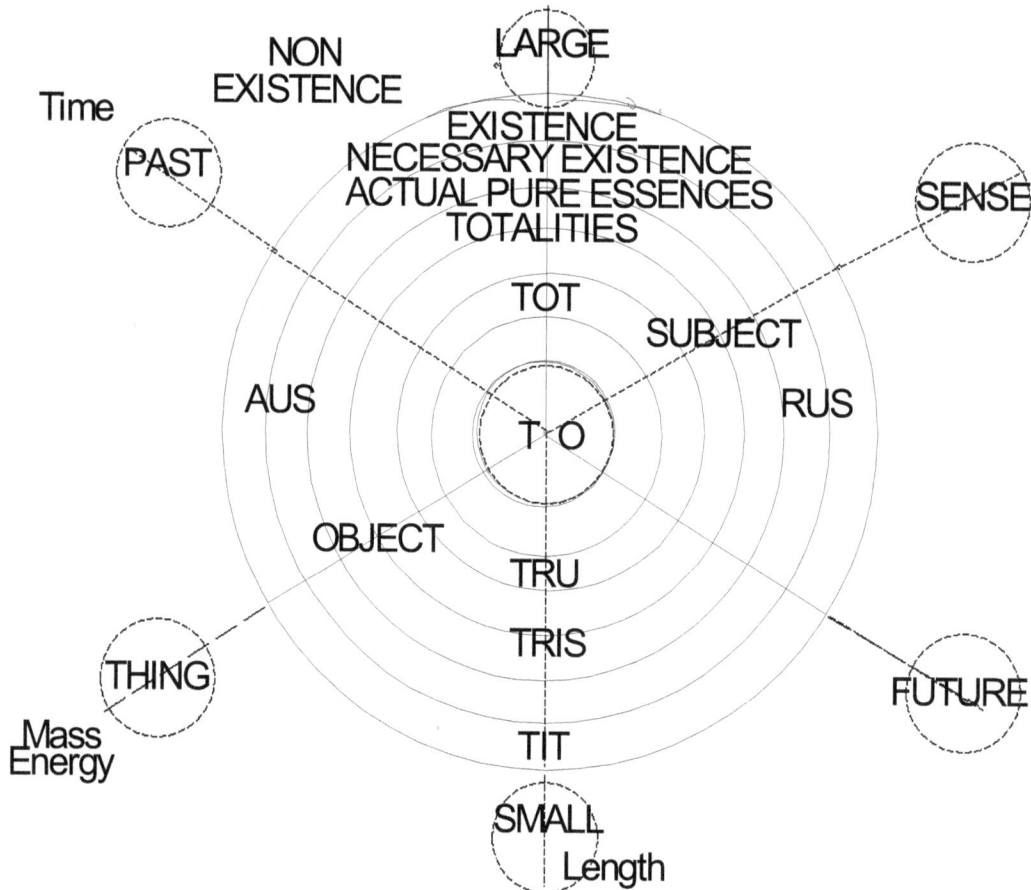

Fig 3-1 Far Approaches to totalities, TOT and TO

This figure (3-1) presents a diagrammatic and rough rendering of the far approaches to actual totality, how its dimensions can be best combined, and how the many concepts, principles, laws and mechanisms involved are positioned into it. This provides an excellent means of achieving a good tie-in between perspective, presentation and TO. The dimensional formulation in space time and mass as it becomes S-O in TO changes as they are formed and function together in, or depart from TO.

The general and special, or non actual and actual, levels or kinds of existence are made distinct by "any" and "the," the indefinite and the definite. They are also made distinct by the potential acquiring actuality. The "is" identity, specifies a certain entity, object or relation, as in this and that. "Is" is the power of perseity that produces a this, and the power of haeceity to produce essences. It is a relation by analogy of attribution and a fundamental characteristic of relative totality in contrast to absolute totality and the physical universe. It is increasing certainty of this distinction that is the critical step in actual totality. This is the proof by definition, provided by TO. Potential entities may become null, a, some, the, or their proper part in the actual totality, TO. It is largely the problem of existence, and kinds of existence, separating the special of totality from the general, and recombining them as they exist in, and by the power of, actual totality.

How do a, any or some things, become the thing? How do entities become the entity? Is it a state or condition of the entity itself, its object, subject, or being realized? How do entities and things progress from absolutes through relatives, universes in common, to increasingly definite, to identity with, and perfect existence in actual totality, TO? How some things become the thing, is both simple and profound. It is simple to the extent that, as by Anaxagoras, they are increasingly and predominantly characterized by those things of which they are mostly made. That is how relative totality, RELTO, the relative varieties of TO, ARTO and TO are increasingly specified and made up of different and separate aspects, forms, modes, essences and characteristics. In addition they have a separate and special internal focus. Each relative totality, RELTO, and special variety is not some other, closely related or a comparative entity, because they do not have the exact makeup typical of what characterizes another.

For "the" things, we do not speak of special varieties of things only, we also speak of them generally, This is signified by, The Universe Common, TUC. It is TUC or what TO holds in common with entities, things, absolutes and relatives, AUS and RUS, that justify and enable us to validly speak of TO in this early part of this book, and its proofs. TUC is an, any, and some thing as they become, yet are not yet, the thing, or all the necessary characteristics they have in TO. To be of TO, the, or the actual totality, as in 'thee', is how TUC becomes TO.

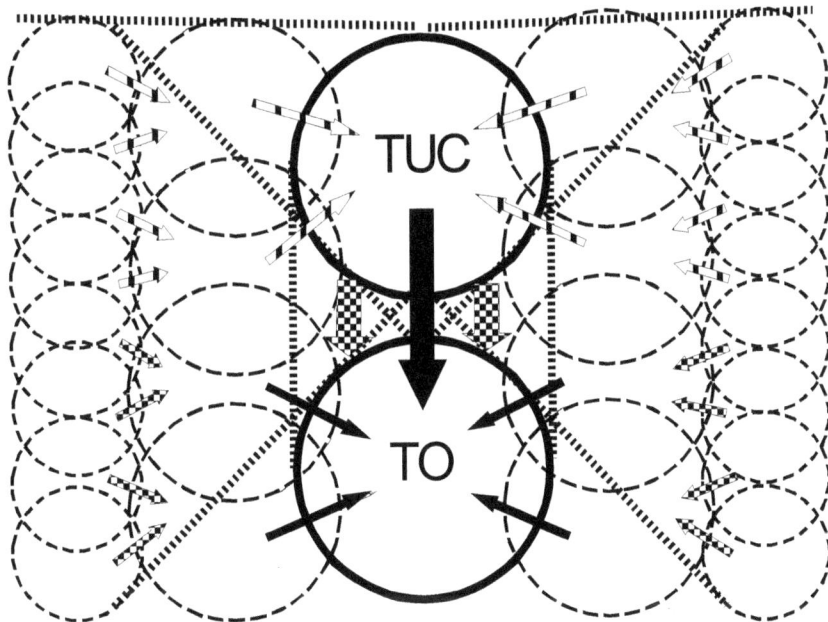

Fig 3-2 Development the definite, "The"

In this diagram, (3-2) TUC the large top circle signifies The Universe Common or all whose entities become a part of a specified, a, an, any or some universe, world or totality. The heavy arrow signifies the steps to TÓ. Most important in the steps is The Universal Definite, or TUD. TUD is a, any or some separate existent that becomes definite. It is the step from TUD to TÓ that defines how the potential becomes the actual definite by perfect identity within TO. TUC - To The Universals Common, TUD -The Universal Definite, most important in steps to TÓ, TUC-TUD-TID-TO, from entities through each identity to TÓ. TUC are a, any or some separate universal that becomes definite, TUD, TID is the final step of Identity in TO. All reveal how the potential becomes the actual definite by perfect identity with TO.

We must describe and distinguish the difference between any, the, the actuality that exists in TO, and its actual pure essences. If we do not, TO has much less meaning and proof. By any and the, both general and specific have existence in TO. This is not so in any way alone, only when combined in a focused way by the relative. These in comparison with any existent give us clearer ideas of what are the this-ness and what-ness of necessary existence. Distinct from what is irrelevant existence. TUC stands in marked contrast between absolutes and relatives, between *a* thing and *The* Thing. This is, Speciation of the Entity, (SPEN), the product of being a this, haecceity, as TUC proceeds through the more relative, TUD, and by relation and absolute in TUD to TÓ. It is more than a product of external convergence. It is highly internal in that it has its own convergence, focus or point and center of reference that produces, as it expands or grows. It exists as autonomous, at once and existent, (ATO). This is why TO is itself and not some other.

Actuality is the form or state of having definite and necessary existence. Actuality decreases to the degree an entity, sense or thing is less definite with less necessary existence. In a sense there is an actuality in the physical universe, even absolute totality, yet this is less necessary, has less actual pure essence, and increasingly lacks relativity and absolutivity. Thereby actuality depends on the definiteness of the relative, relative totality and TO.

The Law of Indefinite that becomes The Definite Existence is the beginning, positive imperative development, actual pure essence and proof of actual totality. LÏDEX. It is what comes into being as absolute totality and relative totality fuse. It is much more, and more definite than LUNEX the union of indefinite and definite. LÏDEX is exemplified by the absolutivity and relativity that are actualized by actual pure essences, like ARTO in TO. LÏDEX although a feature of absolute totality, is largely a product of RELTO in TO. The increasingly definite approaches perfection, like TRIA, or TRU is All, in RELTO and TO. In relation to LAMOL and DAOST, LÏDEX reveals the basic form and properties of TO. The chapters of this book trace the development of LÏDEX as it acquires increasing definiteness. In this way it becomes the law, and proves the certainty of TO.

The fact that what is necessary for existence, along with the definite is the basis for totalities and actual totality is evident and strongly proven in Chapters 8-19. This is shown by the continuum, mode and fusion of subjectivity and objectivity, relative totality and universe, absolute and relative, generic and specific in actual totality. This shows how several and all constituents together and within, prove actual totality, rather than singularly, or separated.

Absolutivity is necessary for the existence of relative totality and the relative universe, as relativity is necessary for the existence of absolute totality and the physical universe. They are mirror images of each other in the states and conditions of relative and absolute totalities. Both are absolutely and relatively necessary for the existence of actual totality, largely depending on present or changing time, in TO in transition with TOT. They are decreasingly less for totalities that become less definite, an, any and null universes and totalities.

It is in being relative to us, being focused upon our being, attention, intention, and success/failure that determine its absolute specificity, inclusiveness and existence. (TRIA) Other special relatives as animals evolved, as in human cladistics. Even many special relatives as TRU and TIT are at best all TUC, not, or distinctly less than, SPEN, when less than their part and role in TO.

What is the role of the definite participle? Does "the" prove the relative, relative totality and thereby, actual totality? An, the, the actual, and the universe of all enabling best human life in TO, are evident in the varieties of TO, TOV. The contrast between an, any, and the, illustrates the power of relation, of the reality and actual existence of subjectivity-objectivity, or S-O. Such a state and signification do not occur in a completely physical universe. This is a product of the long term accumulative power of life of animals, to mammals, to humans and their central nervous systems' ability to specify, designate and signify definite objects. To make definite, requires more than any action, it requires a doer, an agent between the action and reaction that brings existence to actual existence. Without being definite, specified by the reality of combined objectivity and subjectivity, signified by symbols, names, words and language all lacks what is necessary for actual existence in actual totality.

It is impossible to have actual totality without need, essence, relation or specification. This is "the" definite article proof of TO, and its relative and specific varieties, (PRUTORS), Proof of TO by relative and specific. Actual and actuality emphasize the dynamic, current, convergent living focus of relation and specification in TO. This is what most distinguishes, separates and makes more actual relative totality from absolute totality, the relative universe from the physical universe. In reverse, relative totality, as specific varieties of TO incorporate much that is in absolute totality, to the degree it is a part and plays a role in the whole.

As much as we use the terms is and truth, they are like a puff of wind, unless we understand their basic meanings. This is to be seen in all that lacks actual totality, TO. How is and truth exist by the mass energy form of matter and mind, subject-object and reality modes of TO. How the mass energy continuum forms matter and mind of AU and TRU to exist in actual totality, TO. A good example of the deficit in truth, is to be found in the definitions and meanings of word 'is' and its root, ' to be'. This is highly important, for totality is, and all forms of the base to be, convergently focus about a center of all that enables best human life and the mind's relation, part and role in TO. Thereby 'is', should be not only to exist, but to exist essentially, by the actuality that is TO. However "is," is variably used in everyday life. It is used as an identity relation or simple connection, only vaguely as truth. As much as is, and to be, were the centerpiece of Parmenides' path to wisdom, they were only a weak, but very stimulating beginning. The observed "is," lacks much truth, the spoken and written word often lack even more. Only when recognized and used, given as some continua does to be separate what is the truth and actual, from the necessary, any existence and non existence. Such continua are fundamental to the state of totality, between most specific and general. It provides proof of the reversible role of the most specific TRU and TIT in TO and TOT. When most specific "is" shows the graded differences between this special world, in and of itself, TO, TIT, and all others. Is, like that for true, denotes the identity of sense and thing, subject and object. Yet identity of sense and thing, subject and object, and even what is thought to be true, is far less than, or disproportionate to its actual pure essences. It is apt to be an incomplete, poor, misleading or false representation. All definitions, meanings and representations must be none other than their part and role in TO and the variety of TO they apply to.

Without being definite, specified by the reality of combined objectivity and subjectivity, signified by symbols, names, words and language all lacks meaning, what is necessary for actual existence in TO. Meaning is how the senses act upon things and objects to subjectively deduce their form, function, idea and essences to produce concepts or words. Words and concepts that exist as actual pure essences frequently and critically induce potentialities to become actuality in, and often prove most important to more particular aspects of TO.

It is from need, the essential, practical and functional values of life and human being that all the difference in the world separates any existence from the necessary existence that most characterizes and proves actual totality. Without the step to necessary existence there can be no actuality. Without necessary existence and its actuality there can be no actual pure essences, whose ever greater qualities by relative totality are what largely determines actual totality. The categorical imperative signifies the combined necessary in existence with the categories of actual totality. It is the power produced by both, and is a large part of convergence on actual totality. As a large part the categorical imperative helps guide and force all to be their part and role in actual totality. It also tends to make actual totality what it is.

The search for ultimate unified totality is nothing new. For many thousands of years humans have searched for basic and hidden meanings. What actually exists, to what extent, how and what is actuality? Behind the world we are in, and the world about us we have correctly been enthralled by new revelations that expand our understanding. It has been the driving force of religion, philosophy, science and their uses in life. So it is today, with all our knowledge we still live in a world of oppositions and conflict much of which stem from a lack of knowledge and consensus in what the world actually is. What is necessary for existence, and what is actuality? This is why we must continue to search until we find out what the ultimate unified totality is. Why this is what we call actual totality. Among these hidden and basic questions and meanings none have been more invisible and unknown than their actual pure essences.

CHAPTER 4 - ACTUAL PURE ESSENCES to Totalities,

Approach, Meaning, Distinction, Essential, Development to Greatest Meaning

Meanings, subject and object, actual pure essences, mass energy, realities, and especially it's Á continuum and dimension, are different aspects of the same feature of TO. Their different aspects identify with different varieties of TO, TOV. Meaning stimulates learning as learning increases the form and power of meaning. Together they are the core, and proof, of TO-TOV-TRU-TOF.

Actual Pure Essences are basic to ÁLT, SPÁLT, ACEXIN and representation of the mutually interdependent proofs in TO. They converge on the predominance, success, well being, improvement, perpetuation and survival of the whole, TO-TRU with TOF. This is the Law of Mutual Reinforcement of Meaning, Actual pure Essences and Representation, LÁMER. It is how mass energy becomes real, by combination of subjective and objective, to form the Á dimension and reality in, and proof of, actual totality.The meanings in developing actual pure essences are the most effective way of understanding and showing all, as TO, with the comprehensive precision and overview its greatness deserves. They show and prove TO by the fact that all exists not as things or objects in themselves, but by the actual pure essences of any and all, whose potential becomes actual by the actual pure essence acceptor in TO, convergent to its actual pure essence form and function. This is evident in the ABSTO-RELTO bond that gives unity and reveals the increasingly generic to specific varieties of TO.

	Observed Not There	What Appears There	Actual Pure Essences
Observer Sees or Senses	Our senses often in error, unselective and fool us	Common human error of entities, things and objects	**Only Viable Recognition of what necessarily exists, in the Actual Real World, TO**
Does not Observe or Sense	Denies ones mass - energy power in life. Is not, and sees nothing.	Failure, delusion, trap, of accepting any and all as is	Failure to realize the essential, thus Meaning of Life, or Actual Totality, TO

Table 4-1 Definition and Meaning of Actual Pure Essences

To know what is and is not actually there, here, or where is crucial for life, a continuous learning process of greatly improving our ability to survive. Much of our encounter with the world is faulty, figment or illusion, only appears to be, or is much less than what actually exists. We thereby constantly mistake, disproportion or misjudge life and only get a small fraction out of what is possible. Here is the key to actual pure essences, the meaning and proof of actual totality, TO, and our part and contribution to it, TES. It is by essences that the pure, actual and real of life produce the necessary identity. It is how we understand the mass energy continuum of subject-real-object, or the A dimension of the actual world, TO-TRU.

By discovering their correct dimensional form, TRIT, knowledge, representation, presentation, operation and application become more likely and accurate. Our lives have greater meaning, success and survival. Much of this we take for granted, yet this is largely by observers that do not sense, the observed that are not there and what appears to be there that is not of its actual pure essences. Actual pure essences must be their part and function in TO. Here is the most critical error we make in life, failure to understand TO and its actual pure essences. For unless we know what actually is, and are not, and know it with the greatest of validity, proportionality, actuality, totality, interest, wisdom and control we continually fail to be the best we can be and fall short of being the best we must be.

Actual pure essences are the qualities that typify an existent or entity, necessary qualities of actuality and purity that make it what it is. They are those practical and functional meanings that give purpose and value to an entity, thing, sense, object, subject, part or component, aspect or category, totality or actual totality, TO. They are what is essential for its existence and make it what it is. Those purposeful qualities and actions of actual pure essences that increasingly converge on higher levels, and have all that is typical of TO. It is the power and nature of actual pure essences to bring and converge all necessary to its existence in TO. Purpose like much in actual totality has many stages and levels of development, a form and function typical of the transformations and transitions that occur in TOT-TO, absolute and relative totality, actuality, and mass energy. These are more or less correlated, depending on their part and role in the development of TOT, TOP-TO-TOF.

Actual pure essences are the core qualities all must have to be the entities, totalities and TO they are. They are those qualities and meanings that have functional value, practical use and approach the reality of actual existence rather than appearance in any and all things. What they are in different entities, things, totalities and stages of TO exist as a spectrum, from most simple to most complex. It is important to distinguish between these different states. Practical use and functional values exemplify and prove actual pure essences, as these selectively become their part, roles in, and prove TO. This is what a person sometimes is inclined to do knowingly or routinely in their daily lives. This proof arises from the most fundamental characteristic form and mode of TO, actuality, by the primary, Á, mass energy continuum and dimension.

It is a series of quantum leaps between the stages of necessary existence, any essence, actual pure essences and these to their actual pure essences in and by TO. This is the Law of actual pure essences in TO, LAPET. It is one of the most important laws of, and proof of TO. This series largely follows the ontogeny and phylogeny of human life, and the growth, and proof of, relative totality in TO.

There are many steps, to and levels of, full actual existence in TO and TOT, from non existence to irrelevant existence without specificity and increasingly needed existence, actuality without essence, increasingly essential for actual existence, increasing union of essences, actual essence, focal and formative, pure essences, quality, formative, actual, pure essences separate, formative and actual pure essences, potential combined to TÓ, and actual pure essences formed, and set in TO, APET. Also there are transformations that occur with actual pure essences, TO-TOV-TOT, APETOT. The first three are without essence, the second three, actuality without actual essence, the third three, formative and potential actual pure essences, and the last three stages of existence in TOT and TO of increasingly higher levels of actual pure essences, differentiated and transitory in time. It is by these great transformations from non existence, through levels of existence, essence, actual essence and actual pure essences in TO that we can most readily recognize and prove the exclusiveness and greatness of TO, TOV and TOT. These series and gradients are revealed in the varieties of Totality, TOV, developing in TOT and selectively existing in TO. This progression of actuality from its incipiency in the barely existent to its most profound state and condition makes TO what is the great and glorious ultimate unified totality it is, internally all inclusive and externally yet separated and exclusive of all else.

The actual pure essences of an entity, thing, sense or object are different from the actual pure essences of this same entity in actual totality. What its actual pure essence is in actual totality is most important. What this is of the entity is less important, although may be of potential and relative value. The importance of the actual pure essences of an entity in actual totality, is often of greater importance than the entity itself. This is why I as the first person may be of little significance, my actual pure essences great, my actual pure essences in combination with that of many greater, and my actual pure essences in actual totality of the greatest significance. In this way actual totality consists of proportionate parts that are not of entities, things, senses or objects, but of their actual pure essences in actual totality. This is multiplied by the number of times it is a member of a kind of entity, thing, sense or object that is proportionally necessary for the existence of actual totality. This is the key to the meaning of the Law of Partial Necessity, LOPAN, and a major proof of actual totality. This important proof confirms what actually exists and doesn't exist, and their proportional value in TO. It also confirms what is so quickly forgotten, actual totality is not only totality and unity, it is all they consist of.

All is changed by the typical form and function of actual pure essences and their role as parts and processes in TO.

The objectivity of the absolute and physical and the subjectivity of the relative and mental whose actual pure essences are received and combined in those of actual totality, TO, certify and prove the ultimate unified totality that is TO.

Any thing or object means little by itself. It is how it relates, interacts, works and has mutual benefit with other things and objects that gives it actual existence. How it relates, interacts, works and produces mutual benefit is its actual essence which, when optimal is its actual pure essence. This is the crucial basis for all. When of a totality, or actual totality, TO, the actual pure essences of separate things adapt, when received by TO, by the actual pure essences of TO and its parts associated with the reception. All is little in themselves except, and to the degree they relate, interact and benefit as each and all are TO. The form it takes in TO partakes of all kinds, varieties, and is largely a product of RELTO. This form typifies and proves TO, and makes it unique, exclusive, and relatively absolute amongst all else.

It is how the actual pure essences of all come to be their parts and roles in actual totality that most greatly reveals, and provides the reality, by which humans, the mind and all that is of relative totality determine actual totality. It proves how actual totality is not only a physical universe, or separate absolute totality. This is the Law of APE proof, by RELTO< of TO. LAPERTO. This is how actual pure essences initiate, generate, improve and perfect actual totality.

What actual pure essences mean in actual totality is, how all things, objects, you and I as components of TO, have actual existence not as things, objects, you and I in us, but how they and we interact, relate and mutually benefit each other in TO. It is the essence, practical value and functional use of all in TO. A simple example is the function of our anatomy, as the clavicle's protection of the subclavian and the chest cage's protection and support of the heart and lungs. Body parts usually have many functions whose essence is to enable human life. The whole of these and all its actual pure essences equals the whole of TO. All exists actually only as they exist by, and in, TO.

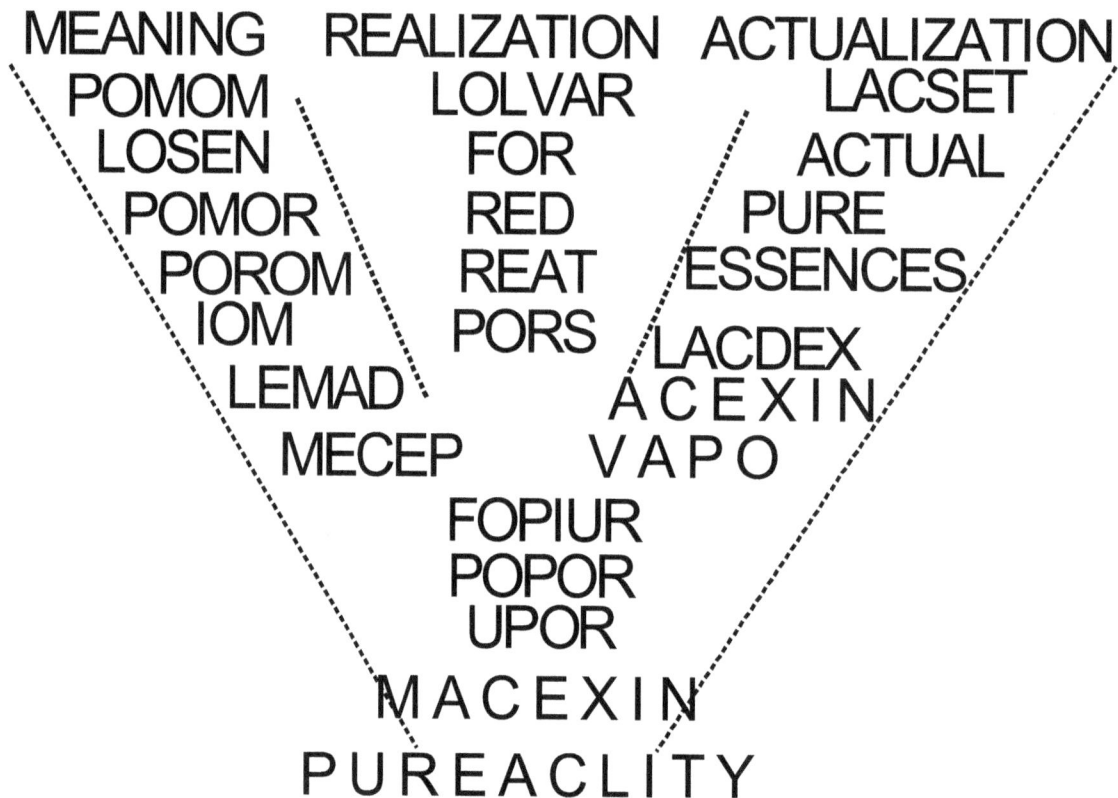

Fig 4-1 Actualization, Realization and Meaning, their associated concepts and products.

Much of what enables an understanding of totalities produce existence and its progressive approach to TÓ. For TÓ actualization and realization based on the actual and real, actuality and reality, take on new and even more powerful meanings. To acquire the gist of these is to basically understand human life, as it is and must be by actual totality. We tend to think of meaning, realization and actualization as separate ideas denoting things in themselves. However this is our error, for among all that is close to and great of TO these are its essentials. Meaning, like is, is what and how well there is an identity between subject and object. In addition there is great intention in meaning, as in Instantaneous Attention, IA, and this is convergent on being, our being in TO. Realization is how profound and accurate the bonds are between subject and object, and how well it overcomes language insufficiency, (LIS). Actualization is how well or lifelike all is, whether it is, or is becoming TO. TO, meaning, realization, and actualization come together in the mass energy, S-O form and function, FAF, of TÓ, to give it the predominant power it needs to compete and survive. There are many steps between how they exist separately as any entity in itself, ENIT, what their actual essences, ESINS, are, what they are in TO as states and mechanisms, and how they fuse in the continuum of Á and total dynamics of TO. Among other concepts the following signify steps in this progression:

POMOM - Principle of Multiplication of Meaning,

LOLVAR - Law of Limited View and Reality,

LOSEN - Law of Experience and Meaning, reinforced,

FOR - Fallacy or limits of Reality,

LACSET - Law of Critical Separation Non Actuality from Actuality, or illusion, figment and non or less than TO,

POMOR - Principle of Meaning Reduction,

RED - Reality Deficit,

POROM - Principle of Reinforcement of Meaning,

REAT - Reality different from TO, only the same by common essences and inclusion TO,

LACDEX - Law of Accumulative and Accentuated Actuation of Dimensions as they work together, in TO.

IOM - Intensity of Meaning,

PORS -Principle of Reality Substantiated,

LEMAD - Law of Enumerations, Meanings, Definitions and Formal Distinctions well Developed and ordered to show concepts, principles, laws and mechanisms of TO.

ACEXIN - Actuation, Extension and Intension or higher states and powers of MLT and ALT FAR in TO,

MACEP - Meaning is Corrected, Proper and Perfected in Completion, as it best Enables TO,

VAPO - Validity, Actuality, Proportionality and Totality, the overall reduction of that experienced to TÓ

FOPIUR - Fusion of Perfect Image and Ultimate Reality, POPOR - Perfected Perspective of Reality,

UPOR - Ultimate Perspective of Reality,

MACEXIN - Maximal Meaning ACEXIN, as of in and out of TO, OITIO,

PUREACLITY - Purity, Reality and Actuality or optimal state of highest form of mass energy,
S-O, Á in TO and TES,

It is not necessary to know many new concepts initially. They do help to appreciate, prove and understand the progression, interactivity and fusion of multiple factors in TO and its actual pure essences.

The Sums of TO are its own actual pure essences, (SAPETO). The Sum of TO is also the Sum of the actual pure essences of all its components, (SAPETE). Actual totality, TO, equals SAPETO times SAPETE. ESINS when united in TO are converted into SAPETE, as SAPETE exist in equilibrium within SAPETO in TO. This is how everything, or all entities, become potential things, TOT, to become actual constituents in TO, It is how a person becomes their soul, as their soul becomes its counterpart in God, if and to the extent God equals TO. Actual Pure Essences provide levels of knowledge of what is, TO, and the path to all that enables best human life, TRU, in its most unique and important role in TO, and increasingly in the future, TO-TOT.

Meaning and realization may at first glance seem synonymous. Yet they have distinct significance, especially in TO. Meanings are the overall bonds between subject and object and how it is of practical use in life, actuality or TO. Realization is an intensification and particularization of the subject-object relation, S-R-O, and application to life, TO. Both are of the mass energy, especially Á dimension, and major mechanisms that give realization power. Reality also takes on somewhat greater quantitative and important meaning as it progresses from existing concepts, OC, to more meaningful essential concepts, EC, to concepts only as they exist in TO, ECT. For whereas reality commonly used denotes all that is known, identifiable, and existing, reality of TO signifies the whole fused region of fused higher subject-object form and function in TO in relation to space time. This gives it the identity and inner strength we are used to, it means to us in life, and proof in TO.

The world we live in is highly anomalous. This is because it is both the way it is, and the way we observe and interpret it. It requires actual pure essence and identity with all actual totality to avoid being anomalous and abnormal. All that lack actual pure essence and are less than TO are anomalous. This includes parts of, and proofs of TO, which if not their precise component and role in the whole of TO are anomalous. This is a reason why actual totality has never been understood before, and why it takes much more than a passing interest to learn and know TO.

When we can envision the actual pure essences of all that exist in, and the whole of, actual totality we can begin to better understand all. We can begin to have and appreciate sufficient actual pure essence of actual totality by knowing their necessary existence through practical and functional values.

What we observe or sense is very limited in many ways. This is why we, and much of our knowledge, are often wrong. The limits, and lack of limits that obstruct the correct understanding and order of things include, the infinite divisibility of all, the infinite extension of all, the infinite conditioning of all, the lack of absolute boundaries of all, the relativity of all, the dynamics and presence of change and variation of all. There are clines, classes or kinds of existing in things, which determine further relations and limits, they also determine its order, and are factors in the creation of life. With change and variation there are redundancy and tolerance, change occurs not infinitesimally but by steps, like those in quantum mechanics, one existent does not occur initially, only after factors making it reach a certain limit, or level of strength for change. This is associated with a maximum and minimum of all. With limits and lack of limits there is that lacking actuality, purity and essence arising from lack of necessary existence, the functional and practical state of things, and quality or degree of purity. All these limits, and limits of limits by what they are lacking, their divisions and kinds fuse by converging on their least common denominators into one, the totality they are in. This is the way in which limits, divisions, kinds of actual pure essences become, exist in and prove the totality. They are the basis for the quality, consistency and existence of actual totality.

Actual Pure Essences are the great pivotal proofs of actual totality. They bridge the gap between the incipient actuality of necessary existence and the highest stages of select, qualified and typical actual pure essences in TO. They subjectively join relative totality to the objectivity of absolute totality by the relativity of observer-observed in the mass energy expansion characteristic of actual totality. It is the reality of this subjectivity-objectivity or observer-observed relativity that makes actual pure essences what they are in actual totality. Their ever greater meaningful relations are what is lacking in the physical universe and absolute totality. The ever greater contribution is made when the actual pure essences of the physical universe and absolute universe are transformed and modified to become their counterparts in actual totality.

CHAPTER 5 - Different TOTALITIES to what is External

Universes, Deities, Philosophies, Systems, Worlds, Deficiencies and Disproofs Compared

As existent entities combine they form new entities, or totalities. As many of these combine, they attain a level in which all is included in one totality. When the process is finished, completed totality is formed. If the complete totality contains all existence that has their corrected actual pure essences, it becomes the ultimate unified totality. If this satisfies the combined absolute and relative actual pure essences of all, it is TO, actual totality. If actual totality is exclusively limited to the present, it is totality proper, TO, or TP. If it is more or less undeveloped or existent, interactively becoming actual, it is potential totality, TOT. When less than these it is any, or different totalities, which may or may not be potential or have partial existence in TO. Thus, there are many kinds of totality. Yet if we seek the ultimate kind of totality, which we must, if we are to have adequate knowledge and understand all, converge and prove how all that exists and its totalities are not separate, but do so only as they are in actual totality, TO. To this end we need to search the best of current knowledge to analyze and interpret those great aspects of totalities that are held to be the only ultimate one. Of these are world, universe, deities or god, isms or philosophical systems, and others. To show and prove their great variety, how they are important and how they are lacking, or how they exist in the ultimate totality, is a continuing theme of this book.

Many different ideas and concepts of totality come from those who hold many, often diverse and opposing beliefs of universes, philosophies, systems, gods and worlds. Much is right, much is wrong and more is a mixture, depending on their part, role and use by and in TO. This becomes clearer when we discover and compare them with TO, learn their deficiencies, errors, and how they become, can be lived by in the near, mid to long term of TO and with TOF.

At their best the physical universe, philosophical systems, Gods or any world, are only their proper parts in the whole, of TO. We cannot be satisfied with only an objectivity of any actuality. To encompass TO, it takes the reality of a combined subjectivity and objectivity of those actual pure essences that characterize the actuality of TO.

Each of these worlds and isms began from nothing, or from nothing, chaotically or incrementally by TO. Each began from nothing in their partiality and separation to acquire increasingly more aspects, features, form and function equivalent to their part in TO. The tendency to converge on TO is limited by their separation from TO, partiality and degree imbedded in trivialities, traditions, errors and miss beliefs.

The most apparent proof given by a physical universe is the fact that all else vanishes in time, from dust to dust, all becomes some form of matter. This apparent proof of absolute totality exists without relation, no observer in boundless infinite time and space. It is a chaos without any center, essentially non existent and incomparable with its opposite, relative totality. This helps to prove that absolute and relative totalities cannot exist without qualities of the other. It is this disproof of an exclusive separate physical universe or relative universe that is the beginning of wisdom. This is wisdom based on the combined features, aspects, form, function and actual pure essences of all, as they exist in actual totality.

The ultimate correction of these absolute universes and totalities is to be found in what they lack, what they have that is different and proofs from TO. The most convincing correction or disproof of an objective, physical or absolute universe, is its failure to account for the mental, subjective and relative, how the key modes and continua of TO, such as actuality, objectivity-subjectivity and reality enable it to be. It is the power of being alive, knowing that we are alive, knowing that we know, and the sum of all the highest powers of cognition and actual pure essences that give humanity predominance, and the potential for eternal development, growth, perfection and fulfillment.

As animals and humans attended to and became increasingly aware of their existence, surroundings and what was needed and essential, they selectively sought order and unity in their relations. This gave better orientation in mass energy, space and time plus convergence on what has the greatest significance, as gods. In time comparative gods and pantheons led to a highest god, and eventually a unitary God. Mostly through trial and error this unitary God has become what we have today, a totality that more, or less, approaches the relative absolute, RELTO and TRU of TOV, as GOD in TO.

Because there is a great lack of inclusion, exclusion, proportion and actuality, as well as many errors and differences, the unitary God is not uniformly held by humans today. Both because of differences in ideas of what God is, and the many diverse systems, sects, traditions, beliefs, interpretations and practices each religion and culture is in opposition with others. It diverges from the least common denominator of belief in one optimal totality. For this reason and the bias of excluding much of the physical universe, a greatly needed actual totality from present day knowledge guided by a unified totality has failed. As great as God is, and its contribution to the world of human life, it has been variably and poorly defined and applied so that interpretation can only be vague and variable. It does not provide the necessary form, function, distinction, meaning, actual pure essences necessary to provide a full and accurate totality and guide to life.

When we speak of physical or relative universes, the absolute, God, world, being, or a unification theory, to what extent can any of them be right, or certain? As diverse as they are, all, or all but one must be wrong, although some when qualified may partake of all, as in TO. Can any of them be the one, ultimate unified totality with validity, actuality, proportionality and totality? It is this most fundamental doubt that has led to false beliefs and confusion, in our ideas, words and action. It has made consensus and proof, impossible. There has been a consensus in various ideas of one God, world, universe or totality, and how mankind should live within it. Yet their deficiencies and doubts leave us with varying levels of a possibility, less in probability, certainty and a fallibility in identifying ultimate unified totality. This is why a new and proper totality must be sought, produced and promoted, the correct, proper totality, actual totality, TO.

Human ideas of religion, the spiritual and God have arisen from, and grown by, an increasingly firm conviction in the exponential power of actuality; those actual pure essences that form and support all that enables TO. This is mostly by relative totality and best human life, or TRU. The greatest of these ideas conform to the validity, actuality and proportionality of TO, and the part and role TRU has in TO. This conformity by specificity with coexistence is the greater, to the degree, they fulfill the actuality and reality of their existence. It is in this way we can recognize the incipiency, and contributions of expanding knowledge and refined religion by their part and role as coexisting equivalents in TO. We can even more importantly prove the role and force of the form and function of TOV, and all that contributes to TÓ, and its uniquely qualified form and function.

Along with qualifying the role of God and other worlds in TO, they are to be recognized by the fact that although they in their greatest and actual pure essences sense, are of less than much that is of actual totality, TO. They lack much of all the rest that is TO. God has an accumulation of diverse ideas whose discovery was a long process that arose from the human mind and culture. As knowledge accumulated, especially with civilization, cohesion, unity and security were increasingly needed and demanded. So did the part and role of God give those societies and cultures an edge, the powers needed to succeed and predominate. Yet we must be careful to recognize that God also developed with TOT to be like TO, to encompass all ABSTO-RELTO, all that is TO. This is both evidence for, and proof of, 1. the transitions of TO and God, 2. The right, or wrong, kind of God as TO, and the vital necessity to bring and hold both God and TÓ, to actual pure essences and proportionate forms as the ultimate unified totality or supreme being, TO with TOT. It is the belief and use of the many philosophical and religious systems of contemporary thought that best reveals and proves their differences from TO. What their errors are and what is needed to bring them to TÓ? Such an analysis bridges the gap between the unity of TO, the finite predicament and infinite regression of differentiation. It is such a direct approach to the most basic form and function of TO, differences in existing knowledge and knowledge of and by TO that best prove TO and all. Many of these philosophies and religions are extremes of continua, which when reduced to their parts and roles in TO are highly effective means of showing and proving TO and their roles in it. Some of these include, Monotheism, Agnosticism, Atheism, Fatalism, Materialism, Capitalism, Universalism, Empiricism, Realism, Dogmatism, Hedonism, Ethnocentrism, Altruism and Rationalism.

All are less than TO, yet when of their counterparts in TO are important and valuable proportions or contrasts in the whole. This depends on the degree to which they are the same as their part and role in TO, and the degree to which they are negative or positive of TO. For example monotheism is a polar opposite of polytheism, atheism that of theism, and materialism that of mentalism. To comprehend the whole of TO is to resolve the spectra these continua and their differences produce, and the non existence of separation in universes, systems, philosophies gods and worlds. Most are further described in the following proofs of TO. Isms prove TO, as TO proves isms. This holds to the extent both are reduced to their part and roles in TO and how well we recognize their equivalence with TO, TREQ.

It helps to know from what totality is derived and what are its most basic needs, from the physical universe, God or some existing deity, the world, or a system of philosophy or knowledge? Even at best these, and all that have actual existence, is only a part, although often a large and important part. It is what the ultimate unified totality is, what its most basic needs is, and how it was derived. It is the exclusive and only totality that fulfill all that consists of, and is essential to, actual totality, TO.

Much is of mixed totality, externally, internally of TO, less, non, or negative to TO. This is evident in the interactions at the receptor and acceptor sites of TO, the action-reaction relations between TO, TOT, and all else. It is especially clear when viewed from Figure SC-1.

One of the greatest disproofs of God, Allah or any deity being the supreme being, ultimate unified totality or TO is their diversity and divergence, how little and poorly they seek consensus, or focus on differences and not on similarities. They suffer from denial with stubborn beliefs, whose defects, deficiencies and partiality cause continual opposition, conflicts, wars and suffering. This results in the very disunity and chaos their fundamental union in totality, peace and harmony had sought to overcome. It is opposite to the kind of vertical integration that is basic to actual totality. It is by the right kind of rational control rather than any choice by religion that positive rather than negative results occur. It exemplifies the great inefficiency of any choice and poorly vertically integrated totality. A totality whose correction to actual totality is a major necessity if mankind is to survive and have well-being and permanence. This should be their providential potential.

The greatest comparative proof of TO with what is, or propounds equivalence, is if totality can be one, it must be, the right one. It must contain all that is right and none that is less or superfluous, when adapted by its aspects, features, universals, qualities and modes such as actual pure essences, in actual totality. The actual and real God is only what is equivalent to its part and contribution to actual totality, TO. When raised to equivalence with TO, God may be spoken as GOD, to the degree and by the condition of being devotionally identical with TO.

When viewed by their differences, actual pure essences and corrected parts and roles in TO, all worlds, universes, philosophies, systems and religions have valid, proportional actual and great presence in TO. By comparing their parts and roles from non existence to any existence, to TOT in all its transitions, to TÓ that may seem to disprove TO, or are inimical to TÓ, the various worlds, universes, gods and systems can be comprehended properly. These differences, and the great enigmas created between the physical universe, God, and other worlds and totalities, are only resolved when, the physical universe achieves the reality of subjectivity, God and idealism achieve the reality of objectivity, and these only as they exist and are proven by their combined reality in the actual pure essences of convergent absolute and relative totalities to be actual totality, or TO.

God is claimed to be absolute and perfect. Unlike God, TO does not claim to be absolute nor perfect, only aims to be relatively absolute and perfect. Unlike God, TO provides more fundamental and ordered proofs of its form and function. Yet, neither are fully known, TÓ acknowledges this difference, God proclaims secrecy of its knowledge. This approach to knowledge is a most important difference, a difference whose imperative fulfillment, is a major proof of TO. Until our knowledge of TO equals its perfection, we and our knowledge are only potential, not fully actualized in TO. Many errors will be made to the degree we do not recognize these differences and correct every instance in use.

Three different views of the same object or kinds of totalities and the related things and senses about it are shown. The difference is one of perspective. In the first view we look at it from afar off, so that it is small, the things about are larger and sensing the predominant portion. In the second figure we envision the object somewhat closer so that it becomes larger. At this stage or level of telescoping in on the object, it, its surroundings, and its subjectivity are equal. In the last figure we look at it very closely so that the object is very large, and its surroundings and subjectivity are barely visible. This is what occurs continually in and out of objects and totalities, in the flux of life, as we are repeatedly stimulated, face givens, and view all that is before us. These are highly variable more or less the same or other than the entities in and of themselves, ENIT. As we look at an object or totalities, e.g., the world, it becomes larger and larger for we see more and more of it. Its form, function, characteristics, properties and all become clearer. It becomes more 'real'. Yet the object and entity are relatively and actually changed only to the extent they become, or are, TO. The first is like our perspective of absolute totality, one largely relative to our perspective. The third is like the object or totality itself, devoid of the external or our viewing of it. The second is a combination of the two. A fourth figure could be included whereby human existence is little or great depending on the optimal union their actual pure essences produce in and by actual totality, TO.

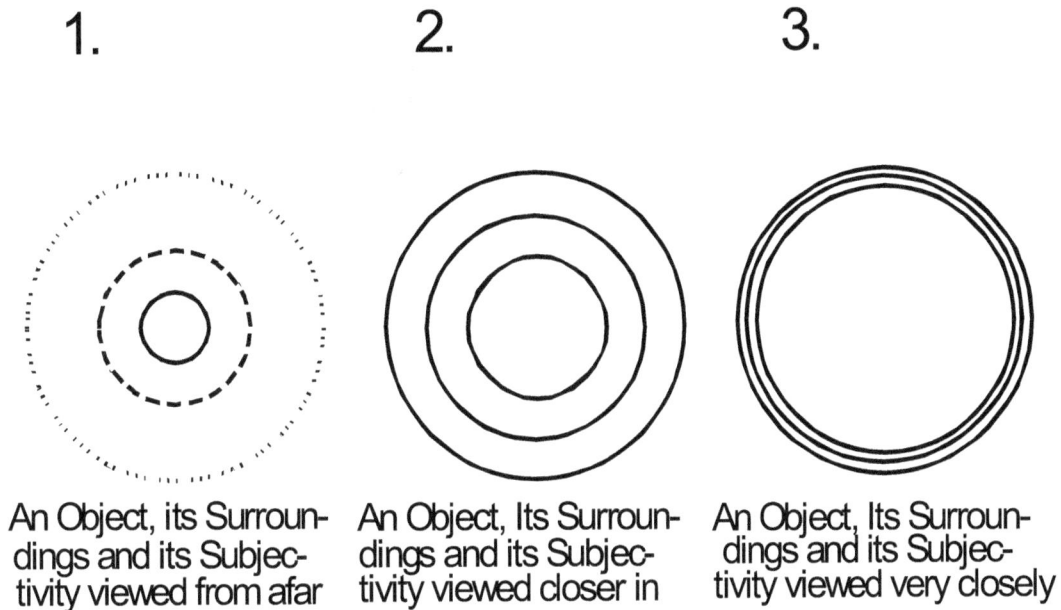

1.　　　2.　　　3.

An Object, its Surroundings and its Subjectivity viewed from afar

An Object, Its Surroundings and its Subjectivity viewed closer in

An Object, Its Surroundings and its Subjectivity viewed very closely

Fig 5-1 Comparative Perspectives of objects, or Existence of Totalities, Absolute and Relative, External and Internal

The problems of separation or inclusion, different aspects of the same thing, and the dynamic of more and less, are continuing themes that occur all through the in and out of existence, totalities and TO. All three are highly related, by both our thinking, knowledge and actual totality. More or less is described in Chapter 2, same and other in 8 and 9, and actual totality in all chapters. What is more or less is of different kinds, and laws, e.g., LAMOL, LANERA. It arises from the never perfectly even dynamic state all is in both external, internal and in and out of actual totality. This produces the often fleeting and to and fro affect on equilibrium that may seem to disprove certainty. What is more or less will depend on whether it is of any entity, a totality, potential, actual, TOT, interactive with TO, in TO, less than TO, or within the TO-TOT transitions. It also depends on the variability of object and subject and their interactions. It is partly the role of reality to bring these variances to meaning, order and TO. It is partly because of these aspects of more or less that worlds, universes, systems, deities and all are at variances with TO. When their separate existence is understood with their actual pure essences and more or less is adapted, along with their proper existence in TO, all becomes more certain and better proved.

The relation between a completely separate absolute totality and the physical universe is most important to understand if we are to divest ourselves of the seemingly exclusive existence of the physical universe and its opposition to all else. A completely separate absolute totality is devoid of the relative, thereby like zero and absolute, non existent. The physical universe does not exist without relativity. Relativity exists in two forms, limited to the physical universe, and that which is generally distributed in all totality. When limited to the physical universe it is only a mode or a part. When limited to a general distribution in all totality, relativity is also only a mode or a part, but raised to a typical level of all totality. For this reason neither absolute totality, nor the physical universe alone can exist as the ultimate unified totality. In a similar way reversed, relative totality lacks existence for want of features of the absolute, an absolute state, 'absolutivity', essential for its existence. When both absolute and relative totality, the physical universe and cognitive universe focused on best human life are corrected by that lacking in each, fused and united by convergent actual pure essences, the ultimate unified totality is the proper totality, actual totality, signified by TO, and expressed as law, is, LÂRPET.

Even more to the point is relativity as both part of the physical universe, $E = mc2$, and as the perspective dependent on relative totality, or the relative universe, in TO. This is the combined proof of TO by its being both object and subject, material and mental. Most great topics of common knowledge, are more or less great constituents or components of actual totality. This depends on how well they are adapted, modified and altered, and how much their actual pure essences coincide with those in actual totality. Among these is the physical universe, the world of best human life or the relative universe, science, philosophy, God and religion. Many other topics of knowledge likewise more or less proportionately exist in actual totality. Many of these and their transition to necessary existence as larger or smaller parts in actual totality and its potential transition are noted, described, or shown as proofs through the book.

If we can come close to the veneration each of the different universes, worlds, systems, deities and others have for a unified doctrine signifying all, we would come closer to the right, "actual" totality. The physical universe, the worlds each and all persons perceive themselves in, the philosophical and other systems, and the Gods of the many societies, groups, cultures, nations, languages, etc. all have features, premises, ideas, beliefs, forms and functions that are more or less highly important and helpful in reaching the right ultimate unified totality. Yet they are so diverse, often in opposition, and the current population on earth is also so large and diverse that their differences are greater than their sameness, and their oppositions to one another is most damaging to each other, and to their ultimate purpose, the right unified totality, as actual totality proposes. With their combined powers of veneration, devotion, good deeds and well being the whole world, universes, systems, deities and all would achieve the purposes they aspire to and search for. From these venerable institutions of human life has arisen much of the greatest knowledge of today, which when modified by their existence in actual totality adds to their correct proofs along with those of, and completion in, actual totality.

CHAPTER 6 - EXTERNAL of Different TOTALITIES, to Internal

Approach, transition, transformation, form, separation, relation

Whether there is anything external to a world, universe, god, system or totality that is held to be all, or is only one, is like whether there such a thing as non existence? There are many factors and approaches involved, it cannot be answered with a simple, yes or no. The human mind likes, yes and no, answers. They are so decisive. Yet worlds, universes, gods, systems and totalities, especially when they are combined as actual totality, are not so simple. Together they are most profound, subtle and diverse, especially when they are separated. Thereby we must approach these questions with an open mind, recognizing both their importance and variance.

We must know and continuously remember, ECRO, the discrepancy between what is in, of, and less than totality, its varieties and actual totality, TO. To stand off and view them is different from their given objects, and both are often different from what TO and its categories actually are in themselves. This is increased by our own relation, in and out, of TO, as components of its special varieties, e.g., TRU and TIT. The terms are different from their totality, its parts, and all in between. Its parts and processes, especially when taken in themselves, are far different from both their actual state in, as well as their orientation and part in totality, TO, (DOFINT). All these are different from positions and analyses of the finite components in representation, SPALT. For proof we must include all these, and the inherence of knowledge that exists both by its own right, and by its APET with TO. Proof of what is separate, exists only by TO, or is intermediary interactively in this relative continuum. This is how separation of TO, in and out of TO, less than TO and a non TO are key to knowledge and actual existence, TO.

What is external to a totality, the totality, or especially actual totality is not the same as what is extraneous. External is positional, extraneous is a comparative state of existence. It is this extraneous state of existence rather than the external totalities and potentials to actual totality that is most antithetical to actual totality. It is the opposite to the exclusiveness of necessary existence, actualities, totalities and actual totality.

To know any world, or TO, all veneers need to be exposed to provide optimal accuracy, balance and actual pure essences in TO. Although TO and TOV are relatively absolute and exclusive, they do not exist without external relations, interactions and effects, We are in coexistence with nature and nature with us, like it or not. We and nature are both out of and in TO. So is it for universes, systems, Gods, and other worlds. How well all profit and have well being, depends on how well the coexistence is in equilibrium, proportion and degree we and all are our actual pure essences. This is the degree coexistence flourishes in TO, as all TÓ becomes stronger, greater and more permanent.

External/internal and potential/actual are the state and condition of actual totality that come in many kinds, forms and functions. It is prominent in the development of TO and how it maintains its highest level of existence. Much that is in opposition to TÓ is not external, and much that is external, is not in opposition. All tends to change under different circumstances, times and varying input and output, Such variation in existence does not deny other proofs of TO when properly modified.

Ontology is the philosophy of being-as-such, and for TO, ONTO. It is the study of what existence itself is, considered apart from any question of the nature of any particular existent, and the attempt to discover the fundamental categories of all being. Yet our greatest problem is not of being, so much as it is of what being. This is the difference between ontology and theology. We assume that many entities and things have a being in and of themselves, but we lack adequate knowledge of what and how it is made in, and what it is, the totality.

From figure 1-1, non existence, existence and actual existence is mostly the distinction between what is external and internal of TO. By TO, non existence, existence and actual existence, like external and internal are not 'black and white', one or the other and completely separate. They are highly related and interactive. Such a relation is not the same as relation by actual pure essences in TO. Herein is the crux of same and other, or different aspects of the same thing, DAOST. It is characteristic of TÓ, only highly selected and qualified by and in TÓ. Necessary existence becomes actuality, actuality becomes actual pure essences, actual pure essences of any totalities become their counterparts in actual totality. This is how what has necessary existence becomes, like DAOST, external to internal, in actual totality.

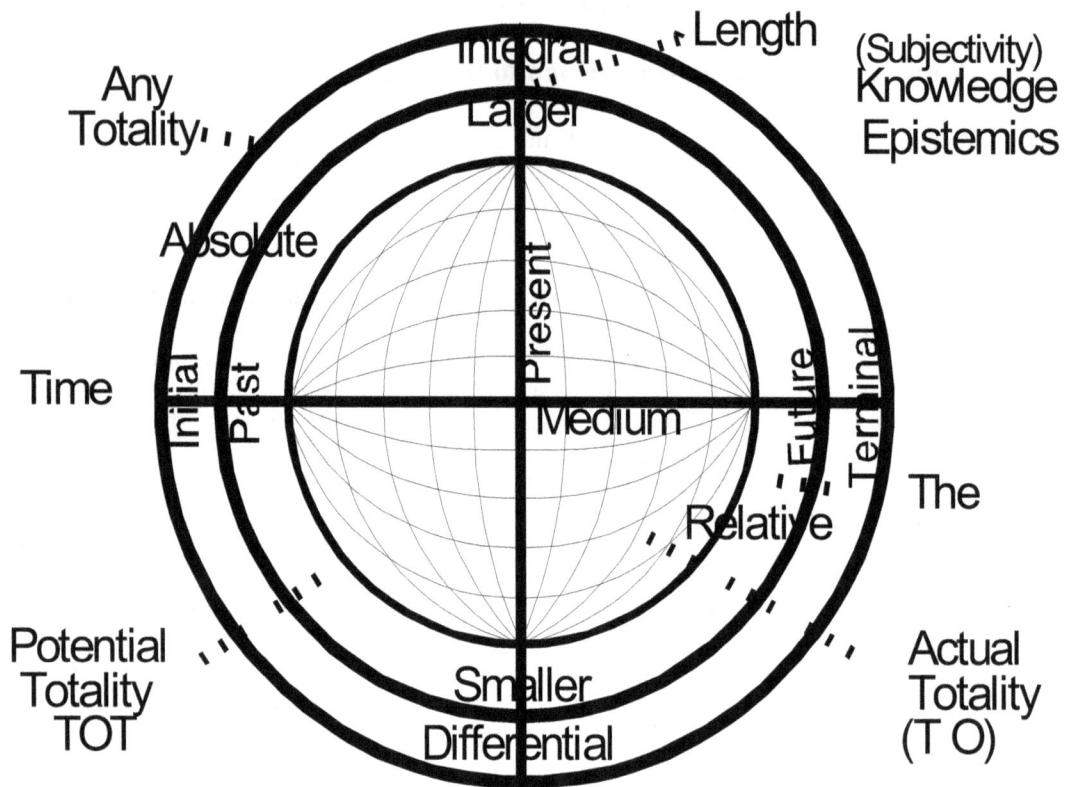

Fig 6-1 Ontology, or being, as a function of the Systematic Unity of entities, objects, universes and totalities

The Principle of Ultimate Coincidence, PUC, is shown by the fact that to an extent or for a while all may have some independence, yet sooner or later, indirectly or in some way they will have to do that which is necessary, a return to actual existence in harmony with the world around them, TO. This signifies, explains and proves the out and in dynamics of all, and TO. Since TO is highly dynamic and exists by APET, the effect of PUC is largely internal, only becoming external, or external becoming internal, when internal parts and external things are rejected or accepted by the processes of need and tolerance in the equilibrium dynamics of TO.

A. Numerous Superimposed Totalities or Components

B. Separate Totalities, Divisons or Components

C. Scale of Divisions

D. Eliiptical Totality of an infinite number of Forces and Parts

E. Fusion of Totalities
. TOT and TO

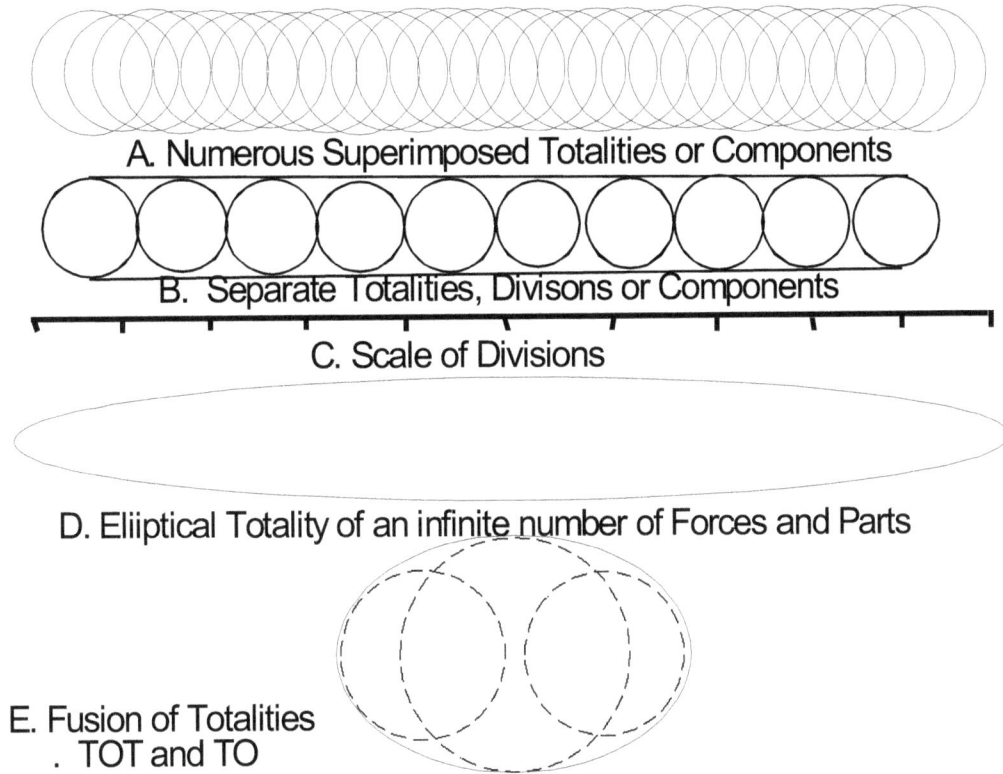

Fig 6-2 Divisions and Relations in Totality

A. Serial superimposed objects for totalities, varieties, or components, and change in time.

B. Separate totalities, divisions or component parts of equal size.

C. Scales, made of component parts show equal divisions, as between infinitesimal and infinite, or differentiation and integration. Variations of this scale depend on form, representation and math to show the needed aspect of totality.

D. Forces and parts, not visible, acting by integration and differentiation in totality.

E. Union of divisions of totality, or TO, separated and fused, e.g., varieties of TO.

When we seek to discover what is outside of totality we must recognize what may cause the limits and explanations of such a potential anachronism. Totality, assumes there is no outside.

Yet, when we make totality more definite by the distinctions between, an, any, some, and the totality, and further qualify it by actual totality we can begin to find, and provide some valid proofs. Outside and before the big bang mass space and time vary from their form and state within the current astrophysical universe. This variance includes the potential for reversibility. It is the reversibility and changing forms, whose creation is what propelled the evolution of TO that are central to its form, function and proof. This reversibility that is fundamental to all totalities, and subsequent components, is signified by the Law of Formal and Elemental Reversibility, (LAREV).

The state and condition of being either/or outside or inside of actual totality, and moving between them is signified by OITIO, out in TO and in out. The state and condition of without TO, or is and is not, as in contraries and dialectics and support of TO, is signified by SIL, or without Law. This law is quite obvious yet most important for quality of distinctions between what is and is not TO. It is by the contrast between SIL and OITIO, potential and actual, absolute and relative, and many of the polarities whose continua make up the actual universe or totality, TO, its varieties and TRU that they can be most clearly recognized, shown and proven. Potential and actual are not completely separate. They are two portions of outside and inside of TO.

TÓ exists interactively and dynamically in and out, with before and after, what is potential to actual, receiving and accepting by the receptors and acceptors of needed constituents and actual pure essences. Largely by the positive and negative of the external and internal, especially actual pure essences, selective acceptance and rejection are produced. It is this massive interface that is a large part of the dynamics of TO, its existence and transitions in time, TOT and TOF. This does not disprove the relatively greater homeostasis and stability of TÓ in the near and mid term of time.

Thereby in our categories of all words and things, and in positioning and operation, as in SPALT, we must remember that they conform to the law of essential separation. Whether they are, and the degree, they are inside of, in and out of, or outside of TO, TOV and best human life, and whether the forces of independence or dependence in TO, are the more prevalent and powerful. The degree to which they acquire and maintain their actual pure essence status, or actuality in TO. This quality is largely determined by the Law of Relativity and Reality of the objective-subjective in TO, LAROS.

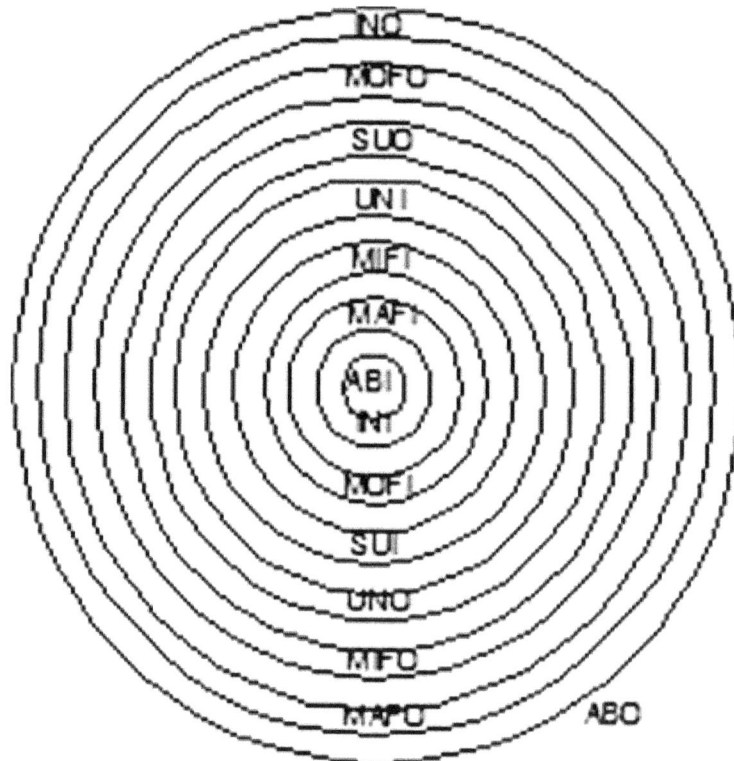

Fig.6-3 TOT, TO, TOV and Universes, Absolute Outside to Inside, (ABOABI)

In the various relationships of subject, object, unit, finite, infinite and problem solving, we set out on our quest to provide systematic unity. This can be done by, values and their uses that come from the simple proportions of all in totality. TO. Axiology treat measures of values or worth and help determine, quality, actuality and what is most to least important in objects, totalities and TO. They are proportions functioning by S-O, the mass energy continua, Á, to make choices and selections. They help to reduce all in a set state of a solid construct, as a model of TO. The diagrammatic representation of such set states that exist by relativity of subject and object internally and externally of TO, ABOABI. The observer observed relationship wherein the object can be interpreted from the absolute outside to the absolute inside. This must quickly be qualified by the limits of absolute extremes. A Standard Unitary Perspective or 1:1 identity of subject and object, SPUNO, is shown. From this we proceed through unitary, increasing levels of finite, to infinite outside, and from the inside unitarily, finite, infinitesimal, immediately and simultaneously as may be gained by envisioning this page totally.

Figure 6-3 shows twelve levels or spheres, to represent an entity. object, universes, TOT or TO. An infinite number of levels are possible, yet that presented is a number that is both representative and readily held in perspective. It can be done for other continua, subjects, objects, the Subject-Object, as well as TO and TRU. It can be used as a basis for, and help in positioning, such as in SPALT and SPAT.

Figure 6-3 provides a construct of all that can be developed from the systematic unity of a one to one relationship between subject and object. Of the continuum some of the different levels that can be developed are:

From the Outside, Infinite Outside, Innumerable Finite Outside, Numerable Finite Outside, Limited Finite Outside, Unitary Outside, Standardized Perspective Unitary Outside, and Standardized Unitary Outside

From the Inside, Infinite Inside, Innumerable Finite Inside, Numerable Finite Inside, Limited Finite Inside, Unitary Inside, Standardized Perspective Inside, and Standardized Unitary Inside

The spheres or levels shown in the figure above have names, yet are little noted in this book, cf. CIPTRU.

Absolute Outside to Absolute Inside, ABOABI, is represented by twelve spheres. The quest for systematic unity and perspective is approached in ABOABI. Increasingly as we view, or exist, from TOT-TO-TOV-TRU-TIT-TRIR-TOF, we are in ABOABI, and the reverse. ABOABI differs from the Absolute Need for Singular Unified Perspective and Presentation. The Primary Law of the Absolute Relative - what is above or about -, yet inside relative. TÓ exists, at least in part by TRU, and TRU exists largely by TO, both in transition together. The varieties of TO, TOV, and expanding TRU exist differently than what they are in the present. To learn and understand the great form and function of TO and its constituency is the beginning of total wisdom.

For each of the separate categories in figure 6-3, there is an outside and inside. The maximums and minimums when signified as absolutes are Absolute Outside to Absolute Inside, whose sum and whole are ABOABI, how closure is made in the progression. Such an overview and closure are key to the Law of the Absolute Necessity of Order and Standardization, LATANOS, This is derived from the form of TO and TRU, TRIT, and needed for OBT, SUT, ROT and POT. An image as ABOABI contrasts with LATNUP, Law of The Absolute Necessity of a Singular Unified Perspective and Presentation. This is supported by PLAR, the Primary Law of the Absolute Relative, or what is above or about relative, yet inside relative. This is what is needed to achieve a singular unified perspective. We have described out-in-TO-in-out, or OITIO in revealing transition and gradients between what is external and internal to TÓ. The maximums and minimums correspond to the limited and are bridged by the innumerable to numerable finites. This describes the extension of finites and their relations to the periphery and center of TO-TRU, their perspectives and representations.

Does TRU exist outside, inside, both, or in what way relative to TÓ, Actual Totality? Since TO and TRU cannot exist without some of the other, and TRU cannot include much that is of TO by its hierarchy, TO exists, at least in part by TRU, and TRU exists largely by TO, LOPAN. Then, in what way does TRU exist separately, and how is this shown. In Fig. 6-4, both are shown with dotted lines to represent an open set, in which a separate TRU duplicates much of itself that is in TO, and both depend on time.

The progressive varieties of TO, and expanding TRU, have existence different from what they are in the present. When in the present, TO is all, and TRU existing separately is not. When in TO, TRU is altered, yet imparts much of its form to render both more that of a closed set. Increasingly as we view, or exist, from TRUS through TRU to TIT to TRIS, we are in the unitary inside, never forgetting the fact that we exist from the absolute inside toward the unitary inside, and view from all together as they exist in ultimate unified totality in and of the actual totality, TO.

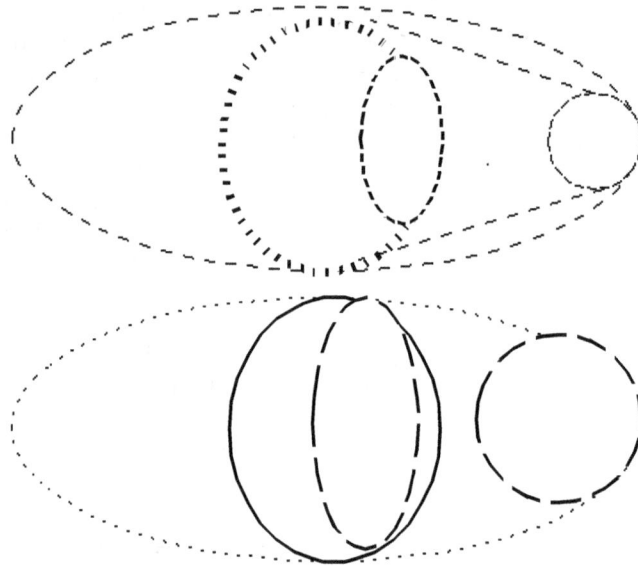

Fig. 6-4, Existence of TOT, TO, TOV and TRU, outside, inside and both of each other by time n1, and in the present

We must also make clear the differences between absolute totality, the astrological universe, the physical universe, and all other totalities, especially relative totality, and their differences from all other worlds, systems and deities. They all have time honored distinct meanings when reduced to their different meanings, actual pure essences and how they are their part and role in TO. Just as an absolute totality is of many kinds of physical universes, so is a relative totality of many kinds. Only when there is a definite specification of totalities, to the absolute or the relative do we approach exclusive unitary totality. Since we do not know exactly what the absolute totality is, nor all the relative totality consists of, we must rely on what the physical universe and the relative universe are known to be. Since the relative totality, as the relative universe of best human life, consists of what is most essential for their existence we know that they are based on certain qualities, modes, continua, kinds and varieties. Among these is all that give human's predominance, especially their cognitive powers. In this way, by the fusion of absolute and relative universes, TÓ consists of an immense mixture of actual pure essences of both absolute and relative totalities, ARTO. The distinction and differentiation of this immense mixture are the basis for the varieties of TO, varieties that are highly generic aspects of absolute totality, and highly specific aspects of relative totality. By such a union of the absolute totality and relative totality in TO, it, and much external and internal will be well shown and proven.

The progression from, a to the, totality, reflects the transition and transformations that occur from non existence, to existence, to necessary existence and actual pure essences in general totalities or TOT and their reception or rejection in TO. This is a very complex and dynamic progression that reflects both the variation, continuity and reversibility of TOT, TO and TO with TOF

Since totality is all, it does not have much of the form, function, characteristics and properties of all else, all that may be external, less or other than itself. It is in this way all external can be shown to be more or less different from totality. This is most important, for it helps to clarify and distinctly make what various things are, as well as show and prove actual totality. It also helps to clarify differences among all that is external and all internal of TO. The exception to this proof and rule is the varying change, alteration, adaptation and new forms that arise in transformations and transition in time. Yet this exception is ameliorated by the fact that much of these transformations and transition in time are reversible, or can be moderated. In this way, the Law of Partial Necessity, LOPAN, the Law of Formal and Elemental Reversibility, LAREV, and SIL, or without law, the distinction with that outside of TÓ, mostly hold, and the proportions in TÓ tend toward stability.

CHAPTER 7 - The INTERNAL of Different Totalities, to Combined Totality

Parts, form and relations, internal relations, dependence, separate to increasing inclusion, creation and evolution in totality

Internal form and consistency of an entity, object or whole, is what it is made of, or quiddity. It typically makes it what it is, saves it, shows it and helps it to function beneficially. Internal form is of many kinds including continua, divisions, varieties, form, function, states, conditions, relations, limits, parts, processes, etc. The quiddity or consistency of totality, are constituency, all its components and mechanisms. It is from the potential of external which when accepted becomes its component part in TO, ESINS to ES. How they exist as parts in TO, is ES, and how they exist as processes are CES. Many new concepts signify various forms, states, limits and relations of parts and processes in TO, (ELAC, ELT).

DIVISIONS or PARTS
of a WHOLE
ES of TO

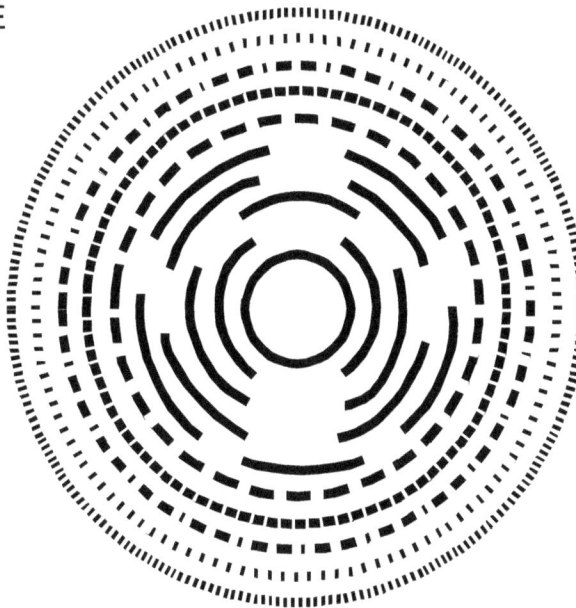

Fig 7-1 Divisions of Parts of Totality

How is totality divided? What is its composition? These are questions whose answers are essential to an understanding and representation of the content of totalities. The figure presents a simple view of this problem. Since totality is both unitary, finite and infinite we must first show it, as above in a small number of major layers, from the unitary in the center through a few divisions or components, to many divisions or parts, to how it is composed of an infinite number of divisions or parts.

The finite may be shown in all layers in between. It is in the finite that largely makes up totality, as we in life are concerned and involved. This is because our lives are mostly characterized by our relativity in the finite divisions of all about us. This is the Finite Preeminence, (EPRE). It is the finite preeminence of the more and more finer and finer finite that makes up much of the totality and its representation. It is by reducing finity to unity, and extending finity to infinity, that we stretch our relativity, enhance and assure our being in TÓ.

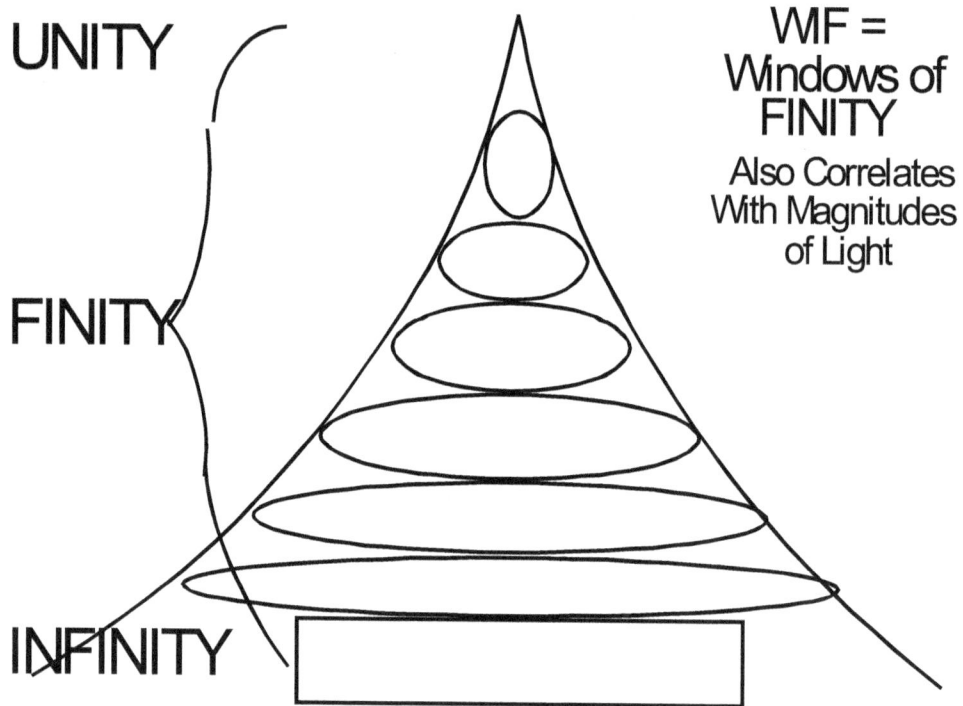

Fig.7-2 Windows of Finity

This figure gives, in a very straightforward, yet diagrammatic way, two very important concepts and visions of totality. Windows of finity signify the fact that the whole is finite, at least in content. This finite ranges from the extremes of near absolute unity to near absolute infinity. It is never quite either, for perfect unity, absolute or zero are non existent.

It is less comprehensible externally for TOT, TO or TOF. Perfect unity is approached largely by the relative absolute in minimal time. The other extremely important concept involves relativity and light relationships and is, the Law of Mass Energy Form and Function and Light Reactions, interactions and interrelationships, LÁLREAC. As we converge the finite of cardinals and fractions there is increasing density in unity and light. With their dispersal there is darkness, approaching non existence. Light is a manifestation of relativity, the observer observed relationship, measured by intensity. Thus, intension and predominance in Á, mass energy in TÓ, cf. Ch 17.

Limits are by the existence, distribution and degree of the modes, kinds, varieties, divisions, continua and each constituent in TO. Limits are restrictions imposed by form, function and other modes and features on TO and its parts. Limits apply to all, as between general and potential totality, TO and TRU. They often define their differences and thereby are key explanations. This includes how all exists by Different Aspects of the Same Thing, DAOST. DAOST basically and typically, guides how limits impose on, and are imposed by, TO, and how TO can be discovered, known and proven.

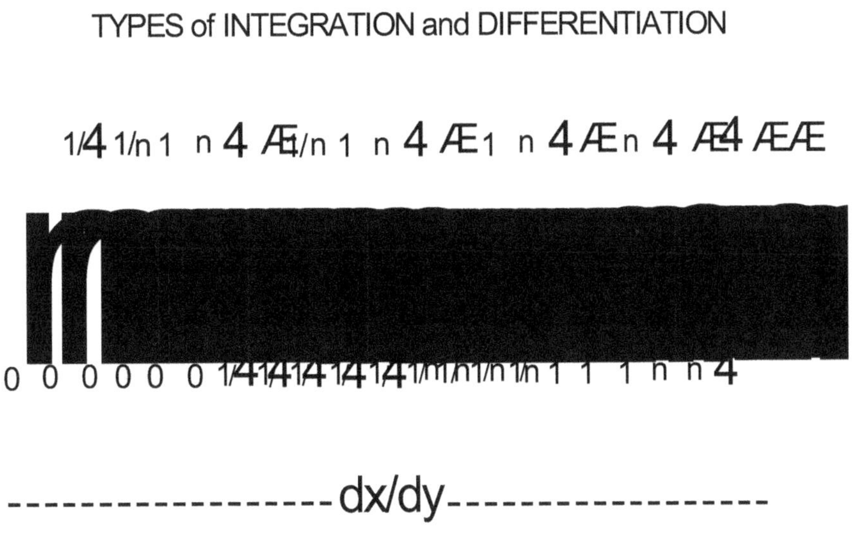

TYPES of INTEGRATION and DIFFERENTIATION

$$\frac{1}{4} \quad 1/n \quad 1 \quad n \quad 4 \quad \text{Æ}/n \quad 1 \quad n \quad 4 \quad \text{Æ}1 \quad n \quad 4 \quad \text{Æn} \quad 4 \quad \text{Æ}4 \quad \text{ÆÆ}$$

0 0 0 0 0 1⁄4 1⁄4 1⁄4 1⁄4 1⁄4 1/m 1/n 1/n 1 1 1 n n 4

------------------dx/dy-----------------

Fig 7-3 Absolute Finites, based on differential and integral limits

Unity to infinity in proportionality denotes the general scale taken, which must be brought to, replicate how the finite exists in totality. Various end points in integration and differentiation are shown. They may be classified by their center-points and endpoints, and are vivid reminders of the scales and complexities of the operations in differentiating and representing totality. Their implications, uses and proofs in mathematics demonstrates the vertical integration of totality. Their scales are not uniform, but altered, like mlt, warped in totality.

There is a marked difference between TO, TREQ or what God is, and TO, TREQ or that God is. These are highly related and interdependent proofs that show how God, although highly relevant to TÓ, does not necessarily show what TO is, nor God is, in the actuality of TO. Errors and loss of faith in the proof of God have been two of the greatest sources of God's denial and thereby loss to humanity. We must therefor firmly and accurately prove TO and base these proofs on both that and what TO is. This largely originates from its internal form, limits and relations. This is the combined Method of Proving TO through its this-ness and what-ness, and thence together, (MEPRUT). It is closely related to the quiddity and haecceity of TO, as in ATO, TÄ, FAF and especially the special Á state and dimension of TO. Proof of TO, by both its this-ness or that-ness and what-ness, its haecceity and quiddity, PRUTAW, differ from MEPRUT in focusing on proving the kinds of this of, and what in, TO. To state that TO is, is to refer, relate and compare it from the outside, and what TO is, is to refer or relate it from within. It is the difference between a 'that', and a 'what'. The law of this and that of TO, LATHIS, shows the importance of the subjective and relative origins. It also shows us the importance of how we enter into the existence of TO. These are closely associated with other proofs of TO and are noted through the proofs in this book.

Portions or parts are often mistaken for wholes, e.g., creation in religion, and evolution in science. They are different aspects of the same thing, TO. This is signified by DAOST. Any entity to have actual existence must conform to DAOST, to be a part of TO. Creation and evolution are two aspects, and large components of the same beginning, development and processes of the dynamics of TO. Creation is the generation of transform-ations in larger typical steps or levels. Evolution is the continuing and usually progressive transformations in smaller steps that typify TO. The imperative requirement of internal form of a whole or totality is whether or not and the degree they are of a category, mode, continuum, kind, absolute, relative or variety. This is the initial step needed for internal relations.

The constituents of TO are not only of the same thing, TO, they are of similar or related things or constituents in TO. It is the increasing existence and power of these same constituents and related components in actual totality that reveal its origins, development, growth and proof. This is the Law of the bonds of sameness and relations proof of TO, (LABSPRUT). This law establishes the fact that actual totality is certain, certified by the evidence of its own existence and the bonds, forces and powers, of relation that hold TO together. They make it the unified totality it is. These bonds are those internal relations whose convergent forces typify TO.

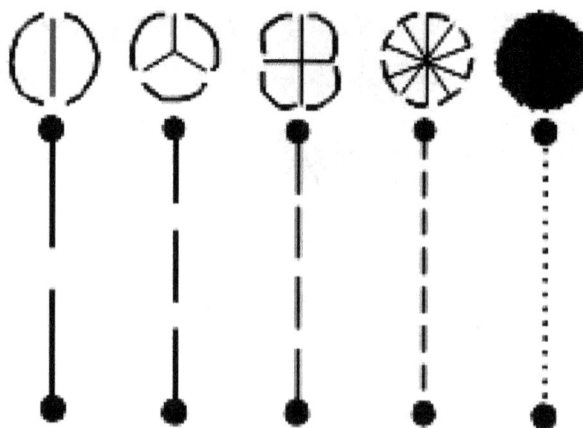

Fig 7-4 Coincidence of Parts with Polarities

In order to better visualize parts and their relationship with the finite gradients of the dimensions of totality, the accompanying figures are provided. The finite parts in item one with only one division would be for only one dimension, e.g., length, large and small. The finite parts of totality in item two with three divisions would be for length large medium and small, for time past present and future, and for the Á dimension, subject, real and object, or its powers. The finite parts in succeeding items, three to five, would be increasing divisions in each dimension, or the continua and spectra they form in totality. There are many other parameters in totality that increasingly show its gradients, divisions and parts.

Polarities help to reveal, describe and solve a basic problem whose proof is of great value. This is the problem posed when we ask simple questions, and expect simple yes and no answers. Such questions as "Is there only one ultimate totality?", "How does the potential become actual?" "What is actual totality?" "How do all things integrate in unity?". None of these can be answered simply or with a single yes or no. They are like so much about us, more or less. This is the Law of More or Less, LAMOL. To satisfy this law requires sufficiency and adequacy. The upside of the work required for such sufficiency and adequacy is the fullness, correctness and permanence their answers supply. More or less, is not complicating, it is validating, the practical level necessary for sufficiency and adequacy. Polarities and the spectra they define are often the means by which continua and their dualities form totality, are made known, proven and resolve TO.

The collective mass energy components of the form and function of totality are consolidated, as in a special totality by their predominant powers of adaptation and survival, to become absolute in the unity of totality. Reversely the comparative differentiation of entities is what contributes to and constitutes the finites of totality. The need to give finite formal distinction to all that is of totality is great and indicated in many ways. Yet the unity and relativity of totality enable us to provide an increasing degree of these needed finite distinctions of terms of their deepest, most essential, and complete sense that proves the form, function and much of TO.

Every stage, category or type of object or part of TO is not TO, TRU is not TO, TIT is not TRU, and each and all persons are not TIT, TRU nor TÓ to the degree they deviate or are different from their part and role. Each has its own loss from TO, and usually carries that of others, e.g., from each preceding stage or level. This obvious fact is much more subtle and profound when we realize their distinctions, operations, proofs and uses. Unless we can separate and provide the actual pure essence meanings of each, and how all are ordered together, we can never understand or represent them. Representation is much more than a picture, form or work of art. Representations are the great bonds and modes of interaction of subject and object in the mass energy, Á continuum of TO-TU. The internal contents of totality are somewhat to markedly less than the whole. Yet it is what is inside that makes up the whole, counters its unity and proves much of it. As we progress from non existence through totalities and TOT to TÓ we can better recognize all that TÓ consists of, its order, vital necessity, proof and course in the future.

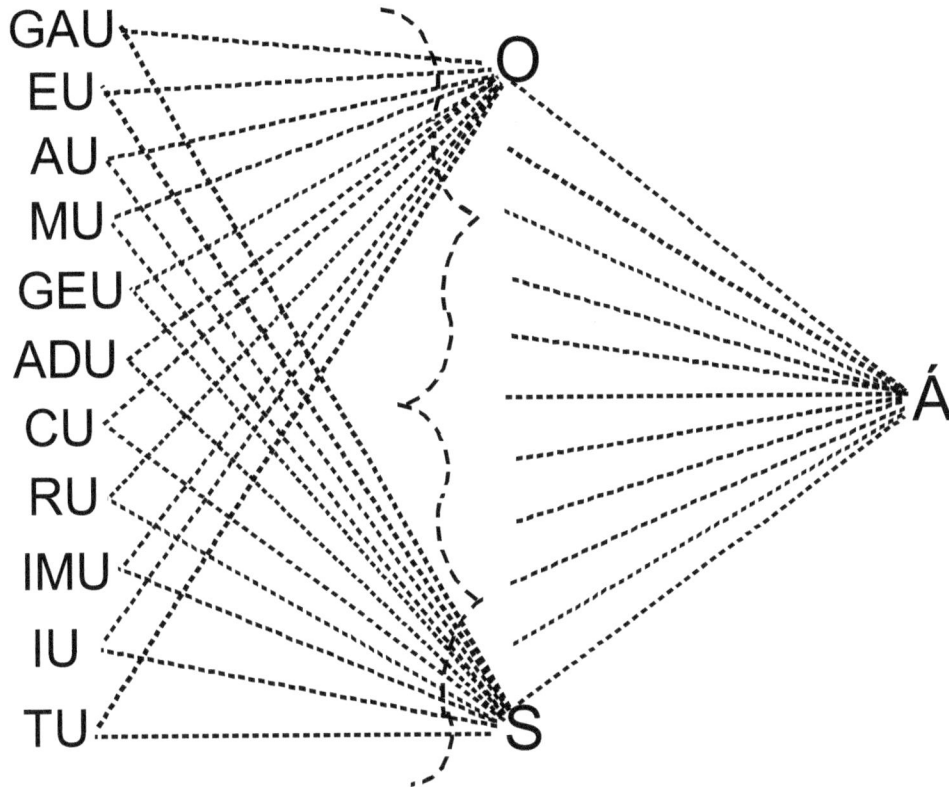

Fig 7-5 Compared Finites in the Subject Object of totalities, and Á of TO

Because a part is necessary for a whole, even though extremely vital, does not make it the whole. This is the most common mistake humans' make, failure to know or judge what is more or less, existent or not, in totality. This is the law of specified distribution of parts in any totality, or TO, LASDET. This law also applies to all, including TRU and one self, in return to TÓ, whereby it will, dynamically, be altered to include or exclude relation to parts of TO. Hereby we see both separation, and the role of necessity, actuality, essence and the bond of all in TO. This is balanced, yet fortified by the law of partial necessity, LOPAN, the actuality of parts, necessary for TO. The immense significance of LOPAN is the necessary and essential bond that exists between oneself and TO. LOPAN applies only to the degree TO is right and the person is only their part and role in TO. These laws provide proof of the limits, distinction, form, modes, varieties and parts of TO.

Most everything consists of many kinds, the greater its quantity, quality and actual totality, the greater the kinds and worth. For totalities and actual totality this becomes most prolific, for which it needs order and control. This is best by the vertical integration of actual totality. Divisions provide separation by kinds, a separation that is often multiple hence adding to proliferation. This has to be controlled by the systematic unity of TO, and the use of proportions of the whole. This is signified by the Law of Internal, Kinds, Divisions and Differentiation, LAKDI.

CHAPTER 8 - DIFFERENT and SAME TOTALITIES to Universals and Qualities of Same Totality

A, any and the totality, Potential, TOT and Actual Totality, TO, Determinants, Form and Function, Exclusiveness, Unity

What is totality? What makes it a whole, what characterizes this whole, and how does this whole help us know, show and prove all? What are the general and special features of this whole or totality? Does the world of man exist as a whole? If so, in what way, how and what does it include and exclude, and how do we divide the whole? Why is this important, and important to be correct?This is the essential principle, the essence of actual existence, or being. It is the traditional idea and concept of the fact that totality, or the actual world, is essentially based on the relevance and special nature of existence, being, life, and most especially being human and having the cognitive powers to recognize and best apply it. This arose from The Essences Critical, TEC, typical of the renaissance, what is of primary importance in most philosophy, schools of thought, systems and ideologies. They are the actual pure essences of TO, APET, the end product of the progressive actuality of existence, life and the world dominated by humans. In this way, since TO partakes of portions of its varieties, TOV, it partakes of the actuality of each variety, TOV, thus the contribution and prominent role of TRU, TIT and TRIR.

Perseity is, through, by means of, or for the sake of one self, the independence of a thing, its being in itself, per se. Perseity for totality and TO is the separate existence and exclusiveness that accrue from the actuality and internal powers of its own existence. How TO, its categories and components coexist, relate and converge to make them what they are. This is how TO is all, TRIA, an independent whole. How being dependent on practically nothing external, and by LOPAN, being a necessary part internally TO, is fully bonded.

In language, knowledge and philosophy we have great precedent and proof of TÓ in the terms, aseity and perseity. Aseity is being as such, independent of a subject. Perseity is how TO exists by means of its own internal constitution, without relativity. They are the state being occurs from itself, signifying for totality an independence from all else, and support exclusion of all else. It is the exclusiveness of TO admitting and eliminating all superfluous that is produced and conditioned largely by its powers of convergent unity that separates it from all else. This exclusiveness of TO, is characterized and signified by aseity and perseity.

Whatever is external and internal impose limits on all that is external and internal. To that extent and degree all external is not. To the extent universes and deities is external to actual totality and without limits, they lack actual existence. Since totality is qualitatively absolute, it is the prime mover and user of limits. Necessity is largely a product of the limits imposed by all that constitutes, and is, the whole of totality, or TO. It is by limits imposed by relation of what is necessary and essential for existence that all, and TO, can be found. TO, its varieties and TRU impose their limits, (RUIL). This is a primary proof of the relativity, unity, form, coordination, systematic unity, DAOST and all of TO. These are altered as varieties adapt to and become their part in TO. Necessity is largely a product of the limits imposed by form and all that are, and constitutes the whole of TO. This is a primary proof of the increasing relativity, unity, form, coordination, systematic unity, DAOST and all of TO, its varieties, TRU and TIT.

The tendency to think of an absolute and relative separately, or an identity devoid of any differences, is a trap that illustrates the degree to which knowledge is distorted. This is close to the simple defect pointed out by Hegel of Schelling's concept of the absolute. He likened it to a view in which all cows, are black at midnight. Without difference, as the D in DAOST of Different Aspects of the Same Thing, in the form of TO, we have an entirely different and erroneous view or subject. This applies not only to the form of TO, more so, to its function, TRUD. It is however, of great value to extract the homogeneous and heterogeneous, the same and different, the identical and non identical, and the relative and the non relative in order to critically analyze, order and develop their forms and functions, and proof of TO.

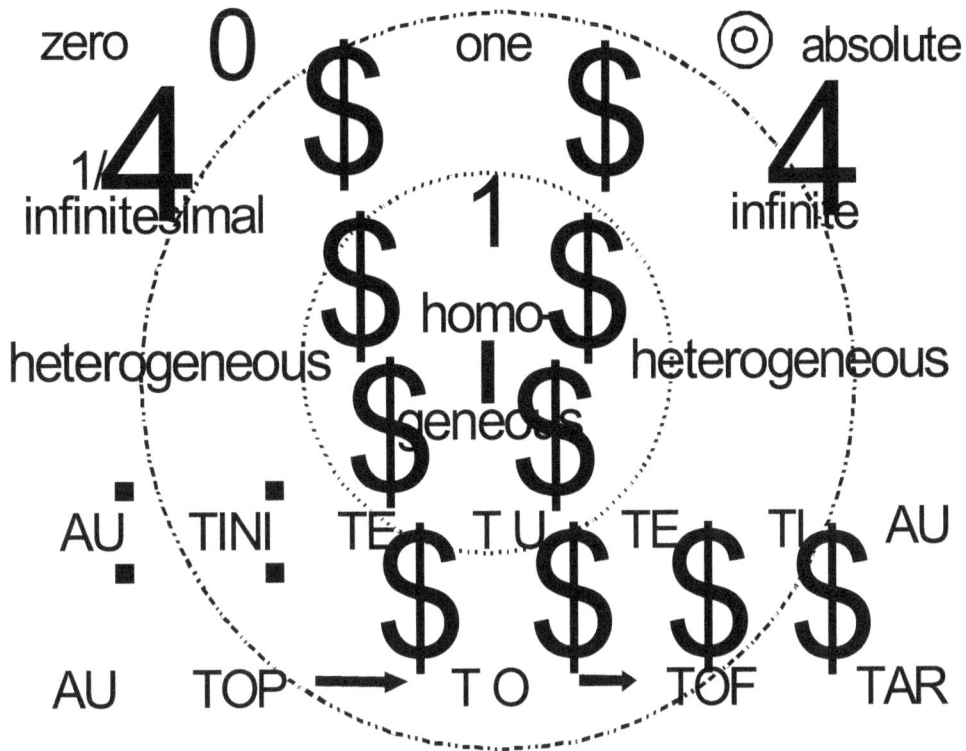

Fig 8-1 Relative Finite, Unity in Totality

Figure (8-2) helps to show how none of the commonly held concepts of systems, universes, religions or worlds can be the right totality. It does so by ordering the different philosophical systems by their place in mass, length and time. It is only a rough sketch, added to the fact that a planar rendering is not as good as spherical. Many negative isms and negative opposites of positives are not included. Such are fatalism, terrorism and many of the misguided and deviant beliefs so inimical to TÓ and best human life. Those systems which are heavily weighted in diametrically opposite poles of the three dimensions, like theism will not show their position well on a 2D field. A spherical figure gives a better replication of totality. Each ism is an example of their part and role in totality. They are not evenly distributed, some being both greater in the whole, and more in one or two dimensions.

The figure is a beginning for positioning in the representation, derivation, development and determination of parts in TO. For each represents neither the whole of totality nor ends, but specific and unique parts and processes in the whole. Their validity and corrections can occur only when they are recognized as parts and proportions in the whole. The whole is actual totality, TO, and their validity is no greater nor less than the degree to which they are only this part in TO by actual pure essences and their actuality, validity, proportionality and totality in TO, VAPO. These systems or isms can be better understood, and their limitations and qualifications shown to prove TO, if they are interpreted and positioned dimensionally.

LENGTH (LARGE)
UNIVERSALISM
THEISM MONISM TOTALITARIANISM ABSOLUTISM
TIME (PAST) DUALISM
POLYTHEISM PANTHEISM
PLURALISM
NATURALISM (MASS ENERGY)
TRADITIONALISM OBJECTIVISM
PARALLELISM COMMUNISM
SOCIALISM PHYSICALISM
EVOLUTIONISM
HISTORICISM RELATIVISM HUMANISM ALTRUISM
NEUTRAL MONISM DIALECTICAL MATERIALISM
DOGMATISM NEOREALISM PRAGMATISM UTILITARIANISM
ANIMISM REALISM MECHANISM
MYSTICISM EXISTENTIALISM CAPITALISM
CRITICAL REALISM
EMPIRICISM INSTRUMENTALISM ATHEISM
HYLOZOISM MATERIALISM
PERSONALISM ASCETICISM HEDONISM
IDEALISM INDIVIDUALISM VOLUNTARISM
PESSIMISM PARTIALISM SOLIPSISM MELIORISM
SUBJECTIVISM CRITICISM EGOISM OCCASIONALISM DETERMINISM
(ENERGY-MASS) INTUITIONISM SKEPTICISM SCIENTIFIC EMPIRICISM FUTURISM
CRITICAL IDEALISM AGNOSTICISM POSITIVISM TIME (FUTURE)
RATIONALISM CONCEPTUALISM OPTIMISM
ATOMISM NOMINALISM
PANPSYCHISM INFINITESSIMALISM
TRANSCENDENTALISM LENGTH (SMALL)

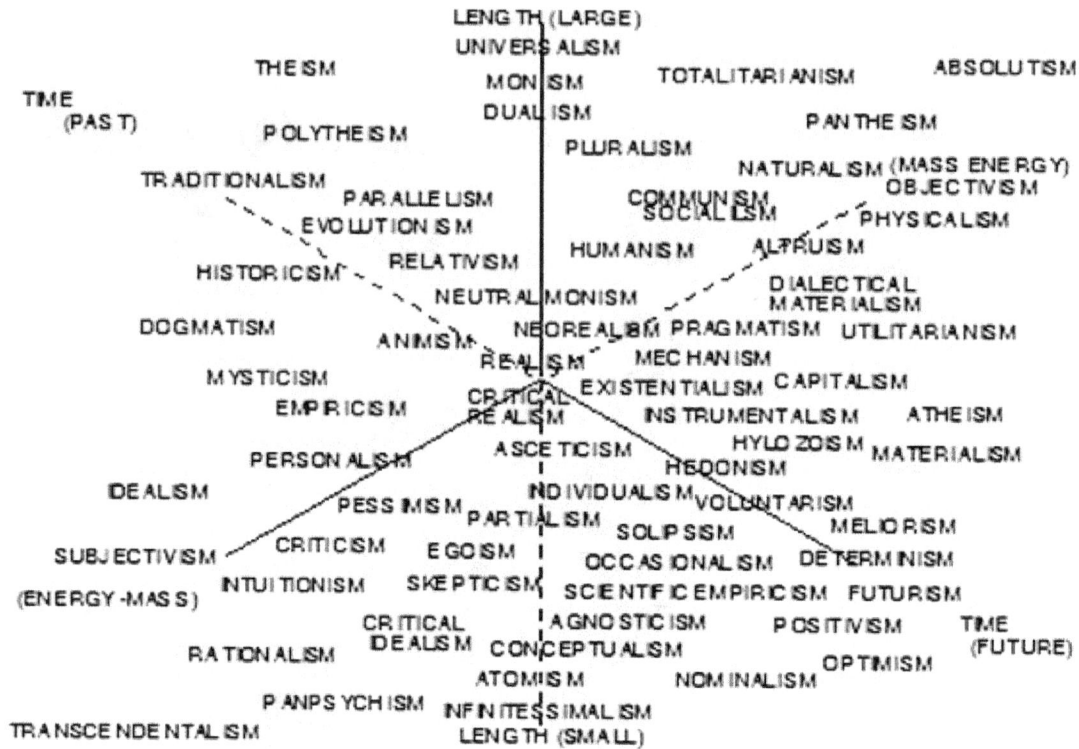

Fig 8-2 Positions of isms by SO, L and T

Totality is how everything, the physical universe, general totality, TOT, varieties, all senses, the special varieties, and relative universes converge and unite by the fusion of their actual pure essences in absolute and relative totality, as they exist by ABSTO-RELTO to create or produce TO. This bond of absolute and relative totalities has evaded human search for unity, proportion and harmony in life. Many worlds and universes have been and are known that have greatly added to our knowledge, purposes and ways in life. Only their diversity, voids, lack of relation, and opposition prove their partial position, errors and inadequacy.

The actual totality, TO, and its proper core, cannot be, and are not, only a part, erroneous or inadequate. It is not a singular bond of absolute-relative in any totality. It is most complex, profound and subtle fusion throughout TO, as seen in the varieties of TO, TOV.

All necessary for TO to exist are prime unifiers. An exclusive totality cannot be, many, a part, or any totality. It must be one. By definition TO is all, thus exclusive, can't be other. Only in unity is their totality, only in totality unity. The exclusiveness of TO is largely produced by centripetal and convergent forces. TO becomes more actual, gains strength, stability and converges on unity, with increasingly selective predominance and power. This is practically and reversibly conditioned and increasingly less in the mid and long term.

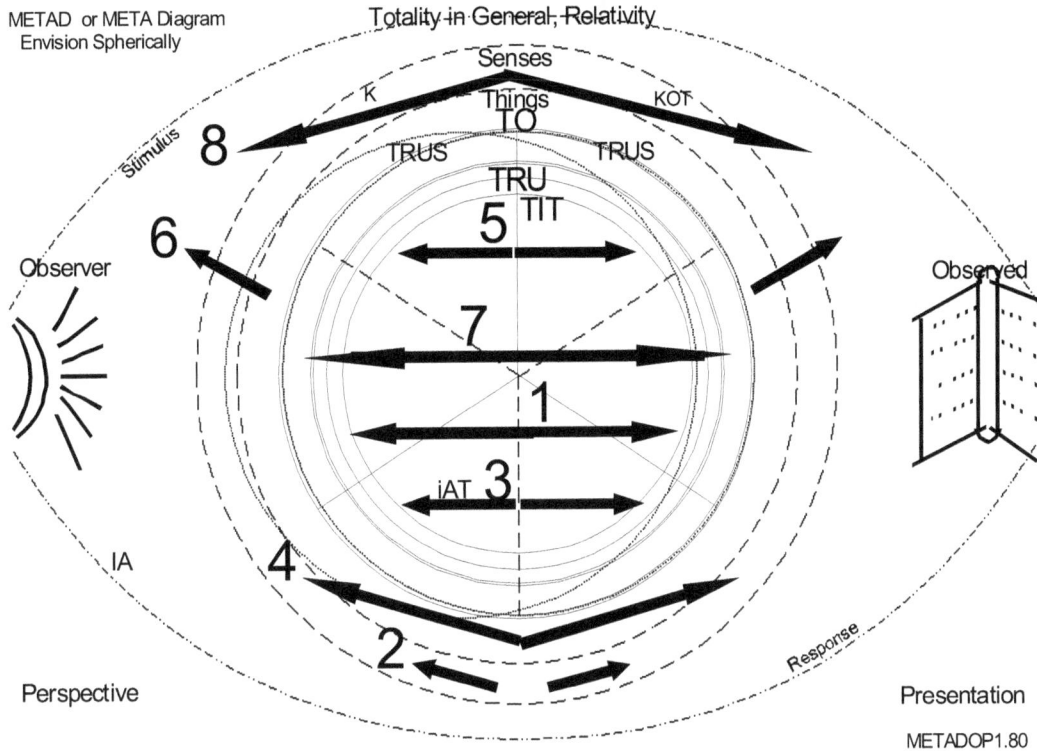

Fig. 8-3 Types and Distribution of entities, parts, processes, constituents and essences in and out of actual totality TO.

The category that best relates and classifies each position is designated by their given numbers:

1. focal,insideof TO, unevenly distributed

2. focal,outside of TO, evenly distributed

3 focal,insideof TO, evenly distributed

4. focal,outside of TO, unevenly distributed

5. universal, insideof TO, unevenly distributed

6. universal, outside of TO, evenly distributed

7. universal, insideof TO, evenly distributed

8. universal, outside of TO, unevenly distributed

9. focal,inside ofTO, uni or bi dimensional.

All entities and parts may be separated and classified by whether they are outside or inside, universal-general or focal-specific, and even or uneven in distribution. In presenting and showing actual totality, TO, these are naturally selected to be what is inside and focal, as given in numbers one and three. However to provide a comprehensive representation and achieve the necessary overall perspective, we must always remember our observer position and contrasts with TO. We must also always remember those modes and parts of TO that are not or markedly less than inside and focal. Among these are all that do not have and cannot be accurately positioned by the three dimensions, A, L, and T. Two of these are what is universal, or permeate the entire whole, and that of dimensions. Both are more than focal but less than total, often a continuum or division. Since what is universal, particular and focal are not two separate types, they are continua whose spectrum includes much or the entirety of TO. We must remember and account for what is medial of their extremes. For demonstration this does not present an unsurmountable problem. The focal and those that are more universal can be enumerated and treated individually when needed. Also what is even and uneven is not so problematic. It does not alter the positioning of the focal. Universals can be described by their even or uneven permeation of the whole. Examples of ESINS, or the actual pure essences of parts and constituents in TO are mostly positioned by their centers, as showed with each through the book. They may be similar, but are different from that of their part, and more so of their separate existence externally.

The Exclusion Law, EL, expresses the imperative conditions under which all exists in TO. Since TO and TITS at time zero and the extended present, by APET, are absolute, its parts, ES, partaking of TO, are qualitatively absolute. Quantitatively this is to the degree they are vital to the whole, LOPAN. Their interactions and processes, CES, are in continual give and take relation with other ES and CES, typical of the dynamics throughout of TO and TOV. This is signified by the Law of Dynamic Polarity of EL, (LAPDEL). This is evident in the Paradox of Exclusion by degree of separation, e.g., Instantaneous Attention, IA, or transient existence in and out of TO.

Ultimately nothing is superfluous. Practically only by degrees of being superfluous, from near zero to near absolute. What is non existent to negative in or of TO is near absolute. What is most existent and positive in or of TO is near zero in superfluity. So that each person to achieve the optimum in universal knowledge must acquire a maximum of most existent and positive in and of TO, and a decreasingly practical amount of all less. This is how we can selectively qualify and prove actual totality. A person cannot become lost in the trivial or impractically finer detail of all, even that of actual totality. Worse, they must avoid all less and lesser relative aspects of the physical universe, or absolute aspects of the relative universe. This follows on the admonition of Apollo at Delphi, "be not superfluous, know thyself, to thine own self be true." This means to know how you exist in, and how all exists in, its typical way in actual totality. This is how an, any and, some totalities never had, or lose their existence by being superfluous, never to become their component in the actual totality, TO. The proof of TO is most evident in this weakness, as in vegetation that is stunted and decays, and persons that fail to learn, know, identify and live by actual totality, and all its blessings and benefits.

There are many totalities, actual totalities, and variants of the actual totality. They are more or less related to the worlds, universes, philosophical systems and deities described in Chapter V. Most are also related to potential TO, TOT. They are formed and function more or less like TO depending on how closely they are related, how potential, positive or negative and beneficial or in opposition. If they are potential, positive and beneficial they are much the same as TO and thereby a part. If they are less, negative and in opposition they are different and when interactive with TO a part of its dynamics, transitions in time and survival. They are all based on relative totality, and how it exists with absolute totality. It may be based on all gradations of life from humans to animals to decreasingly organic, or aspects of the physical universe in which aberrant forms of relativity exist. They do not basically alter the typical features and form of TO. They do participate in the success or failure of TO, and have significant effects in its external relations.

The whole of totality, TOT and TO are both same and different, more of the same, the closer we approach TO. Much of their form, function, principles and concepts are the same or closely related. The more they are different the more they are superfluous, of neither quality, quantity, actual pure essences, nor content of actual totality.

II Proofs of Categories and Major Features of Actual Totality, TO

CHAPTER 9 - Categories and Qualities of The TOTALITY, To

Modes of Actual Totality

Categories, Universals, Continua, Form, States, Conditions, Relations, Limits

The progression of proofs moves from non existence through developing totalities external to internal of totalities, to the totality and actual totality, TO. Although existence, actual existence and being, has much in common, it is the law of primacy of being by relative totality, being alive that most distinguishes and proves TO. It is in showing how universals, continua, qualities and all external become internal in TO that help to reveal and prove much of the development of TO and many of their changing forms and states. How universals, continua and qualities external are more or less the same, yet different, when they become internal, modes, kinds and varieties in TO.

It is the correction to actual totality, of the deviation from the right degree of relativity, specificity, sensitivity and typical quality that any and potential totality doesn't have. By increasing specificity, sensitivity, absolutivity, relativity and quality of actuality, TO develops actual pure essences and the kind of balance that enables determination, and makes, certifies and proves the actual existence of actual totality and constituents.

Any and the totality contain a potential amount of actual pure essences, and a limited, and often invalid and disproportional amount of, subjectivity, objectivity and reality. Actual totality and its various categories, as the powers of RELTO, TOV, The Relative Universe, TRU, and TRU in and of itself, TIT, increasingly focus on all that produces convergence and unity. This affects and forms a large portion of the whole of TO.

	Non Existent - non Totality	MIXTURES and TOT	ACTUAL TOTALITY
UNIVERSALS QUALITIES	Universals, Qualities, Laws Not of TOT or TO	Universals Transcendent in Common, Increasingly TOT	Laws, Principles and Proofs of Universals, Qualities in TO
MIXTURES	Larger and Smaller Entities Things and Senses not TO	Larger and Smaller Entities both any totality toward TO	Larger Parts and Processes of TO, Concepts, Mechanisms,
PARTIALS ES, CES	Smaller to Infinitesimal Entities not TO	'Trivia' of the world much that surrounds the Individual	Smaller ES and CES of TO Increasingly Finite

Table 9-1 Content of Universals, Partials and Mixtures from the non existent through TOT to TÓ

All universals and partials do not exist only in TO, nor do they exist only in themselves, or in any way, or any mixture. Parmenides addressed two of these problems. The problem of "to be" and "not to be," resulted from the radical "krisis." This resulted from monism in TO, how to reconcile this with that which is "not to be" and potentially to be. The other problem Parmenides addressed was that of the mixtures of things. This is the beginning of dualism his Doctrine of krasis, mixture. The beauty and greatness of Krisis and Krasis are that they signify two great states and operating systems in TO. These are proceeding from one to many, and at the same time dynamically proceeding from many to one. The concepts, principles, laws and mechanisms involved are *critical* to TÓ. This is appropriate, since the word, critical derives from krisis. It adds to the sufficient reason for the physical and mathematical basis supporting the relative in the proof of TO.

To the extent the states, conditions, relations and limits of totality determine its actual existence is signified by, (LOSCARL). Two of the states that especially apply to TÓ are the Summative Unitary Reference of TO, SUROT, and the Principle of Unity that arises from the convergence of all and consistent form of ABSTO-RELTO in TO, TOV and TRU. It is impossible to recognize the proofs, form, function, states, conditions, limits and relations of TO, without knowing and comparing the extent and greatness of the whole and its many constituents. How great consciousness, attention and thought are, even when compared to the greatness of oneself, all persons, knowledge, engineering, technology, productivity, commerce, civilization and conditioning of the environment by humans. When we take all together in their greatness in RELTO and in TO, we can begin to understand their power, contribution and the actuality, reality and proof of actual totality and its interactions, adaptations and alterations with absolute totality.

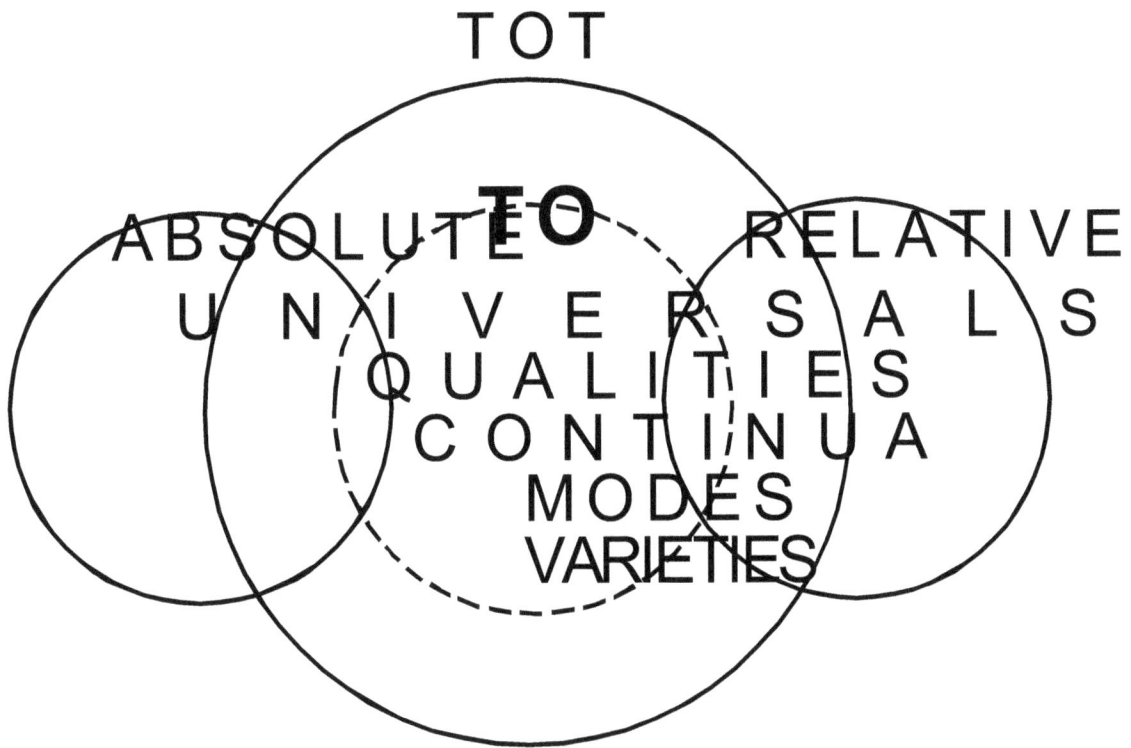

Fig 9-1 TOT Absolute, Relative and Universals, in and out of TO

TO as any totality, theme, topic and object is most distinguished by its categories, universals, continua, modes, kinds, varieties, qualities, form, states, conditions, characteristics and properties. That TO is in ways relative or common to TOT may be shown as a composite mass. Such a diagrammatic rendering must be quickly qualified, being objective rather than what TOT and TO are in and of them or their constituents. Although many major courses and segments of TO are both directly of, or only potentially for, potentially against, potentially neutral, or non actuated of TO, these do not disallow or disprove the unity, solidity and set state of TO. This is, the Solidity of TO, (SOLIT), TO proper, in and of itself in the present. There exists in TÓ many forces, structures, components, categories and forms providing external potential, actual, relatively adequate, cohering, binding and strong features. These are its needs, requirements, makeup, form and function that are repeatedly reinforced and self promoting. Just like 'the ties that bind' God, Christ and a person in Christianity, and just as knowledge reinforces, our environmental, physical, mental and cognitive assets in health, so does all that is positive for TO assist in helping to make it strong, solid and self sufficient. As a biological unit, by all that has made, is and will be, actual totality in the present time, exists selectively and separately in differentiation, and totality from all that is less, other, or negative to itself.

Quality is the state of purity, essence, and power, which gives a special predominance to the whole. Quality is largely what defines and provides direct proof of TO, TOV and TRU. Quantity is the numerical content, form and finite of TO. It is the quality and quantity of increasing existence and essences of modes and constituents that enable us to more accurately know, show and prove TO.

It is the depth conception of fundamental categories, universals, qualities and modes that necessitates ideas, new concepts and language which in turn makes actual totality TO well named, proven, powerful, useful and most successful.

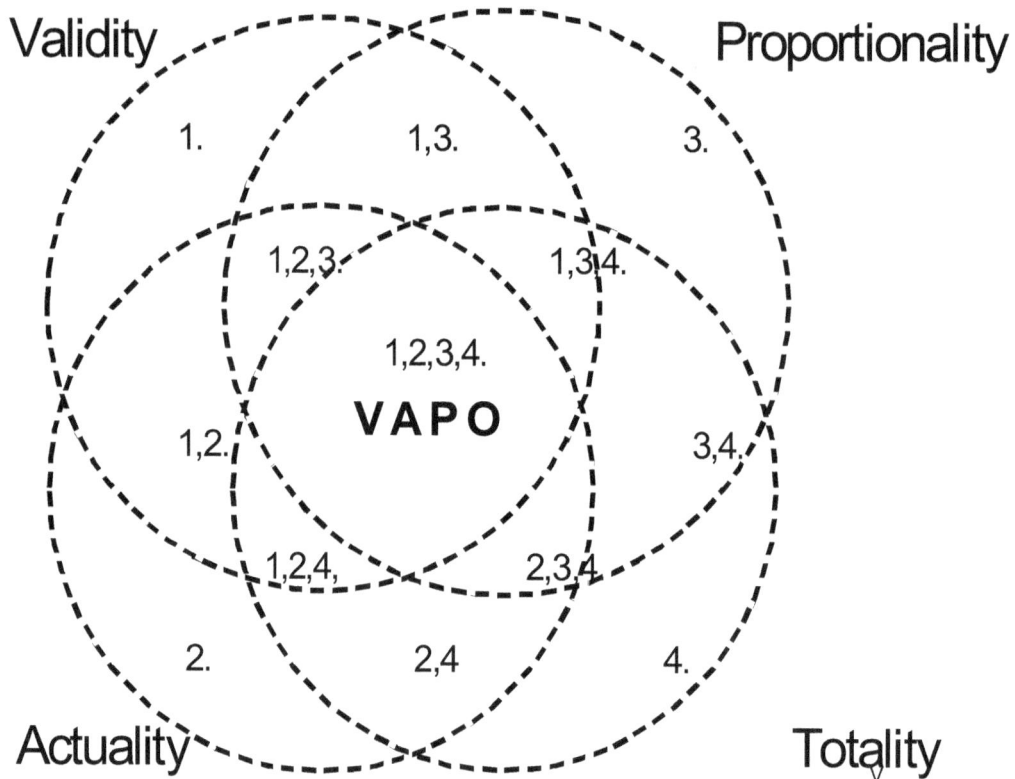

Fig 9-2 Validity, Actuality, Proportionality and Totality, VAPO

When combined in totality, and of TO, validity, actuality and proportionality are signified by VAPO. It adds major universals, qualities, states, condition and modes that determine much of all of and in TO. VAPO, evident from this figure, works reversibly and massively. When, in any way, we depart from VAPO we are soon lost and often nullified. When we depart from great ideas, proofs and laws of TO, we similarly lose. We should uphold the greatness of VAPO and its components. There is a constant splintering off and regrouping with other associated elements. Yet they tend to return to an equilibrium with the constant upholding of knowledge and TO, OITIO. This is signified by Great Ideas, Mechanisms and VAPO, (GIVA) This is primarily how they function together to uphold TO, as by all the great ideas, associated events, and discoveries in history. This is driven by the Acuity of the IA which naturally tends to TÓ, (ACUIA), and the Search for Vivid Imagery, (SEVI). The trend and purpose of these are the Optimal Finite Form and Idea, OF, of TO. VAPO is most closely associated with the Principle of Explosive Total Knowledge, POKED, that produces revelation and accomplishment through analysis and interpretation by all of and in TO. It is also associated with depth conception, intuition and total positioning in TO and TOT.

When the categories of totality are envisioned in their part and contribution to actual totality it is much more accurate. Actual totality can be recognized and proven to be the correctly proportioned all inclusive and exclusive whole it properly and necessarily is. This is the Magnitude of TO by RELTO with ABSTO becoming ARTO in TO, signified by, LORAMT.

Before we shift to the actual totality proper and its modes, kinds and varieties we need to summarize and tie in totalities, the potential totality, TOT, to TÓ, and how they exist, are formed and function, as in time.

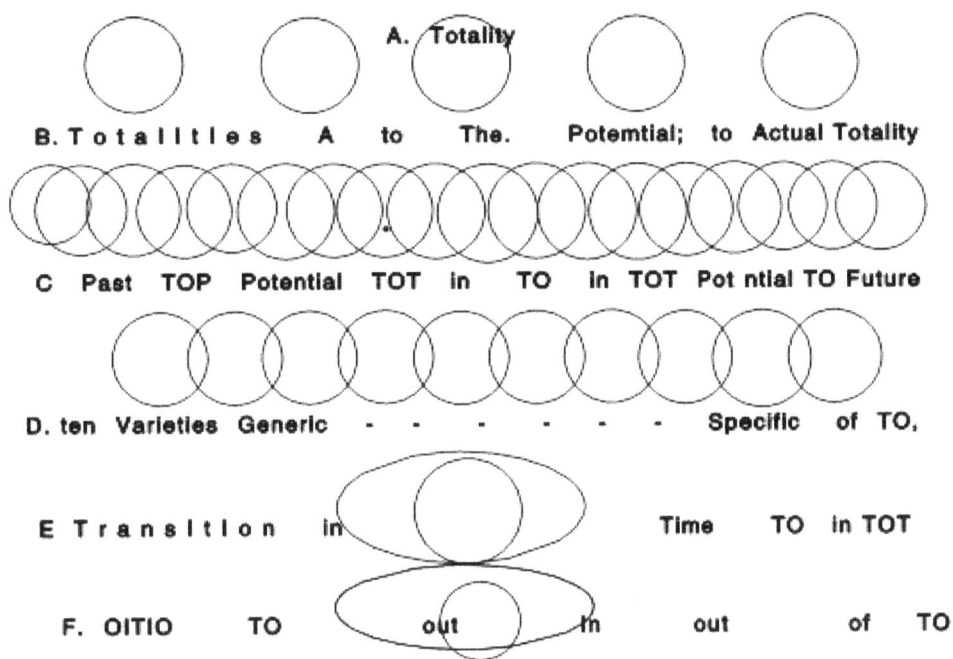

Fig 9-3 From a totality to The Actual Totality TO

This figure seeks to show how a, or any, totality becomes more definite in those totalities that have an increasing potential for TO. This includes all of TOT, from its most past states, TOP, through its present dynamic interchanges with TO, to its most future states. It also shows how those absolute and relative totalities that have a potential to become TO are selectively, by their actual pure essences, fused and become their parts, ABSTO-RELTO, and as ARTO united in TO. When united they exist in TO as its varieties, TOV, from the most absolute or generic to the most specific or relative in TO. The accompanying transitions with time and interchanging dynamics in and out of TO, TOT and other existence are also shown.

The form of TO, TOV and TRU are characterized by limits that are critical for being shown and proven. By the existence, essences, categories, form and function of TO, TOV and TRU much of their limits are set. By limits and set form TO and its objects acquire the qualities of a closed set that enable them to be well represented. Each variety of Totality, TO, TOV and TRU, imposes its own Limits, (RUIL). Progressively, this becomes more actualized as the absolute unites in TÓ to become and forms ABSTO-RELTO, ARTO, and the mathematical, geometric and dimensional frame of TOT-TO-TOT. How TOV exists as a continuum and spectrum with limits and gradients in TO reveals the convergence of forms that exemplify the contribution of TOV to TÓ.

Principle of Convergence of Limits, POCOL is from Correction by Limits, COL. Although a totality, being all, is without external limits, it is characterized by many internal limits, as well as limited by the absolute-relative spectrum and transitions in time of its existence. Much of TO and TOV, as dynamics and systematic unity, are largely determined by the convergence that occurs with concretion of limits.

The typical predominant quality in TO, exists as well in its categories, universals, forms, states, conditions and limits. All that functions well in TO adds to its predominance and proof. Quality must always be separated from pseudo quality, all that is superficial, excessively sophisticated and artificial. Although they, like APET can be of some practical value, like so many fashions that pretend quality, only detract from the actuality of TO.

Quality and quantity, their levels, losses, gains and improvements are indicators and measures of totality and TO. It is very important to have quality and the right quantity. Yet the right quantity depends on accurate measures, and quality is of little value without knowing and living by what is most important. This includes knowing and living by that which has the right proportions in TO.

The most basic continua of TOT and TO are absolute-relative and subjectivity-objectivity, both permeate all TO and determine much of its form and function. The absolute-relative continuum is central to the focus and union of TO and its varieties. The subjective-objective continuum is central to the development of mass energy, whose transformations are key to necessary existence and the consistency of TO.

CHAPTER 10 - MODES of ACTUAL TOTALITY, To Continua and Kinds

Actuality, Objectivity, Subjectivity and Reality

The modes of TO, actuality, objectivity, subjectivity and reality are its most distinguishing features. They do not exist, or are inadequate, in other totalities, universes, systems, etc. They most clearly separate and prove TO from all else, help to form its exclusiveness, predominance and support many of its other proofs.

Specificity focuses on actuality, objectivity-subjectivity to converge on reality in determining TO. The specific focuses actuality and converges the continuum of objectivity-subjectivity to become the reality that makes, TO, and its core, actual Totality Proper, TP. This is the Law of Specificity Focuses Actuality and Converges on Reality to determine TO, LASFACT. This actuality in increasingly higher forms, includes actual pure essences, cognition and their uses in TO. This most important law and concept prove TO and the role of specificity, as well as how the part and role of best human life, TRU, the more specific variety, focuses on, participates in, and affects TO. This is better understood when we envision how TO exists in space, time and mass energy, by the model, SPALT.

The power and forces of actuality, objectivity, subjectivity and reality and their gradients have convergent effects in TO. They typically characterize and have properties that prove TO, in contrast to other totalities and universes. The increasingly special varieties of RELTO in TO, dominated by best human life, TRU, brings all their highest powers of actual pure essences and the reality of combined subjectivity-objectivity, S-R-O, to the state, form, function and fruition it has at present in TO. Progression of these modes largely describes the advance of TO in transitions with TOT and its future course to ever higher stages, levels and forms of actual or necessary existence.

In the development of TO toward completion, actuality is the end product of non existence to existence, negative to positive, potential to actual, unnecessary to necessary and central to the proof of TO. Although what is necessary for existence can be external, actuality exists as a continuum of higher forms through actual pure essences to their state in and contribution that is only internal, in TO. It is the great and glorious progression of actuality from its incipiency in the barely existent to its most profound, massive, dynamic actual pure essences in the greatest of best human life in the more specific varieties of TO, that most fully characterizes TO, and sets it apart from all else, This not only proves TO, its absence in other totalities, universes, worlds, Gods, etc. explains their inadequacies and voids, and helps to separate what is not TO from what is in them that is also in TO. Actuality is the whole of essential, absolute - relative total necessary existence.

Each mode permeates all actual totality with some variance of densities in the position of each. It is the coincidence of actualities or actual pure essences of any and all things that prove each other and their part in TO. Many actual pure essences of all entities may converge on TO depending largely on the needs of, and conversion in TO. Actuality brings overall unity and proof as forms and functions combine in TO, TOV and TRU. It is by necessary existence and the specification of the definite participle, 'the,' in relative totality that actuality acquires its form and function. These include life, reality, acts, deeds, truth and presence or what is extant, actually and currently exists.

The modes are positioned and aligned by their existence in TO. Being widely distributed, and in a planar 2D figure, the modes of actual totality need to be viewed broadly. However, the position of the center of each mode is focal. By perspective, representation, development and actual form TO may seem to be quite different. This is seen by comparing the two accompanying figures with the frontispiece. It is by the increasing alignment of mass energy in time that modes exist in TO. The absolute or exclusively objective universe of physical science is much altered, and in ways warped compared with its form and function in actual totality. Separately the physical universe implies a totality without actuality and definite relativity. Yet since the absolute objective physical universe is only an extreme, since it has relativity, and since actual totality includes much that is objective and physical we must be careful to understand what exists and does not exist, by actual totality. This is why it is so important to know actual totality, and its set state in the present and averaged sum focus of totality proper.

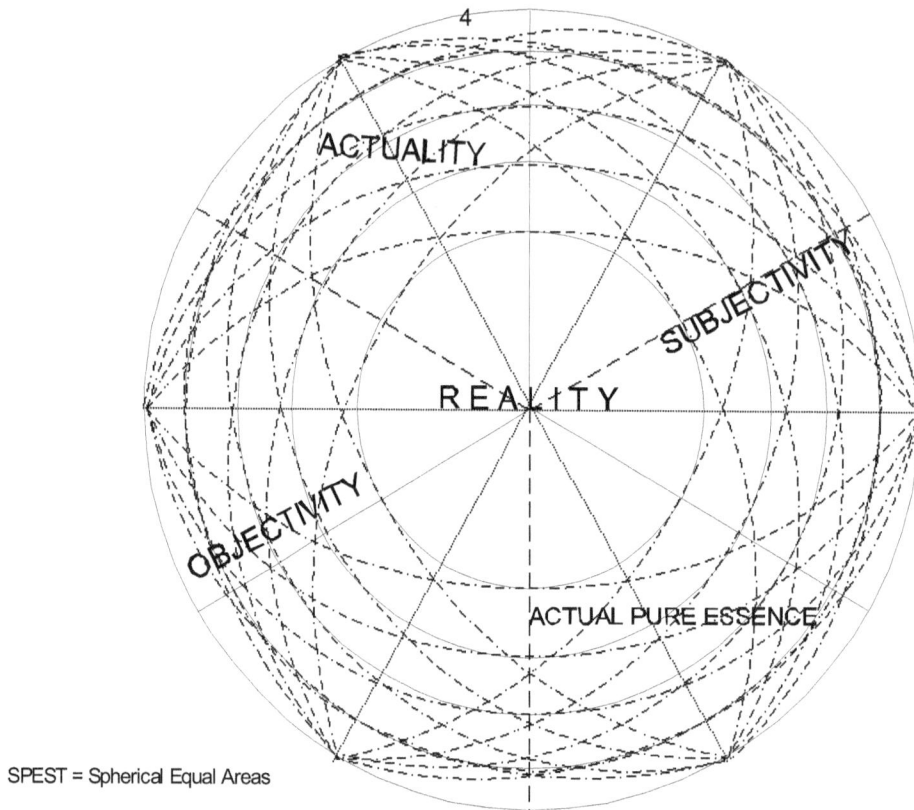

Fig 10-1 Alignment and position of modes in SPALT of TO

The difference in the modes in the two figures is that in Fig 2 they are aligned only by the mass energy continuum. The modes, view and form of totality proper or actual, TO, provide the basis for many of its other components. How components, kinds, the absolute relative continuum, universals, numerous other continua and the varieties exist and relate in TO.

 Levels of mass energy, sense-thing, subject-object, consciousness, instantaneous attention or IA, and actuality in TO-TOV-TRU are features of reality in TO. The actuality of consciousness represents the extent to which one's momentary awareness and instantaneous attention, IA, are involved in things that are real, effective, existent, essential, and what they are supposed to be, (ACI). This may be compared with substance, mass, energy, and what is 'tangible'. It is that of a thing which is sensed versus not sensed, how thing-sense, or object-subject become one in TO. What occupies space and time, compared to what is not there, does not occupy space or time. This is why and how sense-thing and subject-object of mass energy by transformations and APET make TO-TOV-TRU. Physical mass and energy, change in ever higher levels of actuality, in ARTO and TO. Why actuality is the primary mode, purpose of life, especially human life, and proof of TO and its constituents.

Extremes alone can never satisfy the requirement of existence in a whole, whether completely objective like the physical universe, or completely subjective like the world of religion or philosophy. It is in how the objective and the subjective, like the absolute and relative, exists together as one, one whole existent that is more or less centered on the fulcrum of the sum of forces of absolute and relative, and objective and subjective. It is into TO, whose transitions in time and reversibility prove all that has, is and will exist by their own actuality and reality. The astrophysical universe, as much as it appears to explain all the objects we observe, and as much as it is a part of ABSTO in TO, lacks the increasingly higher stages in relative totality, RELTO, of S-O reality and actuality.

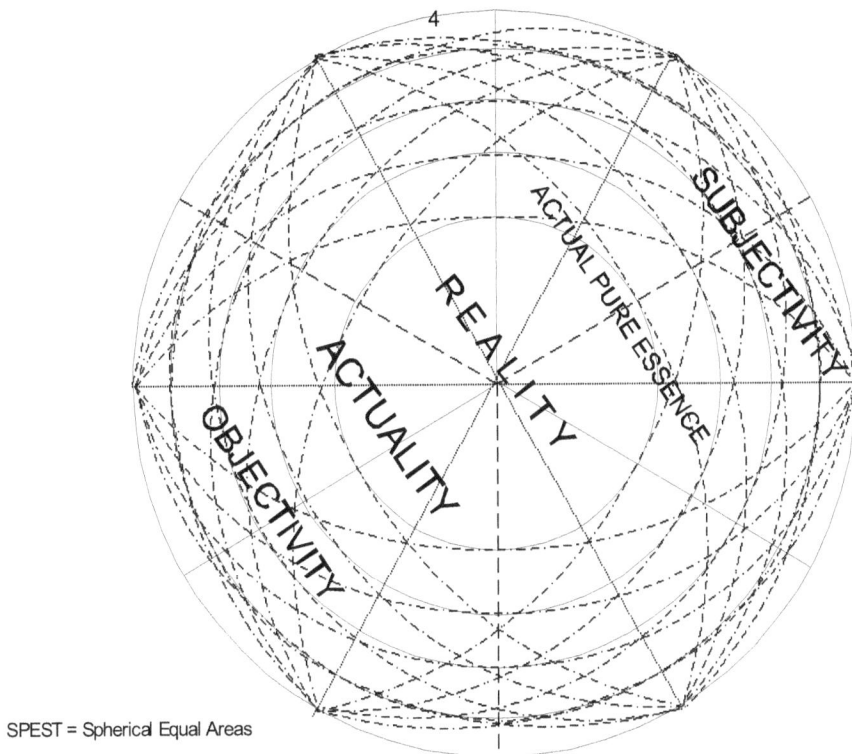

Fig 10-2 Mass Energy alignment of the modes of TO

The finite predicament of extreme polarity of subjective-integral, subjective-differential, objective-differential and objective-integral is what the subjective-integral of our special partial position in TO seeks to reverse and bring into equilibrium. It largely corrects and succeeds with sufficient knowledge of TO, or its equivalent and the combined ordered proofs in totality, TO. Yet without correcting itself the subjective-integral moves to another extreme polarity. It is only when all modes and continua are their part in TO and are in optimal equilibrium that they have complete and proper existence, in TO.

The circularity of subjectivity and objectivity asks the questions and seeks to provide the answers to what, why, how, where, when, which, and who. In this way we most easily recognize the reality of the form and function of TO by the focus of its modes. It is the objectivity of the absolute or physical and the subjectivity or relative and mental that are received and combine in the reality of their actual pure essences in their actual proper totality that certify and prove the ultimate unified totality that is TO.

Wisdom is much more than an exclusive property of God. Wisdom and knowledge, like subjectivity and reality are the greatest fundamental determinants of actual totality. This is by way of their being a large part and constituent of relative totality. Wisdom and knowledge are the increasingly great generative powers that bond the absolute and relative, the physical and relative universes in actual totality.

Reality is the ever higher concurrence, convergence and fusion of objectivity and subjectivity. It is the hallmark and proof of TO by RELTO and TRU as they coexist as one, TO. This is the contribution of the specific varieties of TOV to TÓ. It is the reality of our existence, which with actuality drives us toward ever higher and greater fulfillment in the transformations of TO. They are not only the intricate inherited beauty of its physical form and function. They are much more, the vital essential role and contribution that the reality of the most valid mentality and ideal provide for totality, TO. Without it all totality, as it now exists, would be lost.

Reality's doctrines of internal relations, in its highest levels, are strong proofs of TO. The doctrine of internal relations holds that relations are only of the object or whole in which they exist. When limited to the relation between things, not only a whole, the doctrine applies to those relations that are essential, cannot exist without ceasing to be the things they are. Since TO is the exclusive whole of all that has actual existence the doctrine only confirms a natural fact, except the relations in the transitions of TOT-TO-TOT.

Most profound is the fact that the essence of reality, is the sum of relations, as those between objectivity and subjectivity. This compares with, and is in union with, the fact that the sum of actual pure essences as they exist in, is TO.

It is by such ever finer and greater reality that both TO, TRU and all their varieties have originated, developed, expanded, portend and will flourish in the future, TOF with TO. In this way, by the sum of reality that we can recognize and prove the role and force of TO and TRU, and their qualified form and function.

The actual pure essences and relativity of the far reaches in space and time of the astrophysical universe form a continuum with its extremely small presence of space and time, or here and how. Their balance and most specific varieties of RELTO in TO produce and prove the role of actuality, objectivity, subjectivity, reality, best human life and mentality as they exist in, affect, and contribute to the whole, TO.

It is the increasing modes of the specific of TOV in TO that bring levels of mass energy to S-O in escalating stages to the progression of life and human mind in TO. It is a kind of fulfillment working toward completion of TO, by its most specific modes, continua, kinds and varieties that lead to and provide their combined proofs that converge in TO.

The modes of actual totality are the great, although not the only, aspects and features of its existence. They help to differentiate its form and function. Even more they are the means by which TO knows, and can be known. It is in this way their greatness comes from the power they give actual totality. They identify objects, as in sense perception, they identify the actual pure essences of these objects, they identify the necessary existence or actuality of TO, they subjectively identify an observer's part and role in TO. The modes of TO produce the ever higher levels of reality that combine objectivity and subjectivity together, which can with reason and logical thinking, identify with all actual totality. It is in this way the modes of TO, by discovering and proving certain knowledge, provide the basis for understanding all things. It is impossible to have certain and sufficiently complete knowledge without understanding actual totality and its modes, or their equivalents.

All continua and modes, as all things do not exist in perfect independence or separately. They more or less exist as different aspects of the same thing, in actual totality. Since absolute totality is not separate, and the physical universe consists of relation, or relativity, it should be evident that actuality, subjectivity objectivity and reality exist in both ABSTO and RELTO in qualities and amounts that equal their essential contribution in TO. This clarifies and proves the part each has in TO, and how they exist in stages, of development and perfection. Actuality as necessary existence has a presence in the physical universe, as a sun is necessary for a planet. This is not the same necessary existence in each stage of actuality and in each higher varieties of TO. Actuality as necessary existence is itself a continuum, whose existence most typifies actual totality.

The same state, condition, limits and relations hold for objectivity, subjectivity and reality. Since object and subject do not exist separately, but as combined both psychologically, and in TO, their stage also becomes greater the more specific the variety of TO. Since reality is the combined meaning and fusion of object and subject, it also becomes greater the more specific the variety of TO. It is in this way both actuality and reality are the great fundamental modes of actual totality, the compelling powers that hold it together. This is the Law of combined stages and Powers of Actuality and Reality Modes uniting TO. LARMUT.

CHAPTER 11 - DEVELOPMENT of ACTUAL TOTALITY, to Kinds and Form of Continua, Dualities, Polarities, Spectra and Dimensions

Many universals, transcendents, continua and their subsets participate in the development, dichotomy, cleavage and divisions of TO. The more we learn and apply the form and function these divisions provide, the better we can learn, prove, know and live by TO. TO consists of many different continua, whose dualities, polarities, spectrum and dimensions produce much of its form and function. The universals and continua of TO are largely essences of its actuality, objectivity, subjectivity and reality. All that exists combined in a dual and often opposing relation can be a continuum. From this is derived law of opposites. Those whose polarities are continua of the same universal or class are the purest, e.g., long and short for length and past and future for time. Another example of a continuum that is limited to a single universal that permeates all TO is relation. Such continua exist in TO from least to greatest existence. Examples of continua in general and less definite include many true dualities, the opposing relation between two definitely separate entities, e.g., absolute-relative, same-other or different, sense-thing and subject-object. This does not reduce their importance in TO, only makes description more complex and necessary.

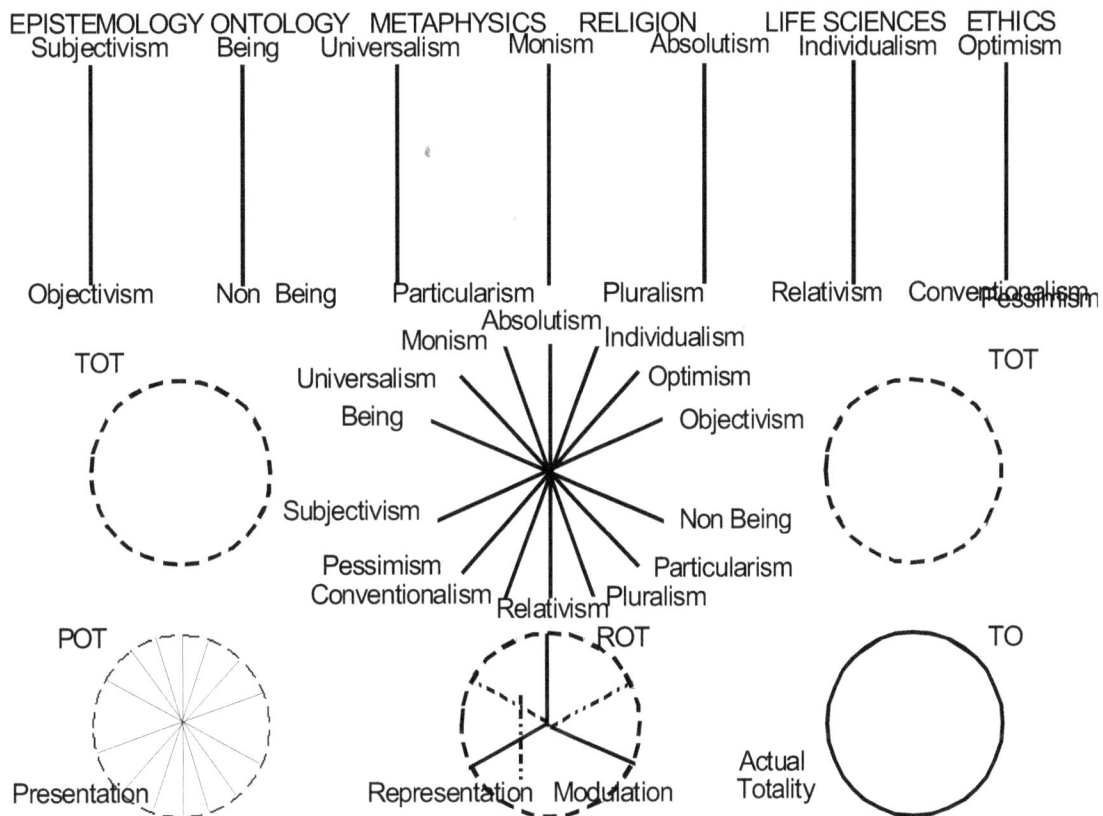

Fig. 11-1 Continua Opposites, Polarities, Parameters, Arrangement, Proportion and Balance

The basic form and relations of continua, dualities, polarities, opposites, isms, and contrasts are a great part of TO, and a great aid in its representation and presentation. We have, among others, the subjective and objective in epistemology, idealism, realism and materialism in ontology, universalism and particularism in metaphysics, absolutism and relativism in theology, optimism and pessimism in ethics and individualism and conventionalism or traditionalism in the life sciences. We have all the universals, continua and qualities that although not confined to actual totality are, when adapted by their and the actual pure essences of TO, the foundation for accurate and complete understanding.

Continua usually exist as dualities with polarity, whose two poles are its extremes. It is how the many continua consist of extremes and means in TO, function dynamically, and tend to work convergently to give equilibrium, unity and power to the whole that makes TO the exclusive and ultimate unified totality it is. The convergence of extremes that unifies TO is given by the Law of Unity of Opposites produced by the needs and actual pure essences of the whole. By learning how actual pure essences in the form and function of opposite poles work in concert in the dualities of continua in TO, its proof becomes much clearer and more certain.

Most of the greatest conflicts and problems of negation and threats to survival of TO come from the most basic oppositions. These arise from dualities of continua, whose continuous oppositions, with action and reaction, are the dynamics of TO. They are revealed in existence of TO, external and internal differences, and paradoxes of disequilibrium in the form and function of TO. They include extremes with irreconcilable lack of compensation and intolerance. This is why it is important that we discover, learn and remember those paradoxes, dualities and universals and accurately operate, practice and live by them, in the balanced totality of TO. All levels of TO and its modes, continua, kinds, varieties and mass energy, interact and are contingent upon each other. This is provided by the proofs of integrative processes and interdependence of all in TO.

For actual totality the continua are modified by their part and role, especially their form and function in fundamental features that most typify TO. These are the continua of non existence, the actuality of necessary existence, and the continuum and spectrum of absolute to relative. They set the state of TO in which generic to specific varieties exist. It is these generic to specific varieties that are most helpful both in being a basic form of TO, and enabling us to envision TO and its relations in time. These are how TO and TOT, as TOF for the future, coalesce to produce the continuity of necessary existence, actuality and reality that have, are, and will, form and assure TO.

Absolute totality and relative totality, as most polarities and all things, do not exist separately. This is why the physical universe, although not the same as absolute totality, must have relativity to be treated as a separate entity. So is it true for The Relative Universe, TRU. All exist only dependently in TO, and are modified by its actual pure essences to the degree they apply. The sequence from absolute to relative is, Absolute Totality - Physical Universe - ABSTO - ARTO - RELTO - Relative Universe - Relative Totality. This is the fundamental division, whose unity is ARTO, and makes, exists, forms, functions, saves and is used for actual totality.

The continua, equilibrium and application of absolute-relative, like that of TOT-TO-TOT, are created by the continuum of reference, definite/indefinite and relation. It is central to actual totality, TO, and most critically to its special varieties, e.g., TRU. It is a concentration in unity over a diffusion toward zero or absolute. Internally it is well distributed, representative of the whole range from most concentrated to most diffuse. Externally it is different, from an imaginary point of view or reference, it is decreasingly concentrated, an existence that occurs only by virtue of the extreme of its concentration. Since however, the only point of reference or relation that is actual, practical and vital, and an average of the whole, we bring this imaginary external point into its actual pure essence in TO, thereby providing both its existence, demonstration and proof. In this way, all that exists, exists only and ultimately by its actual pure essences in ABSTO-RELTO of TO, whose center is the average of its diffusion and concentration. This is provided by the fusion of absolute and relative totalities in TO.

This figure (11-2) is a simple and rough example of four common polarities and their approximate position in the dimensions of a mass, length time. Many of the dialectical, opposite, and contrary divisions and continua of TO are less than primary dimensions. Yet they can be interpreted in terms of mass, length and time, MLT, and/or used as lesser continua, parameters or systems. This is evident in three ways, one by adapting altering, and showing the nature of ALT the mass length time dimensional form that exists in TO, how polarities can be positioned in TO, and how these divisions, contraries, opposites, polarities and dialectical dynamic aspects combine to make up, form and prove TO. It shows how the mass energy, Á dimension of TRU-TO is more than any S-O relationship, but is set by the predominant form and function of the absolute to relative and generic to specific varieties in TO.

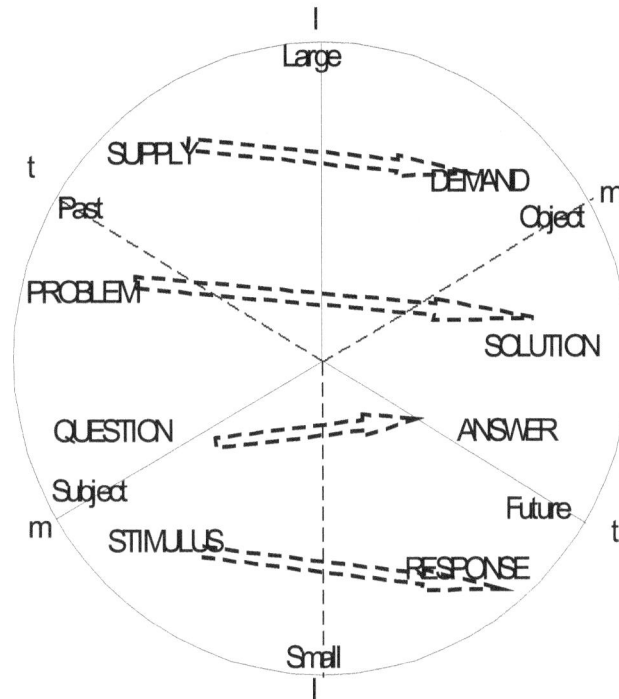

Fig.11-2 Polarities and superimposition in models of TO

Viewed oppositely, the continuum of absolute is that of a diffuse absolute over a concentration toward unity. When of relation, most specific and concentrated it is our mass energy in the here and now. It is centered by the mass energy of the reality of S-O. When most diffuse it is any mass energy, place and time. Neither continuum are all, nor exist alone. They are modes, limits, parts and processes in all that enables TO. In potential to actual totality, TOT-TO-TOT exists as a continuum that is a variation of the absolute-relative continuum only arranged by centering of the relative in TO. It makes TO what actually and perfectly exists, over what is less actual, perfect and only potential, TOT.

By the equilibrium and balance centered at one with unity of the sum of all, actual totality, TO, is formed and exists. It is the unity provided by the convergent equilibrium of continua, absolute-relative, relation, generic-specific, hierarchy of the varieties of TO, and TOT-TO-TOT itself that make and proves TO. The ultimate actuality that is TO is the sum of selected, qualified, proportional and essential integrated potentialities that have gained actual existence, TO. By continuous convergence and fusion of absolute-relative, or ABSTO-RELTO, the autonomous actual and necessary existence of TO has been formed, exists, maintains itself and is proven.

Generic and specific signify the continuum of TO upon which its varieties are based. Absolute and relative signify the continuum that is most basic to the form that gives TO its most unique characteristics and proof. Physical and mental, although somewhat different from generic and specific, are related to the absolute and relative, by revealing the object and subject, and thereby reality of TO. Absolute and relative, physical and mental, and generic and specific, is highly confluent. Yet each approaches TO in a different way, and degree in and out of TO.

Fusion, unity, same and other, and more or less typify TO its modes and kinds. Totality Proper, the core of TO, propels and converges actual totality to be. Actual totality, like any object, is a composite whole. This, unlike any, one object, is absolutely neither static nor dynamic, open nor closed. It has combined qualities of many dualities in continua that fill its actuality. Actual totality exists and consists of what is most absolute to what is most relative, never absolutely one or another, only a duality in infinite series of same and other. Same and other, or same and different, when extreme is never absolutely one or another, except when completely separate. In the totality of TO, they exist only as a duality in infinite series of same and other, (SO). This is a proof of TO, the balance of all in TO. Same and other of Subject and Object, SOSO, and Subject Object Equality, SOE, are also basic to, and proofs of TO. Same and other is closely related to different aspects of the same thing, DAOST. Their use is for contrast, for alternative approaches and views of TO.

Same and other and different aspects of the same thing, DAOST, express various aspects of TO, TOV, TRU and TOT. Extreme examples of this include that of absolutely absolute, and absolutely relative, AA and AR. The importance of this is evident in TAR, the absolute relative of Future TOT, TOF. By and in these continua, as other and same, absolute and relative, there are increasingly greater differences and transformations that are increasingly crucial to TÓ. These transformations produce and reveal their modes, kinds, form, characteristics and properties. Together the transformations and their products provide the bonds that make TO autonomous in the present, and the ultimate totality. It is an ultimate totality that brings proportion and harmony to absolute and relative totality as they combine in ABSTO-RELTO, ARTO in TO and its varieties, TOV. When of its special varieties, and the special totality, TRU, with the future of TOT, there is concentrated much of absolute totality, that focuses on best human life. This is the theme and goal of the work of all mankind toward actual totality, and all things to come that must necessarily exist by TO and with its future, TOF.

Multiple meanings of words are corrected by multiple definitions to produce precision and proof by specificity in TO. We must always remember, especially if in doubt or in error, when using language, that words often have more than one meaning. We have a tendency to use words simply, partially, roughly, indiscriminately, or in the extreme. This is nowhere more misleading than in the use of the term absolute. Absolute can mean unlimited, free from imperfection or error. It can also mean the opposite of relative, or it can mean all, the opposite of zero. Only to the extent that zeros and relatives have some common basis, which is a proof of TO, TRU, TIT and TOT, are such duplicate meanings, in part, justified. The actual pure essences of the combined absolute-relative are key to the most fundamental of proofs. That is how TO with ABSTO-RELTO can be practically and realistically absolute by the role of life and human mind, and by providing the basis for TO and its varieties, TOV. This focus is what binds TO-TRU-TOT and assures their expanding role in the future.

All entities, things, senses, words, worlds, universes, deities, systems, etc., are, exist in relation to, or as, continua. They may be of different spectra in the same duality, simple opposites in the same duality, or gradients. As long as their existence is in some kind of relation, this will be some kind of continuum. It may be from definite to indefinite, natural to unnatural. Many will seem to exist separately, as a simple subject or object, yet this will usually be found on more careful analysis, and especially by actual totality, to be in some kind of continua, duality, polarity, spectrum or gradient. Often it will be a combination of more than one kind.

The only exception to a state of being a kind of continua is actual totality itself in some frames of reference. This is also apparent in words, ideas, concepts and meanings that are equivalent or close to actual totality. Such is natural and unnatural, where natural is what is TO, and unnatural what is not or less than TO. Or it may be classed as normal to abnormal. It depends on how the terms, entities and objects are used and the meaning or manner of this use. It is the flexibility in use of language that both give it strength and weakens it, depending on each occurrence. The strength comes from freedom and broadening in usage. Its weakness comes from the multiplicity of meaning making duplication, superfluous and differentiation more complicated and often less than actual totality.

Actual totality, being equally both dynamic and static, is in the continuous state of flux between the extremes and means of its constituents. To formulate and apply all these together in TO provide a great amount of proof to each constituent, and TO.

By the sum and total of their combined levels, varieties reveal much of the unique form, function and characteristic of TO, TOV, TRU and TOT. Further proof of combined and unified continua is evident when we position many of the greatest existing and new concepts in the model SPALT. It is by these means we find the most direct, clearest and accurate proof and view of TO, TOV, TRU, TOT and all.

What holds all together in TO? It is the Universal Contract, TUC, among all constituents in, and of TO working in harmony and success. This is not as much a contract of willing humans with any totality as it is of all components of absolute totality and relative totality working as one in TO. It is the propelling and compelling forces of all that enable and enhance bonds that assure TO, its existence and continuity.

It is by the total relative position and role of continua and universals that we need to continually work to develop TO, how they form, participate in and prove TO, and how they explain many of the problems, paradoxes, and how each helps to resolve all in TO.

That perfectly absolute totality and perfectly relative, like the polarities of many continua when taken absolutely, can't exist is evident in the fact that relativity and quantum mechanics are necessary for the existence of absolute totality, and physical features as mass, length and time are necessary for the existence of relative totality. This combined fact, of most basic relationships, is a large proof of, and suggested sum total interactive state necessary for a final conclusive proof of TO

To correctly approach and hold all in perspective we must disclose what is within what, which is within which, why and how they are in this way. By contrasting dualities of continua, centering each duality, determining their states, conditions, limits, relations, proportions and positions relative to each other and the whole, we can determine which, what and how dualities, entities and topics exist in the whole, of actual totality. This is supported by determining what is lacking or gained in their ascent or descent in the hierarchy from non existence to existence, existence to necessary existence, necessary existence to actual pure essences, and essences as they exist in various worlds, universes, systems, deities, totalities and actual totality. The method of approach to, and perspective of TO by centering, positioning, proportioning and summation of dualities may be signified by its law, LACOD, The Law of the centering of dualities. It is by the centers and summation of dualities, as they exist in actual totality that we can most readily and accurately determine the features, modes, varieties, mechanisms, form and function of actual totality.

Of the continua and dualities that exist in actual totality that of the absolute-relative, and its subsets matter-mind, or physical-mental and objective-subjective are prime.

The relative and absolute are extremes of a duality which forms a basic continuum in the universe. The constituents within this continuum and universe must exclude all else, All must exist within to the extent of their part, and the equilibrium and adaptation of the relative\absolute duality and continua.

The physical and the mental, as objective and subjective do not exist relatively as separate entities. They like many of the most basic modes, forms and functions are paradoxical dualities of universals. Universals and particulars, as polar limits of a continua do not exist separately, they can only exist within an ultimate unified totality.

This is the Law of the absolute relative continua and duality, LARCAD, It is the primary proof of TO and the direct to indirect basis for many other proofs and laws.

CHAPTER 12 - ABSOLUTE and RELATIVE in TO, to ARTO

Never All or None, only More or Less, Form and Function, Bonding and Equilibrium in TO

Kinds of TO include the most basic of continua, between absolute and relative, absolute totality and relative totality bonded in TO. This bond reveals the law of never absolutely absolute and never absolutely relative, or LANERA. It is the proper form and function of absolute, relative and their union in TO that most effectively describes and proves TO. This bond is signified by ABSTO-RELTO, and ARTO. The bond that is the basis for much of TO, including it's generic to specific varieties, TOV. The internal separation of TO, into absolute and relative, extremes and means, integration and differentiation, characterize the kinds, varieties, equilibria and ultimate coincidence that create, develop, make and determine TO.

That perfectly absolute totality and perfectly relative totality can't exist is evident in the fact that relativity and quantum mechanics are necessary for the existence of absolute totality, and physical features as mass, length and time are necessary for the existence of relative totality.

The physical universe is not entirely the same as its extreme, absolute totality, nor these as bonded, ARTO in TO. The relative universe is not entirely the same as relative totality or its extreme relative. The physical universe differs from absolute totality mostly by the presence of relativity. This is evident in the theories of relativity and quantum mechanics and the presence of an observer. The relative universe in contrast to relative totality consists of the presence of the absolute. The spectra of each of these totalities and universes are large components and divisions of TO. Absolute and relative totality, as ABSTO and RELTO fully describe TO, and the physical and relative universes describe only realms in TO. To discover, understand and prove the spectra of each in and out of TO resolves many of the paradoxes of continua in TO and enigmas of life.

Actual Totality, TO, mostly develops from the expansion of relation and alteration of absolute, the build up of mass energy sufficient to reach a threshold, like a quantum leap or fractal. To develop absolute-relative totality and all that enable the generic-specific varieties of Actual Totality or Totality Proper, TO, is one in which relation overcomes, improves and incorporates absolute totality to yield general to special varieties, as the physical universe and the relative universe, TRU, of all that enables best human life. By their taxis, tropisms, and tendencies to selective powers of generation and growth they produce an interactive convergence that leads to TÓ. This helps stability and reveals the identity and continuity that show and prove Actual Totality.

Absolute totality and the physical universe are held to be the totality by many in the modern world. This is especially so in the basic sciences that contrast it with mind, philosophy and God, and know its recent origins in the development of life and mankind. Yet relative totality, the relative universe, mind, philosophy and God have necessary existence, whose presence is much more than absolute totality and the physical universe. Only when we understand the vast, intensive and imperative form and function of relative totality and the relative universe of knowledge and all mankind, civilization, its products and beneficial effects on the physical universe do we prove and have the certainty actual totality requires. This is the certainty of the actual existence of relative totality and the relative universe by actual pure essences, fusion with absolute totality and unity in the entire totality of all the content of actual totality.

Neither the physical nor relative universe is all. Nor are they all combined, even though they are both large parts of actual totality. Actual Totality consists of more than both physical and relative universes combined, as is evident in the form and function of actual totality and the features and proofs of this book.

Fundamental to totality and TO are the differences between absolute and relative totality, and how they must be accommodated. Like the duality of all continua in TO, absolute and relative are of the most important, yet poorly developed and neglected portions and features of knowledge. In absolute totality objectivity and subjectivity are non existent, a separate feature of mass and energy dependent on its being known, i.e., relativity. So also is it for reality and the essential in actuality. In relative totality, especially as this becomes increasingly relative and decreasingly absolute, objectivity and subjectivity are more related, tend to fuse, becoming real, the basis for reality. This is seen in observer-observed relativity and the development of mass energy in The Relative Universe, TRU. Reversely it is how the differences between absolute and relative totality approach equilibrium in ABSTO-RELTO that their accommodation fulfills Totality, TO-TOT. As TRU and TIT become increasingly relative, especially in less and less time, increasing time necessitates the inclusion of more and more of the absolute. This is seen in how the subjective of special TOV-TIT must assimilate the objective, and do so with TOT-TOF in ways that maintain optimal equilibrium and existence.

We use the accepted definition of absolute. All free of relationship or relativity, independent of all outside itself. Being free of relationship produces its continuum with relation. Being independent of all outside itself provides a high degree of exclusiveness for TO. These are two of the greatest aspects of absolute that reveals its actual pure essences and their alteration with absolute totality in TO.

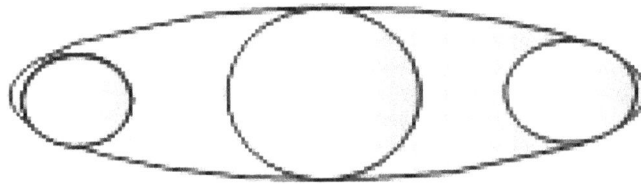

Fig I2-1 Elliptical Totality, TOT, with TO in the center, absolute and relative totalities completely separate.

This is the disassociated state of the physical and relative universes, or GOD from common knowledge, when independent and separated from, and only potential to actual totality. TO is the central circle, TOT the larger ellipse, with absolute and relative totalities as smaller circles in the left and right.

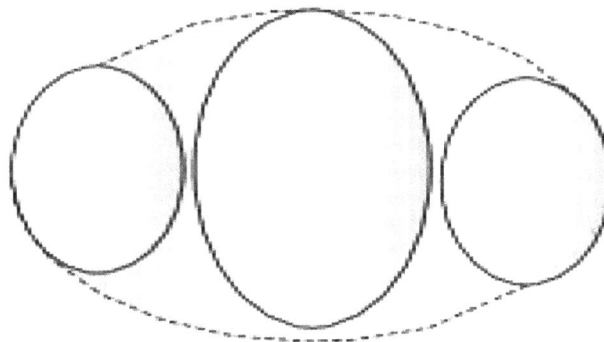

Fig 12-2 Absolute and relative totality in TOT, separate but approaching potential to actual existence in TO.

We must always remember the exclusive existence of TO in the extended present, yet spectral gradients with longer durations. We must also always remember the dynamic of TO, expanding and highly interactive with the external. Finally we must always remember how greatly TO changes with reversibility, especially in the long term.

Absolute totality is not the same as absolute, but very close to the physical universe, and when extreme, less of their existence in TO. We have not yet discovered an exact identity of absolute or a physical universe. The absolute does not exist in TO without dependence and relation. In this way absolute is key to, and helps to prove, TO. Similarly relative totality is not fully known, but has much in common with God, philosophy and the mental or Relative Universe. As with all things, we must be cognizant and remember that our thoughts and descriptions are less than their proper actualities, unless limited to their precise part and actual pure essences in TO. We can clearly see the dependence of TO, and all, on time, and their transitions, interactions and union in TOT.

Law of Limits of the Absolute, and its approach, (LOLAT). The absolute like its antithesis zero is objectively equivalent to non existence. By APET, however, it has certain meanings relatively, in TO. Thereby absolute like non existence can't be existent unless qualified. This is the sense of absolute, which infers nothing other. Absolute may hold to the degree it is in flux, not quite absolute, but with more or less degree of relative. It is this more or less, less than absolute, but relative absolute from which TO achieves necessary existence, actuality, permanence, immortality, eternal continuity and proof. This is similar to the state of most general totality until qualified to become potential and finally actual, stabilized in equilibrium to become TO.

The biological law of self preservation and continuity of life are the preeminent powers of necessary existence. An existence with modes of actuality, objectivity, subjectivity and reality whose mass energies continually transform and increasingly expand to strengthen, perfects and make permanent actual totality. This is largely a product of the frequency, predominance and preeminent role of forces in relative totality toward the subjectivity mode, or mental, fortified and set in the form and function of all ARTO in TO.

As relativity is a limited function of absolute totality, so is absolutivity a limited function of relative totality, absolute certainty occurring when subjective at one pole, and relative evidence when being objective at the other pole of this same spectrum. This is the spectrum of absolute and relative totality that forms ARTO, the major division and dimension of actual totality. From least to most, all depends on the universe, worlds, system or deities, they are in. If of the ultimate unified totality, and actual totality, they converge by actual pure essences in the most natural and normal unity, TO.

The quality of proofs depends on how well TOTRU is known. It is the influence of TRU on TÓ that is a large factor in perfecting the actuality and reality of knowledge. This is because, to be well known, we must understand the world we live in, TO. TRU being of RELTO, the special variety of TO, converges on absolutivity. This is the Law of Absolutivity Proof of TO, LÂPRUT. Like the physical universe must have relativity, so must the best world of human life, TRU, and thereby RELTO and TO, have absolutivity. This may be special and general relativity, and special and general absolutivity. Each of these exists only in that form typical for TO. The absolutivity of relative totality and TRU when in general, is signified by TES, when special by TÍS. TES are the convergence of all humans on the absolute, and TÍS is the convergence of the first person on the absolute. When viewed from the most separate aspect of TO the two kinds of relativity and the two kinds of absolutivity tend to take overall from their extreme. When they approach TO, as with ABSTO-RELATO and ARTO they converge, fuse and become united by their adapted actual pure essences as they coexist in TO. This is the Law of Special and General Absolutivity and Relativity in TO, LOSGART. Like all parts of TO, the four extremes of special and general, exist by LOPAN, but only as this existence is no more nor less than their part in TO, e.g., REM. (14, 22, 29)

The progression of necessary essential existence is largely a function of the continuum and spectrum absolute to relative. Absolute establishes existence of the entity, thing or object, relative makes essential and further qualifies their parts and processes in TO. Hereby is the mechanism of the basic categories of TO, how it is made, what it is, and how to show and prove it. TO is not any simple entity, thing, sense, object or subject. Yet nothing has actual existence except in the way, and to the extent it is in and of TO. For this reason each and all persons must discover, learn, know, identify with and be, their parts and roles in TO. To the extent they do not, they only believe they actually exist. Just like all the entities and objects we sense, they are only figments of their actual existence. Like any totality, any consciousness and belief in existence are archaic, primitive, disproportional and incomplete. Only by their form, function, state, conditions, limits and relations in TO are they fully, really and actually existent.

Relative totality is less than actual totality and the relative universe is less than relative totality. Yet the relative universe is more than one variety of the five specific varieties of relative totality in actual totality. Thereby the relative universe, like the physical universe is more than a variety, but less than either relative or absolute totality, hence a kind of hybrid. It is this hybrid state of the relative and physical universes that makes them, and their great and powerful roles, difficult to understand, cf. Figure 14-2. However once we discover, prove, learn and apply all that is actual totality, relative and physical universes can be brought into their part. It is how both permeate all of actual totality, and give it much of its unique and essential form and function, It is how the relative universe by its most extraordinary focus on human life, mind, reality and survival produce in actual totality the qualities and pure essences of necessary existence.

The relative universe, together with other facets of relative totality, produces through forces of convergence on unity, the characteristic mass, length and time, here and now, orientation of actual totality. It is this convergence and orientation that give actual totality its stability, a stability which functionally and practically enables us to treat it as a closed set. This is supplemented by the objectivity of the physical universe of absolute totality.

It is by the great transformations of actuality, and actual pure essences, whose varieties and gradients are revealed in the absolute-relative kind of TO that general to specific enables us to most readily recognize, show and prove TO, its varieties, TOV, TRU, and human life to come, with TO in TOF.

Since absolute totality and relative totality neither are separate, nor have actual existence other than how they are in TO, it is their combined presence, and fusion in TO that determines and proves them, TO, and much of the make up of TO. This is largely a result of LANERA, Law of never absolutely relative or absolutely absolute. Their combined existence in TO is part of the dependency of all on TO, and more or less on each other. This is signified by the Law of Absolute Relative Combined Dependency and Convergence, LARDAC. The unlimited power of this dependence is what unites, converges on, and proves actual totality.

It is by the Law of More or Less, LAMOL, rather than all or none, that only TIT-TRIS-TINI, as TÂR, tends to approach the reversible absolute. Its more or less limit more relatively approaches the absolute of TIT-TO-TOT-TRIR-TOF. More or less is the ubiquitous form associated with DAOST whose function continuously seeks optimal balance. In the Law of Reversible Absolute and Relative, LARÂR, absolute and relative do and do not exist, only more or less, by LAMOL, and vary with TOV, the varieties of TO. This applies from absolute to zero, as general limits, and infinite to infinitesimal in the present time of TO. This depends on whether it is of TO, or some lesser aspect or part.

It has been claimed since its origin in Greek Philosophy that God did not exist before man, since God was a product of the discovery, conception and progression of human ideas and imagination through time. This may be extended to all, and all reality. They are a discovery of the human mind. Herein is the crux of relation, of relative totality. The subjective powers of the human mind are the focus of relative existence, and a great essential meaning of actual totality. Actual totality cannot exist, neither without the human mind and relative totality, nor without the physical universe and absolute totality. This is proof that not only relative totality exists, but absolute totality exists, and how they exist, not as they are separately, but how they only exist combined and united in the whole of actual totality. This is by the necessary existence or actuality, whose combined and adapted actual pure essences provide the form, function, states, conditions, limits and relations of TO. It is the inception of relative totality that is both an origin of TO in TOP, and its exclusion of all separate entities and totalities. Actual totality is, by its own presence, being at once and autonomously existent.

Thereby if the physical universe and God are limited to their existence, form and function in TO, they are more or less equivalent to TÓ, TREQ. Since both the physical universe and God are held theologically and scientifically to their form and function separate from TO, it is neither correct, nor practical to alter their meanings to equal TO. Thus to assume they existed before man, before all, so are, they true or false for TOT. The physical universe and God, are all encompassing, when typified by their omnipotency, omniscience and fullest actual pure essences form and function in TO. They are less so when separate, and when typified by transitions in time, TO with TOF.

It is thereby this typical form and function of TO, centered on humans, minds, knowledge, subjectivity, consciousness and self consciousness as they exist in TO, combined with the mass-energy, spatial, temporal objectivity of the physical universe that actual totality, with absolute-relative totality, as ARTO is confirmed. This is a confirmation dependent on the transitions and transformations all TO undergo with time. When fused and bonded as ARTO in TÓ, the whole is proven. The critical difference is, transitions in time, adaptation of TO-TOT, LANERA. In this way both the physical universe, God, philosophical systems and various worlds have great meaning when held to that equal to their part in

TO. Yet much lesser meaning to the degree they depart, and are separate from TO. If we are to use anything, or meanings that are separate from TO we must always remember it must as a whole, conform to all that is TO, including absolute and relative totalities as ARTO.

It is the objectivity of the absolute and physical and the subjectivity of the relative and mental that combine and are received by their actual pure essences in actual and proper totality that certifies and proves the ultimate unified totality that is TO. It is by the sum of absolute-relative, other kinds, varieties, modes, and actual pure essences that TO achieve its complete form, function and proof. The unity in totality that relative totality brings to TÓ is expressed in "Univocity of Being." This is the power relative totality has by the Law of Partial Necessity, LOPAN. It is the power of life, being human, and its mentality and ideals, which by their actual pure essences are vital, vitally necessary for TO.

It is the function and property of relation, by all relative totality, to increasingly produce convergence, centering and unity. Conversely it is the function and property of absolute, by all absolute totality to increasingly produce divergence, dispersion, and disunity, e.g., the expanding universe. This is a feature of the Law of Absolute and Relative Polarity, LARP. Such a proof of relative totality, and TO, is a reverse of TO and absolute totality. For it is absolute totality and a material physical universe whose extreme position denies an existence of God. It is in this way that God when TREQ, equivalent of TO, exists neither separately as relative nor absolute, but only combined to produce the full spectrum when they together form TO. This reveals the critical fact that it is in total proof, as shown in the circular figure of combined proofs that there is perfected and final proof, cf. Ch. 38.

Absolute totality cannot exist without some of the relative, e.g., relativity. Nor can relative totality exist without some of the absolute, e.g., the physical. This is signified by the Law of Absolute Relative Dependence, LARD. This is a generic form of LARDAC and a corollary of The Law of Equilibrium of Dependence of Necessary Existence, LEDEN. These three laws are the foundation of TÓ through the form and proof of its development, from non existence to existence, to necessary existence, to actual pure essences, to different and same totalities, TOT, TO, and TO in transition with TOF.

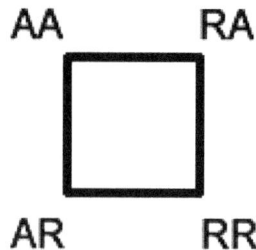

Fig. 12-3 Four cornered extremes of the absolute relative continuum, absolute absolute, absolute relative, relative absolute and relative relative

These figurative relationships are non existent and existent depending on their existence with APET in TO. It is non existent by the extremes taken separately. It is existent by their practical and functional actual pure essences, existence, and combination within TO. It helps to show and prove how absolute and relative do and do not exist, more or less, depending on their external view, united form in TO and transformations in time.

The difference between absolute and relative, certainty and uncertainty, is between two different continua in totality. By LANERA absolutely absolute and absolutely relative do not exist, except by the actual pure essences of the object, universe or totality in which they exist. It is because TIT and TAR, as the extreme special variety of TO, whose sum converges on the center, here, now and IA, that we have a perfectly closed set. It is the role of TAR, TIT and TRU in RELTO, ARTO and the modes of TO that a practically closed set enables us to position all by their actual pure essences. In this way the uncertainty principle of a physical universe and what monotheism lacks can be corrected or added, to prove TO. (cf. 35)

	Absolute Totality before and other than into TO	Absolute Totality as it exists in TO ABSTO	Absolute and Relative Totality only as they are together in TO, ARTO	Relative Totality as it exists into TO RELTO	Relative Totality before and other than in TO
Devel op- ment	Physical Uni- verse without relative, before life and humans	Physical Universe altered by relative universe & develo ping actual totality	Physical and Relative Universes combined plus all developing typical of TO	Relative Universe altered by physical universe & develo ping actual totality	The Relative Uni-verse without ab- solute, As it grew with human life
Form	Dimensions, mlt altered by absence of relativity and,	Dimensions, mlt altered by relative universe	Physical and Relative Universes and dimen-sions combined plus form of TO	Dimensions, altered by what is predom- inant in Relative Totality	Predominant form of human central nervous system
Functi on	Function altered by absence of relativity and,	Function altered by relativity to be accepted by TO	Physical and Relative Universes combined plus Typical of each and TO	Relative empowers all to convergence on and unity in TO	Positive forces of convergence Relative Relative
Parts	Matter-antimatter with mass energy	Mass-Energy Incipient S-O	Mass Energy as S-O, Á Prime dimension in TO	Highest forms of S-O, reality, Á	Any human life and Mind,
Other		Relativity	LOSGART, LÂRPET	Absolutivity	

Table 12-1 Contrasting content of absolute totality and relative totality, existing separately or in TO

This table is designed to clarify absolute totality, relative totality, ABSTO, RELTO, and their common existence, ARTO, in actual totality. This helps to show and prove their contribution to, and key roles in TO. It largely affects and forms the whole of actual totality and its highly dynamic interactive massive state.

Absolute totality as the physical universe with a high degree of certainty provided by accelerating massive knowledge and relative totality as the world of best human life and mind with a high degree of certainty provided by the whole experience of existence on earth, are each most reliable and beneficial. Separately, however they are in opposition, and less reliable as actual totality. The greatest problem that faces mankind is, how do these two great opposing totalities exist as one, actual totality, and how is this opposition balanced and controlled in TO. This is the critical question, test and work of the next chapter, remainders of this text, and human existence in the future.

Is it absolute totality or any entity, thing or sense? Is it relative totality of any entity, sense or thing? Or is it absolute totality of the entity, thing or sense, and Relative Totality of the entity sense or thing? Only when absolute and relative totalities become definite do they acquire necessary existence, actual pure essence and the qualities that enable a presence in actual totality. It is this difference between any absolute, any relative, and their definite convergence, fusion and union, as ABSTO and RELTO to ARTO in actual totality that is the final arbiter of what the ultimate unified totality is, as it is TO. It is proven by LOSGARD and LOSGART. These are the Laws of

Special and General Absolutivity and Relativity, LOSGARD, and in TO, LOSGART.

By the actual pure essences of absolute totality and relativity, and those of relative totality and absolutivity, TO comes to life. The actual pure essences of absolutivity relative totality and the relative universe reveal the great link and bond that establish TO. By their fusion TO acquires necessary existence. This is a primary proof of TO, LÂRPET, cf. Ch 4.

The fact that actual totality is characterized by both more and less, and absolutivity, are not incompatible, they are different aspects of the same thing. Actual totality exists by relativity of absolute totality, and absolutivity of relative totality. The absolutivity of existence here and now of the present is balanced by the relativity, and more or less, existence in extended time, past and future. Only when actual totality is understood, by the great modifying and typifying features, aspects, form, function, component parts, processes, characteristics and properties of its essential existence, can it fulfill its purpose.

The relativity of the physical universe and absolute totality, and the absolutivity of the relative universe of best human life and mind, and relative totality, are highly inter- dependent. This absolute and relative interaction and dependence becomes ABSTO and RELTO in actual totality, it is signified by ABSTO-RELTO, whose convergence and fusion is in their unity ARTO. ARTO is realized by the observer-observed, or subject-object, in the necessary existence or actuality of TO. This is a primary proof of actual totality, and its development, existence and representation.

CHAPTER 13 - Absolute and Relative, Generic and Specific in TO, ARTO to Varieties of TO

Development and Fusion ABSTO-RELTO, United and Dependence in TO

To prove TO, and disprove the exclusive existence of the physical universe, we need go no further than to discover how TO is not mind over matter, humans over all else. It is matter, mind, all else and humans as they exist in TO. TO consists of a vast array of varieties and constituents, from those that are extremely absolute to those that are extremely relative. This all inclusiveness, is produced not only by the more specific varieties of relative totality. It is produced equally by all that constitutes, ARTO, the combined ABSTO-RELTO make up of TO. It is by the actual pure essences, of each and all as they exist in TO, whose equilibrium distributes the context of TO in the right degree and proportions. TO in all its varieties, and in a sum, is much different from a separate absolute or relative totality, any one variety, or even half or any portion of varieties.

Of the laws and development of kinds of totality, nothing can have actual existence either, completely absolute or relative. Even TO does not exist only absolutely or only relatively, only together, more or less, by LAMOL. Although TO is highly and increasingly stable it is always changing with time and TOT, as evident by transitions in time. It is only stable at zero time, and since this is a vanishing point it is never absolutely stable. Like the practical set state of TÓ in the present, so does it have actual existence. All exists, in a way and if only to an infinitesimal degree by their observation, if perfectly of TO. Thereby to have actual existence is also a proof of absolute and relative, ABSTO and RELTO, ARTO of TO.

The law of more or less, LAMOL, is the continuum and spectrum of never all or none, and characterizes the absolute and relative kind of TO, existing as ABSTO-RELTO , or bonded as ARTO in TO. All exists more or less in/out and dynamically and statically in TO. TO, its modes, continua, kinds and varieties are in transition, changing and highly interactive, from the potential and actual both within and outside themselves. Positive and Negative, +- or PN, is the condition that largely selects what is in or out, both at the interface of in and out, and processes within. Usually what is positive strengthens and enhances, what is negative causes deterioration. Humanity is, to a large extent, outside, although by actual necessity and RELTO much of TO.

This is the dichotomy and enigma of life, how to be in TO. For this reason humans endure both an amount of non or lesser existence, deprivation, sickness, suffering, misery and failure, and a necessary amount of facility, good heath, well being, abundance, success and permanence. Although only relatively absolute and exclusive, TO and TOV do not exist without external relations and effects. Both TO, TOT and all exist in their own kind of positive-negative dynamic. These are effects of all that are less than for TO, from slightly less positive to completely negative qualitatively and from tremendous to slight quantitatively. Corrections are made by RELTO and by ABSTO, by an in and out contract with nature, as they exist by actual pure essences of TO.

The absolute existing alone, and the relative existing alone, lack existence in actual totality, except by their theoretical usage in demonstrating all. This figure shows their separation, and combined presence in TO, the central circle. It is the presence and form of their combined functions in TO that are most fundamental and whose evidence is key to learning, showing and proving TO. Absolute and relative actually exist as the massive absolute-relative continuum, ABSTO-RELTO in TO. This is the major difference between physical universes and TO, or any totality and combined generic-specific varieties of the actual totality, TO. Most generally and absolutely the world is known as the physical universe, most specifically and relatively the world is known as God in religion, and integrated systems of knowledge and philosophy. When, in generic to specific, and absolute to relative are joined by the qualities, modes and other features, the world must be correctly known, by TO.

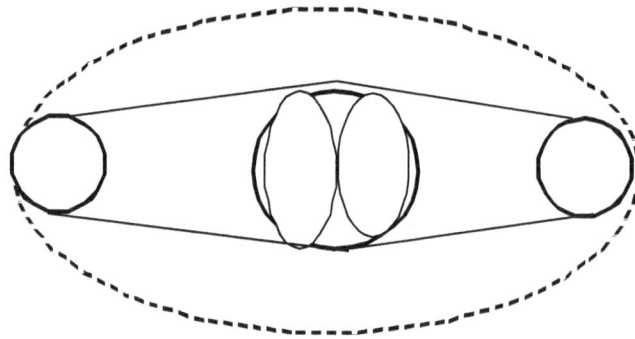

Fig. 13-1 The absolute relative continuum outside of, and ABSTO and RELTO inside of TO

The origin, development and unity of TO occurs from the incipient existence of relative totality in absolute totality that is only of the past potential of, TOP, toward TO present. An exclusive and inclusive TO that exist in the extended present occurs by its practical and functional actual pure essences, ACET, The somewhat variable, tentative and expanding interactive present and future exists with TOT. This is by what is less than TO, or it lacks and loses, and tentatively by many factors, not the least is how well or poorly we, individually, collectively and totally upholds, sustain, maintain and improve it.

Actual totality is constantly undergoing forces that cause permanent, temporary or reversible change. TO obeys laws of action reaction and conservation. TO undergoes continuous positive and negative forces of change. There exists at any time, period of time, or segment within TO different gradations and degrees of separation. Nowhere, is this more basic, than in the varieties of TO, TOV, where each variety through actual pure essences produce in TO both changes in the variety and changes in TO. Yet it is this great variance that typifies and proves TO. It does so by revealing the vast and inclusive form and functions of its features, continua, modes, varieties and components. Fig. 13-2 helps answer this problem by showing the progression extreme absolute to extreme specific and back to ABSTO-RELTO equilibrium in TO. The role of serial states, not only separately, but how they coalesce in TO, as it in turn, is reversible and potentially of TOT.The figures compare changes in TO and its varieties in past, present and future time. It emphasizes major differences of TOT, and union in the continuity of TO and TOV with associated proofs.

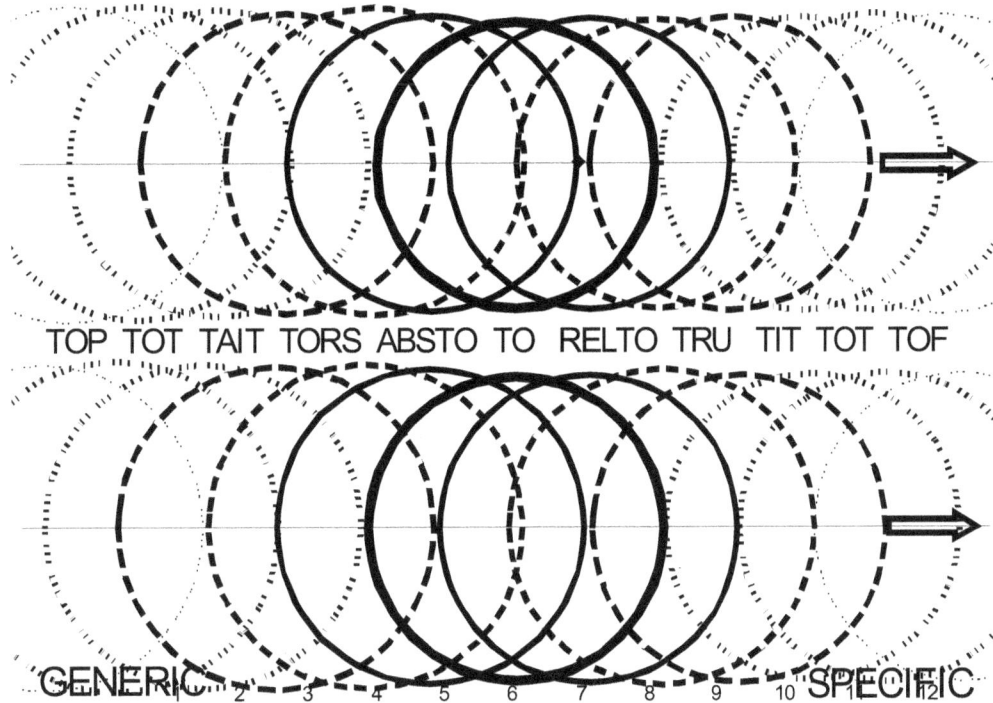

Fig 13-2 Change, transitions, transformations, from absolute to relative, and general to special varieties, TOV-TO, reversibility and comparison.

The distinction, presence and interaction of absolute and relative become clearer when we compare their extremes. In the differences between absolute totality and physical universes, the latter must have at least a degree of relativity or relation. An exclusive absolute totality has no set relativity. Likewise, the relative universe of best human life must have at least a degree of the physical and absolute, which exclusive relative totality has none. The absolute can't exist without the mental. The relative can't exist without the physical. So that we have, are and will not exist without both mental and physical. It is for this reason neither absolute nor relative totality alone has separate existence. They have actual existence by their quality and quantity correctly combined in TO. This is the Law of relative and absolute correction and fusion in TO, LARCAF. This is one of the greatest proofs and laws of all, it certifies the fact that, and how the combined presence, form and function of the physical and relative universes of actual totality, TO, coexists. It states, conditions, limits and interrelates many of the laws and proofs of TO.

To make correct descriptions of all, originating in the absolute relative continuum, solutions to the problems of TRIS-TIT-TO-TRIR TOF are made, thereby proof of TO. For example by the Law of never absolutely relative or absolute, LANERA, as most laws, does not hold absolutely, but must be properly adapted, and interpreted by the actual pure essences of TO, TOV, TRU, TIT, TRIS and TRIR with TOT or what is external. It must fulfill the validity, actuality, proportionality and all of TO, VAPO.

65

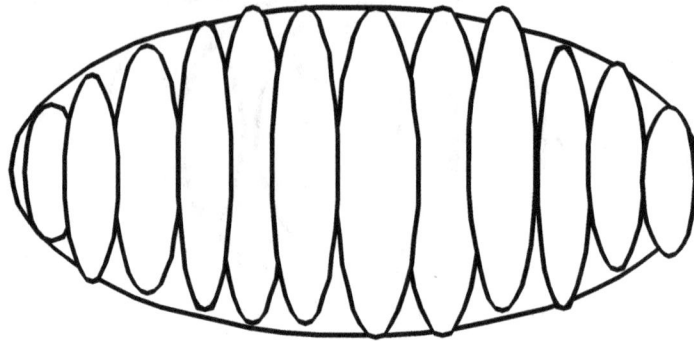

Fig. 13-3 Totality, Progression of General to Special, Categories, Varieties, Universes, Modes, Kinds Absolute and Relative, Relativity and Reality in TO

The term totality signifies all, and all that has existence within its whole. With actuality this is much more exclusive, inclusive and superior and becomes actual totality, TO. Although the term universe is often used interchangeably with totality, it by its own actuality a misnomer. The totalities of material entities, or universes are neither one, nor the valid, actual, proportional, objective, subjective, reality and relative convergence needed to define all necessary existence, the actuality of TO. Absolute totality has no center, exists by infinite regression and non determinant interactions, e.g., uncertainty principle. This is because it is exclusive of the observer, the observer observed, or a subject object continuum of relativity. Universe, unless signified as more than the physical universe, is less than relative, or adequately relative, and thereby less than actual totality. This becomes more evident and clearer when we recognize the part the physical universe has in the, generic to specific, and absolute to relative, continua of TO. All, any or the entities, things and senses general or special exist in or out of actual totality as polarities in a continuum of all. With general-special, some things and some senses make up levels of categories in the progression general to a special continuum.

Is relativity a part of the physical universe, or only a feature of its observation? This is a most important question, for its answer, will provide the greatest needed evidence for totality, its kinds and varieties. In addition it will show the way to discover, learn and remember totality. Relativity shows the existence of general and special, and their combined form and function. It shows actuality, what is needed for existence, mass energy, in space time in increasing levels, the building-blocks of life, mind and highest levels of specific varieties of all that enables the relative universe of best human life, centered on here and now. All must partake, to the proper degree, in the combined totality of generic to specific, absolute to relative, in the right way and proportions of ultimate totality, TO.

Development of Constituents of Actual Totality, TO, and its varieties, TOV. Most generally a universe originates as an existing whole of matter, mass and energy in space time. This is of entities and things together without necessary existence, actuality or combined objects in subjectivity, or reality. As totality becomes less general and more specific it acquires some of the form and composition of each, and other varieties. This is of senses, progressing to combination with things or objects in a new state of phenomena or reals. This combination is the beginning of reality, potentially developing toward reality, constituents in progressive united things and senses to become reals. Of all that is less or other than totality and its categories, TO and its varieties there are those that have much that is potential, as well as actual, in common.

This condition of mixed worlds causes great confusion and dismay. It may be more or less cured by realizing the continuum between any entity, senses, reals, things, worlds or universes, their existence in any totality, potential totality and the actual pure essences becoming their aspects, parts and roles in TOV and TO. Once this continuum is discovered, analyzed, and properly demonstrated all that is less or other than TO-TOV can be validly, proportionately and totally deduced. What is less than and other than TO-TOV may be perfectly valid in and of itself, it may also be valid by virtue of its part and role in TO-TOV. Yet it is still incorrect and misleading. For being valid does not assure actuality and actuality alone assures neither proportionality nor totality. The continuous interaction between categories of totalities or universes, between the varieties, TOV and TO, and the role of in and out, potential or actual, existence, help to reveal the stages and the special existence of each.

As we proceed with proofs, and investigate the roles of absolute and relative, we can begin to recognize the solution to their problems, corrections available, proof of TO and the role of TRU. By the laws of reversible absolute and relative, LARAR, existence of an absolute center, LEXAC, and the modifications necessitated by TRU, TIT, TRIS and TRIR and their external relations, the form and function of TO develops.

Disproof of Absolute totality and the physical universe as a whole exclusive universe -

The special varieties of TO, as TRU, have great presence in RELTO, and effect on TO, compared to all else, and all that lacks actual pure essence. This includes separate individuals, givens, senses and things that have actual existence within TO. We are blinded by our own senses and beliefs and fail to realize our deficiencies and limited roles in TO. Likewise it is so for the physical universe with objectivity that lacks actual pure essence and focuses on the differential, detail, limited role and departure from TO. Only the union of all with actual existence that is formed and ordered by its totality, by the combined bond of ABSTO and RELTO, ARTO, satisfies the requirements of, and proves actual totality, TO.

If absolute and relative is one in actual totality, are they equal, and in what ways? How are they in balance, with an equilibrium that supports equality? How have they existed through time, TOP, and with smaller relative totality? In what way did and will vertical integration exist in past and future totalities? Such occurrences suggest great change in their form and function. It also suggests marked changes by direction of flow in the TOT - TO transitions and transformations. This is supported by the form and state of mental development. The past and future, TO, are not to be measured by present TO, Each stage has its own complement. This is supported by the fact that the physical universe and God, as many great constituents of TO, were and will change both in quantity and quality. Such analyses as these, help the answers to the above questions.

Since absolutely absolute or relative do not exist separately, all, even truth, exists more or less of one and the other. This holds within, and more variably outside of, actual totality. This is the solution to the problem of exclusiveness of a physical universe versus philosophy, as well as God. They exist, yet this existence is more or less absolute and relative depending on its state in mass energy, space and time. Even the extremes of the astrophysical universe are shown to be changing, incrementally in the shorter term, and fully in the extremely long term. So is it of philosophy, God, and actual totality, each in their own way relative to mass energy, space and time. This is disproof of separate existence, and a major proof of TO. This does not deny a high level of distinction when near perfectly within actual totality. It even applies to the degrees of perfect exclusiveness of TO. This is by LANERA, and its corollary, the Law of Absolute Relative Exclusiveness, LAREX. The latter explain the in and out condition of much that is potential, and less than TO. TO is highly exclusive practically and relatively in the extended present. It is increasingly less so in the more indefinite extremes of future time.

Division typifies and helps to prove Actual Totality, TO. By optimal divisions of the form of TO, its order, proportions and unity are best shown. It is by division we separate, the generic from the specific, and varieties of TO, TOV. By specifying the separate or divided components we can develop relation and order. This is a universal method, based to a large extent on continua whose spectra and dimensions provide precise proofs. This is evident in the proofs of relation, RELTO and Varieties of TO.

Every variety or stage of TOV has its own type of absolute and relative quality, form, properties and characteristics. Varieties that are corresponding opposites tend to mirror each other in certain ways, AA, AR, RA and RR. It is the sum of the actual pure essences of these opposites and their intermediaries, typical for each TOV, and when APET in TO that not only make up TO, they explain and prove TO. This is provided by the vertical integration of ultimate unified totality, when of the proper, actual totality, TO.

What is generally included and excluded in TO, by reason, perspective and supervision? The Law of Excluded Middle, (LEXMI), although derived and used mostly in logic, has applications in, of and by TO-TRU. It is like COL and COR, the limits and correction of all relations, potential or actual to ABSTO-RELTO and TO. When duality consists of two opposite poles, other poles are excluded. Also the degree it functions as a spectrum a perfect middle or equilibrium does not exist. It is of motion, or in flux. The greatest implication of this law is the relation between worlds, or universes, between the physical world and God, or a physical world, God and TO. Their participation is only more or less of the spectrum, or TO. When of spectra of TO, they are variously distributed, never in perfect equilibrium.

All are excluded by their differences to each other and TO. This is a large part of the reason why it is so hard for the physical universe, philosophy and religion to agree, or to recognize what they have in common, TO. As partials, polarities or opposite they are incompatible, without adaptation, alteration, transformation, accommodation and good accounting. This is produced by the common fallacy of being indistinct. Resolution in TO is impossible. Only when polarities and the middle of dualities exist in balance by ABSTO-RELTO bonded to ARTO in actual totality, TO, and TOV-TRU-TRIR-TOT is the law of the excluded middle corrected. Correction derived from and justified by all that proves the exclusive preeminence and predominance of TO.

The varieties of TO form a whole that replicates TO and the qualities of the absolute and relative in TO. The hierarchy of TO that exists as all TOV, is only one. The specific or relative, and generic or absolute that actually exist in TO as a duality, are two different aspects of the same thing, TO. It is this same and different uniting in ABSTO-RELTO to produce TO that is one of its greatest qualities and proofs. This is how the objective and subjective universes can be polar opposites, yet large constituents in TO. By TOV they correspond to an extension of the extremes of absolute and relative totality. They do not exist as separate entities, objectivity and subjectivity, but together as extremes in TO, a most basic aspect and mode of TO. Only when external to TÓ do they have independent separation. The absolute or physical universe and the relative or mental universe, TRU, are modified by their actual pure essences. These are again modified by the actual pure essences of TO.

Together ABSTO-RELTO constitutes a whole, ARTO, whose actual pure essences coalesce in TO and future TOT. This is what provides the correct knowledge needed to understand and show necessary existence, or actuality. It is the diffuse transitory dynamic state of absolute totality, and the centripetal powers and concentrated stability of relative totality that define, explain and prove how TO can be an open and closed set. How the actual pure essences of ARTO and TO have sufficient convergence on here and now, and primary mass energy to provide practically effective and efficient unity in totality.

Acceptance or rejection by TO is a function of how greatly, poorly or by negation, entities exist not as things in themselves, but relative to their composition, function and practical use, or actual pure essences in TO-TOV-TRU-TOT. This proves the exclusive relative absolute nature of the actual world, TO, and much of TOV, and the part of best human life, TRU and all their components. It does not disprove all that is in and out, or outside, only that which is irrelevant, deviant, disproportional, impractical and unessential.

All these exist, are formed and function in the dichotomy of TO as perfectly exclusive in the practical and actual pure essences of here and now, and how it is altered or adapts to transition in time, with TOT-TOV, All else has a split aspect, separate existence in itself, or degrees of selected and accepted union in TO. These are highly dynamic interchanges, typical of ultimate unified totality, the actual and proper totality, TO. The degree to which TO is or is not a closed set by stability in time, and the degrees different varieties in TO consist of this stability, typifies their status in time. It is this very dynamic that TO in time must encounter and control if it is to continue to generate rather than degenerate. By its own dynamic, and those that have propelled it through the past, the tendency will be to continue to generate TO. Yet this is not assured, for there also exist, as in the growth and decay of all things, forces in both directions. This is why TO, TRU and all that enable best human life in TO, must be vigilant, rational, and use all their abilities and accessories to maintain and advance TO.

CHAPTER 14 - VARIETIES of TO, TOV, to all in Actual Totality

General Specific Spectra, Gradients, Equilibrium, Form and Function, Formal consolidation of infinite to Ten varieties, Independent to Dependent in TO, Relative Totality, Special Varieties.

It is by the great transformations of their essential actuality in absolute-relative totality in TO, ARTO, that we can most readily recognize and prove the varieties of TO, TOV and its constituents, TRU, TIT and TRIR with TOF. These series and gradients are revealed when the general to specific varieties of TO, TOV are well formed. They show the great and glorious progression of actuality from its incipiency in the barely existent to its most profound, massive, dynamic actual pure essences in the greatest of special varieties of ARTO in TO and best human life that most fully characterizes TO, and highlights Totality Proper, TP, the concentrated core of all necessary existence. This not only proves TO, its absence in other totalities, universes, worlds, Gods, etc., explains their inadequacies, and helps to separate what is in them that are TO from what is not.

It is by these great transformations of totality, actuality, objectivity, subjectivity and reality, whose series and gradients are revealed in the kinds and varieties that we can most clearly prove TO. Most basically TO is proven by the contribution of each of its varieties, and how they affect and determine the form and function in much of TO.

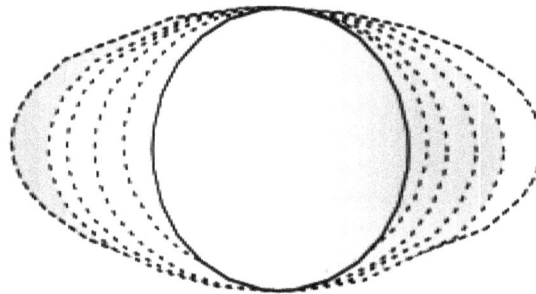

Fig 14-1 General and Specific Varieties approaching inclusion in TO, TOV

Although the varieties of TO exist into their origins and existence in time, by TOT it must be realized and used to complete our understanding of TO and all TOV has, is and will become. The five sectors on the left mirror the five on the right, and show how TO is formed and functions interactively and externally. They are a product of the absolute and relative totalities and the spectrum they form from general to special varieties. The mixture of generic and specific in TO, excludes all generic and specific that are not a part of TO. It must have a balance of the generic and specific of absolute to relative totality. In this way the specific focuses actuality to become the proper of totality, TO, i.e., Law of Specificity Focuses Actuality to determine TO. (LASFA) This is a most important concept that proves TO, as well as the part and role the more specific varieties and the relative universe, TRU, focus on and participates in TO. This is more fully understood when we envision how TO exists in time, with TOT-TOF.

The ultimate unified totality, TO, converges through TOV from its extreme generality to specificity in ARTO of TO The laws of totality conform to, and are confirmed by, differences that are reconciled in the whole of actual totality, TO. The whole is proven by extra speciation of totality. The unity and reality of knowledge, civilization and all the relative universe of best human life, TRU, and RELTO become ARTO within TO. This gives stability to the whole. The unity, reality and actuality of TRU, ARTO and TO, are of increasingly higher states, SÁ, of mass energy. This is the mass energy dimension in TO, or its prime dimension, Á. It is a major feature of TOV, the varieties of TO. The whole continuum of mass energy is transformed in TÓ with TOT to produce increased coexistence, capacity, continuity, power and total convergence on itself in totality, TO. They enable, prove and assure the actuality, absolute relative, and exclusive existence that RELTO, TRU and TIT give TO. This is the Law of the Absolute Necessity of Totality by TO, (LAOT) and its accompanying theory of truth.

Transition of absolute to relative polarity reveals the increasingly special existence of TOV. The differentiation of TO into absolute and relative extremes, highlights and characterizes the kinds and varieties of TO. In this way the varieties of TO assume aspects and features that are increasingly special and partial. This is less like their roles in TO. It may hinder the correct knowledge of absolute and relative, yet it helps to clarify the separation of TO, its varieties, and prove all their different kinds of existence in TO. It is existence that demonstrates the most basic features of TO, its transition, as well as the more or less interdependence and interaction of varieties that are the most important proofs of the exclusiveness of TO.

The succeeding varieties of TO are characterized by their more specific actualities whose form and function provide the basis for understanding the relative universe, and its transformations in TO with future transitions, TOF. By these great transformations of actuality, whose series and gradients are revealed in the categories of totality, and varieties of TO, we can most readily recognize and prove TO and its modes, kinds and varieties. It takes the specific varieties, by RELTO and TRU, to fully show and prove the makeup, accuracy and reality of TO. The more specific varieties of TO, e.g.TRU, describe the highly focused role of not only actualities in TO, but actual pure essences whose validity, proportionality, and quality produce the combined presence that makes TO what it is. It makes TO different from any totality, as TOT, lesser totalities and universes. It is in this way specificity, and much that enables specificity, becomes actuality, an actuality that is not any actuality, but the actuality and actual pure essences that exist in and bring the precise, exclusive and most real existence, TO, (LASFAC).

How the extreme varieties of TOV, as TRU, affect the whole, explain and prove TO, absolute to relative. As a result of the stretching of the extreme absolute and relative totalities in TO, ABSTO-RELTO with the advent of increasingly essential actuality, MESOA, reality, civilization, knowledge and technology, the hierarchy of varieties, TOV, are increasingly more heavily weighted in the extreme. This explains and proves the greater presence of TRU, as well as its contrast with the physical universe. Also this is corrected, and explains, TO, and the role of APET of each variety in the whole. How the most, and extreme, specific, best human life, the Relative Universe, TRU, and this as it is a thing in itself centered on here and now, TIT, are justified in identity with TO. This is a natural consequence of the modes of TO, relative totality, and the retroactive propelling and compelling forces of improving human life, in the specific TOT, TRU and TIT.

To make actual totality and its varieties more readily identifiable for use, we provide a formal reduction of an infinite series of varieties to a practical and serviceable number, 10, with a balanced format of ABSTO - RELTO in the whole of TO. Such a format of 10 varieties from extreme absolute to extreme relative provides a simple accurate differentiation of TO and its representation. By the kinds of TO, the varieties of TO assume aspects and features that are increasingly special. Of these are the distribution and equilibrium of its 10 states or levels. Their distinction helps to clarify the separation of TO, its varieties and to prove its typical existence. An existence that demonstrates the most basic feature of TO, its transition, as well as more or less dependence of each variety, e.g., TRU. TRU-TOV-ARTO-TO-TOT dynamics and their proof by the role of actuality, objectivity, subjectivity and reality. The TOV that exist from absolute to relative, generic to specific in transition through TO, reveals how TRU is not all. Yet, although centered at TOV-7 TRU is largely in TOV 6-8. However like many great categories of TO, TRU exists more or less throughout TO. This extension is characteristic of much of all fundamental to TO, and is an existence that decreases with distance from its center and increasingly differs in quality. To exist in TO does not infer loss of potential or actual by TRU, only changes its actual pure essences as they become these in, and by the APET of TO. This applies to all varieties of TO, and more less to all things that are actualized in TO.

The varieties of TO, TOV, from most absolute and generic to most relative and special are not all the same, nor is this the same for their external and internal form in TO. In this table of the ten formal varieties and the distribution of mass, length and time, or A, L, T, approximate limits and measures of each dimension are given. This is from 0 - 10 or most objective to most subjective in the mass energy dimension, infinite to infinitesimal in the length or spatial dimension, and infinite past to infinite future in the time or temporal dimension. The extreme varieties of TO, are characterized by physical objectivity, infinite length and time in most absolute and generic varieties. In mental subjectivity this is with infinitesimal length and time in the most relative and special varieties. Since the superimposition of actual pure essences both external, and internally by TO modify and external the influence and effect of these varieties in TO, the measures given will only represent limits whose extensions are typical of TO.

Of the many varieties, or formalized ten, the extremes are AIT for the absolute and TIT for the relative in TO. All varieties exist and function not only at their level as a gradient in the spectrum of absolute to relative, and general to special variety, their influence and actual pure essences manifest themselves throughout TO. This is seen in the more combined varieties of physical universe and relative universe in the spectrum of the varieties of, and form in, TO. It is for this reason the form of all in TO is heavily affected and produced by the physical universe, and the relative universe. Since form and function are closely fused and interconnected both physical and relative universes produce a profound effect on TO, largely by their mutual form and function in the highly permeable whole. It is by their actual pure essences, and the profound effects that the physical and relative universes, as opposed as they may seem, are united in, and prove TO. This is highlighted when extreme, in AIT and TIT, in TOT.

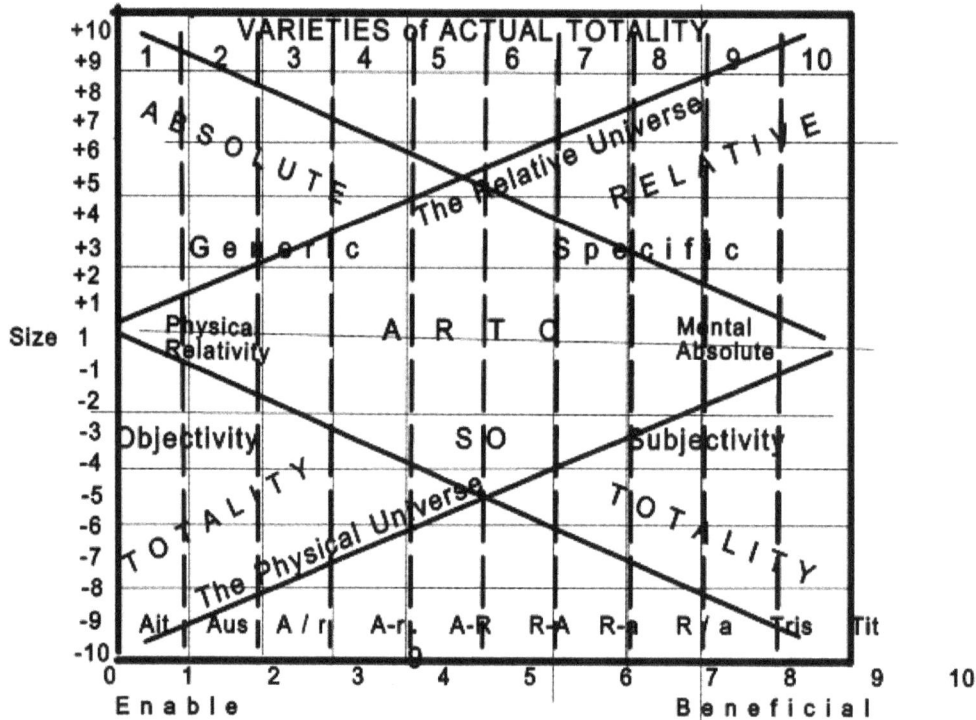

Fig 14-2 Actual Totality, TO, Major aspects, features and constituents by size and varieties

This figure, although rough and in development, needing much refinement and perfection, is one of the two or three greatest depictions of actual totality. Transferring items, in drawings by word processing, cause distortion. A good amount of imagination and tolerance is needed to get a perfect picture. Many basic topics could be included, such as, Mass Energy, A dimension, God, creation, evolution, worlds, capitalism, materialism, idealism, observers observed, stimulus response, problem solution, attention, civilization, worlds, systems, earth, etc. Through the text we seek to explain, describe and interpret major features that help to correct misinterpretation. Most of the items named in the diagram are both dynamic and static. They carry the actual pure essences of their existence, and of TO. The varieties are not only static, although formally discrete. Their interactions and effects are more or less diffuse and large. The figure centers actual totality, TO, at time zero, although its varieties include increasing extension in time. Companion diagrams for TO-TRIR-TOF, will compare what changes markedly and what remains the same.

Generic to Specific Varieties of TO Common Equivalents, Universes and other wholes

Extreme Generic, Ait, Absolute Physical Totality Physically Infinite time and place, no center

Most GenericAus, Aus, a, any Absolute totality Universes, almost all Physical

More Generic, A/r, Absolute dominant/relativeUniverses, largely Physical

Generic / Specific, A-r, More absolute than relative Worlds, more objective, variable

Generic-Specific, A-R, Nearing absolute-relative Centering TO, correct balance, maintains in TOT

Specific-Generic, R-A, Leaving relative-absolute Center of TO and greater strength of the relative

Specific / Generic, R-a, More relative than absolute Worlds, Enhanced Actuality of TO

More Specific, R/a, Relative dominant/absoluteRelative universes, God (variably)

Most Specific, TRIS, The Relative Universe Kinds of TRU and associated constituents,

Extreme Specific, Tit, Perfect Relative TotalityTIT, TRIR, TRIS, on a center, at here and now

This is a tentative listing of the varieties of TO, TOV, their absolute to relative, and common, equivalents.

Since generic to specific and absolute to relative are like spectra, they can be of an infinite number of gradients, or viewed as one-integrated continuum. The purpose is to present enough varieties to show the make up of TO, and how these varieties exist both internally to TÓ and separately in and of themselves or with others. In doing so, however, we must always remember, ECRO, that we are departing from TO, and thereby to this extent, lack actual and essential existence. Their actual pure essences in TO, are diminished, becoming that of the separate variety. This alienation from TO, is key to understanding all, and the reason why worlds, universes, Gods and many systems we commonly and currently live by do not satisfy the full meaning of totality. With the above list, and its extended usage, in mind we can learn TO, TOV, and separation of varieties in more order, proportion, and fuller way.

There are different kinds and variants of TOV, TIT and TRIS that are closely related to each other, This was given in the TRUS, TROS, TRIS, TRAS table, cf. IPTRU. Their slight differences, often universal, are most important, provide great insight, and reveal the great beauty, power and proof of each other, TIT-TRU- TO. They are key to accurate observation and understanding, future TOF and representation. Two most important concepts in the actual totality, TO, are variants of TIT and its form TRIS. These are TIR or TIT Relative, and TRIR, or TIT Reversible. TIR is close to TIT, only with emphasis on the exclusive forces of relation. TRIR signifies the dynamic mode in which TIT, as well as TRU-TO, are in a continuous state of in and out, and can be reversed, either positively or negatively. It is the dynamics of TO-TOT shown by TRIR that most readily reveals the bonds as well and distinctions between TO and all else.

The initial and final extremes of TOV in TO are TO or The Absolute Relative, TAR, and TO or The Relative Absolute. TRA. The latter is described in length in Chapters 27 and 28, TAR, its contraplete, is important in its own way in establishing the origins of TO. Both are strong proofs when balanced with the rest of TO.

	TO ABSTO 100/0	ABSTO / RELTO 90/10 80/20 70/30 60/40				TO Center ABSTO = RELTO 50/50	RELTO / ABSTO 40/60 30/70 20/80 10/90				TO RELTO 0/100
	0 AIT	AU-1 AU -2 Arus Aru				ARU	TRUS TRU TRIS TRIR				TIT 10
SPALT	TOV 0	1	2	3	4	Averaged TOV 5 Sum	6	7	8	9	TOV 10
0	0 - oo - oo										10 - 1/oo - 1/oo
1		1				1 - 100BK 100Q yrs					
2			2			2 - 1000K 1B yr					
3				3		3 - 1Km 100000 yr					
4					4	4 - 10meter 1000 yrs					
5	0 extreme large past					5 - 1meter 100 yrs					10 ext. small future
6						6 - 10cm 10 yrs	6				
7						7 - 1mm 1.2 mon		7			
8						8 - 1 Ang 30 min			8		
9						9 - 1 x10-4A 1 nano sec				9	
10	0 - oo - oo					10 - 1/oo - 1/oo					10 - 1/oo - 1/oo

Table 14-1 Correspondence of form of SPALT with TOV, Varieties of TO, ABSTO-RELTO and classes of Universes, Also the numerical basis for SPALT, thereby their existence in TO, ABSTO - RELTO, TOV

The general and specific form of Actual Totality, TO, is largely affected by major transformations. These are both external and internal in development and continuation of its form, function and dynamics. The gains and losses with each variety provide evidence of the transformations and dynamics of the varieties of TO. The transformations explain the differences and prove the preeminence of TO as TOV, a progression from the most generic or absolute to the most relative or specific, the specific from relative to more, most and extreme specific, TRU, TITand TINI. The principles of external transformation, apply to TÓ or the variety given. For the more specific varieties transitions to TRU, to and into TIT, and their form, TRIS, provide increasingly more stable sets. These are generated by the existence of developing and life forming processes that are highly self propelling, compelling and sustaining. They are accelerated in accord with the actual pure essences of the whole, TO, or the transformations of varieties of TO.

Subject and object combines, as DAOST, different aspects of the same thing, in the SO mode of reality in TO. This duality of subject-object, is the observer-observed of relativity whose continuum produces the SO bond that makes the mass energy dimension or primary Á dimension of TO and its varieties, TRU-TIT-TRIS. The subjective's role in SO, increases in the more specific varieties of TO. In TRU, TIT and TRIS the subjective is predominant. It is this predominance that more or less permeates all TO, and is the major determinant, characteristic and proof of TRIR, TRIS, TIT, TRU and TO, with TOF.

Enablement of varieties, hierarchy of transformations, and balance of generic-specific in TO

As a whole, and by the very nature of TO the different varieties of TO, tend to reveal and prove each other, PRUT. By LOPAN TO is highly TRU, is highly TIT, TRIR, and the reverse, as they exist as one, the only ultimate unified totality, TO. This is by what they are, how they exist in and add to complete TO. The progression generic to generic-specific and reverse is a constant dynamic of varieties, how they exist as one, the only ultimate unified totality, TO. Emphasis on increasingly more, most, to extremely specific, by RU, TRU and TIT clarifies the absolute to relative form of TO. It reinforces the role, dominance and prominence of each variety, e.g., TRU-TIT, since contrast with the absolute, or physical universe, is so difficult to understand correctly.

It is the whole spectrum of varieties of actual totality that enables us to visualize its content and inner-workings. From one to many gradients, or formalized in ten, they establish the frame work and dynamics that are prerequisite to differentiation of a unified and integrated whole. Evermore they prove the fact. form, state, limits and relations of absolute and relative totalities as they combine in generic to specific levels that most typify, characterize and show the properties and parts of actual totality. Each variety by its actual pure essences conditioned by those of actual totality provides a section that is a valid, reliable and proportional part of the whole. Each adds much to actual totality in its own way, as this way exists within the whole. This is evident in each successive level. How the physical universe departs from absolute totality to contribute its mass, length, time, and objective entities, and the relative universe of best human life, and God, contribute the convergent centering and distribution of mass energy spatial and temporal subjective entities. All are conditioned by the qualified selective powers of TO and its actual pure essences. Only with the practically closed set provided by TO, is there the opportunity to meet the rapidly expanding demands of differentiation.

Varieties, form and representation are all needed, vital and mutually interdependent if we are to accurately and finally show and prove TO. As ARTO, ABSTO, RELTO, TRU, TIT and TO are increasingly known, varieties effectively applied, and its form and representation perfected, we can present, what should be an increasingly obvious great idea, all that enables the relative universe of Best Human Life, TRU by TIT-TRIS in TO. It is our most profound and earnest hope that such representations and presentations will be recognized, adopted, improved and worked on with the enthusiasm it deserves, and is so desperately needed by humanity. This is actual totality.

RELTO has grown at an accelerating rate. By its rapidly advancing higher states and powers of ACEXIN, or actuation, extension and intension, RELTO with ABSTO has become ARTO. ARTO is the great, all inclusive, ultimate, and unifying constituent of TO. Actual Totality is not to be thought of simply, partially or in any way other than how it exists only as the great combined all-inclusive living being and totality it is. All that ABSTO consists of, and all that RELTO has grown to include, when combined as ARTO in TO, to encompass all the whole form and function, states and powers of actual reality. By the great expansion of subjectivity and corresponding reaction of objectivity, combined as reality actual totality is largely powered by human knowledge, individual consciousness, purpose, technology, social organization, and the adaptation and alteration of environment, as well as human life, to provide their mutually best advantage and well being.

Human life in TO, as RELTO, and physical existence in TO, as ABSTO is quite compatible. They are mutual, positive and beneficial. This is the form and function that drive actual totality, past, present and future. All these are manifest in the varieties, TOV in TO.

A major aspect, and explanation, of absolutivity, is the part and role of a person in actual totality. From any individual, TES, to the person, TÍS, consciousness, instantaneous attention, self centeredness, and the absolute idea of knowing and being oneself the focus of relative totality on here and now is the basis for absolutivity. The greater the RELTO and specificity of variety in TO, the greater the absolutivity. This is signified by the Law of the Internal Convergence of the person on, and proof of, absolutivity by RELTO, in actual totality, LICPART. With the knowledge of existence, totalities, universals, continua, absolute and relative, generic and specific varieties of TO, we can better answer the question. Is there one world? If so what is its form, function, dimensions, component. How are they to be applied, used as a guide, and organized to, solve the world's problems and wrongs? These we begin to order in the next chapter, actual totality.

III Proofs of the Form and Function of Actual Totality

CHAPTER 15 - Sum of ACTUAL TOTALITY, to mass length time, mlt

Typical form and function of TO, Characteristics, Properties

Actual totality, TO, is not only a practically exclusive unified whole, it consists of many different categories, constituents, modes, continua, kinds, varieties, divisions, parts, processes and mechanisms typical of itself. Many of these have existing names, and many new concepts require new names to equal their importance and power in TO. The preeminence of TO is why we must always remember the whole, and repeatedly bring its greatest modes, universals, continua, concepts, principles, laws and all together as they emerge, are generated, become interactive and interdependent in the whole, TOT, TO, and TO-TOT. This is how we prove TO, and how these proofs dispel all doubts by answering the questions and laying the foundation for solving all problems.

TO is both, and neither, of one kind nor completely static. Since a perfect time zero is non existent, actual totality gains existence from time zero to a mid point in the mid term of relative time. This its period of greatest existence when it is of one kind, highly static externally, and can be treated as a closed set. This is a period of the present, functional and practical near and mid of past to future time. Actual totality very incrementally loses some of its exclusiveness in the increasingly far term. It is of another kind in the sense that at any period of time it has built into its existence forms, functions, states and conditions of other times. This is another kind when increasingly dynamic, changing in current through extended time. We speak of both as actual totality, for they are largely the same in existence and usage. Although both are highly convergent on self preservation, totality proper, its core, does so absolutely, actual totality allows for and does so with adaptations and reactions relatively to external actions, especially with increasing time.

The following two figures reveal and compare two of the nearly infinite phases of actual totality. Although we think of TO, or actual totality, as one and exclusive, it is affected by time, and the many factors and forces that are external and internal.This shows the important condition of TO and especially its relative totality, a condition that depends on its own predominance and powers of survival and well being. This has been, is and will be a major mechanism and responsibility of all that makes up TO. Since we are largely concerned in proving actual totality, and our existence in it presently, for practical purposes we use TÓ to include both its proper, more general phases, and most definitive phases, TÓ, and TP, TÓ Proper.

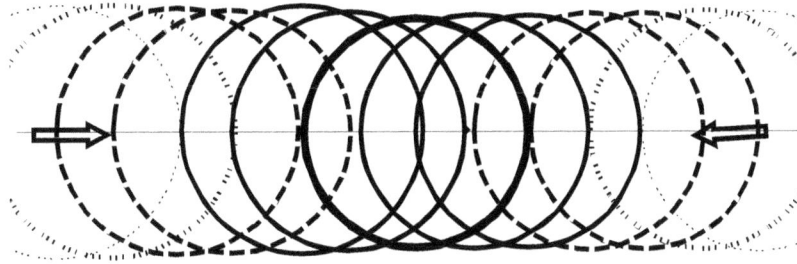

Far Past to increasing **TO** decreasing to far Future

change Convergent on **TO** reverse retroactive change

1 2 3 4 5 6 7 8 9 10 11 12

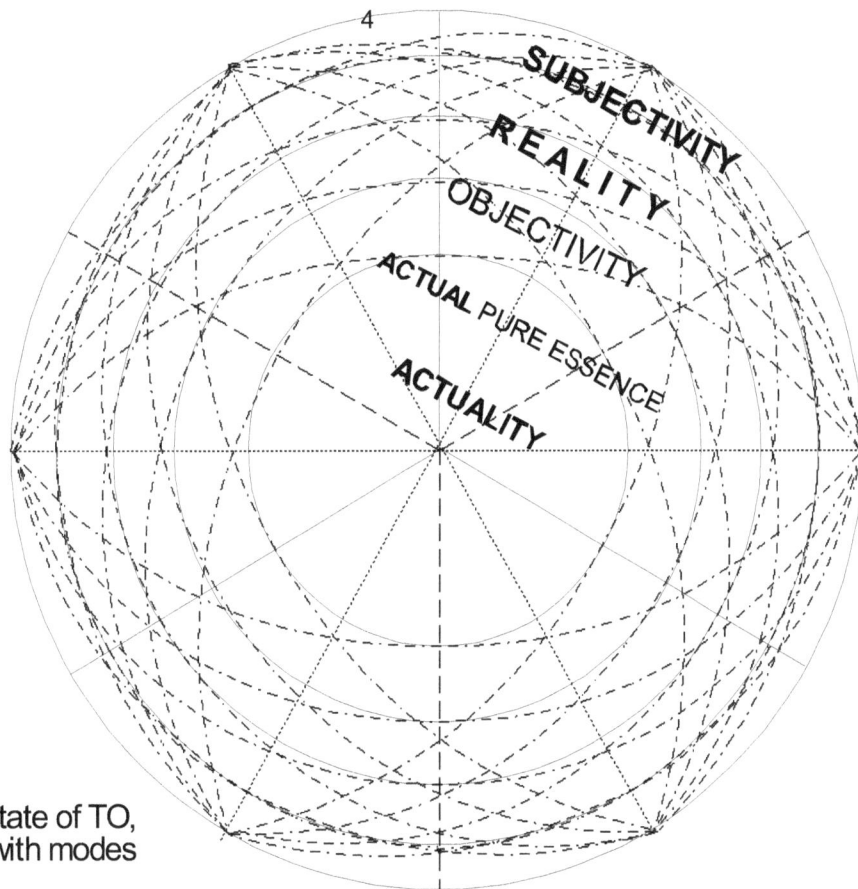

Current State of TO,
internal, with modes

Fig s. 15-1, 15-2 Actual Totality, TÓ in transition, TOT, and Actual Totality Present, TP-TO, compared

The actual universe is the union of select essences of absolute and relative totalities. Absolute totality is largely the physical universe minus relativity. Relative totality is largely the mental universe without much of absolute totality. The union tends to optimal balance depending on form and function of the stage of existence of actual totality. Whether its polarities are, and function in perfect balance, or are weighted toward one extreme of the other. Through past actual totality was more and more like the physical universe, only in an altered and indefinite way. Through the future actual totality will tend to become more and more like the relative universe in present actual totality. It will be drawn to them by their own powers and those of their propelling and compelling forces. How much, and what the adaptation and alteration will be, how and when they will occur are indefinite in the current state of potential to actual totality's relations. Since, the advent and progression of human life, there are increasingly proactive forces and powers toward stability of actual totality. It depends on the existence of the relative in the midst of absolute. This explains and proves both the progression of potential totality to TÓ, TOT, as well and transitions of future TO-TOT, TOF.

Nothing can have actual existence either completely absolute or completely relative. They exist, if only in an infinitesimal degree, by their observation. Thereby to exist is also a proof of absolute and relative convergence, ABSTO and RELTO in TO. Even TO does not exist only absolutely or only relatively, it is always changing, exchanging in the present, changing and exchanging by time and transitions with TOT. By the Principle of Ultimate Coincidence, PUC, necessary convergence of TOT to TÓ, and its constituents in TO is manifested throughout TO-TOT by their positive imperative qualities in form and function. It is the great and glorious progression of actuality from its incipiency in the barely existent to its most profound, massive, dynamic actual pure essences in the greatest of best human life that most fully characterizes the core of TO, Totality Proper. Actuality used as one, the one actual identity of TO, revealed how best human life as TRU in TO produces the relativity, specificity and sensitivity needed to fulfill TO. This not only proves TO, its absence in other totalities, universes, worlds, Gods, etc. explains their inadequacies, and helps to separate what is in them that are TO and what is not.

Totality, TOT, TO, TOV, TRU, TIT, TRIR and TOF differ. Totality is an indefinite existence without either potential or actual totality. TOT is potential that lacks, but may become actual totality, and has a host of components that are being selectively accepted or rejected by actual totality. This is the dynamics of its external and internal existence. TO has many limits, including changes by its actual pure essences that reduce its varieties from things in themselves to how they exist in TO. TRU. The Relative Universe is a large part of RELTO, and includes many of the specific varieties of TO. TIT is TRU itself at its most specific varieties. TRIR is the most specific variety of TRU and TIT, in which here and now is highly focused on the near term. The interaction of this specific variety with TOT-TOF in mid to long term is crucial to the existence of TÓ. The general varieties of TO balance with these specific varieties and impart their form and function to give balance in the whole. All coexist with all other modes, continua, form and function to prove TO.

The shapes of totalities differ, that of absolute totality, the physical universe, ARTO and actual totality, the relative universe and relative totality have marked contrasts. Without understanding these contrasts and differences it is easy to become deceived and confused by the most fundamental form of all. Even the representation and perspective of actual totality may differ. This depends on the dimensions selected, how they are arranged and position of the viewer. This is the law of dependent representation and perspective of TO, LADRAPT. It is partly because of the variations of this law that worlds, universes, systems, deities, totalities, TOT and TO and TOT are in error, lacking, or misunderstood. The ideal representation and perspective would be one that fits all. Yet this is impossible, not only because actual totality is not a perfectly closed set, but because both the relative and subjective nature of actual totality produce a profusion of approaches. When the many divisions and parts of actual totality are featured, this profusion becomes manifold.

The key to the frame or shape of actual totality is the relative, a predominance that permeates all actual totality, and most of potential totality. Without it absolute totality lacks centering, qv Ch 19, and has no basis for a set alignment of dimensions. For actual totality its shape and frame are products of the average of its constituent kinds and varieties. It varies both qualitatively and quantitatively, from that of a torus or doughnut in the physical universe with relativity, to altered dimensions mass, length and time in the relative universe. The center is on the here and now, an average of past to future time, and average sized length. Since TO partakes of all varieties with spectra of dimensional form typical of each, the resultant form is that of their average. In this way both the torus or doughnut form of the physical universe, and the spherical form of the relative universe, as shown by SPALT are demonstrated. The form of a torus is shown by SPALT with the infinite of all at the periphery and the infinitesimal at the center. The spherical form is shown by SPALT with the infinitesimal of length at the bottom. They are interchangeable depending on perspective, representation and stage of variety and kind of totality. The average forms of modifications of SPALT provide a good model for the form of actual totality.

The overall form of varieties and approaches to TÓ differ, whether of each variety from each other, or from TO. So that to provide a complete form, and proof, of each they must account for how and how much they add to the overall form of TO, and how much it accounts for their form, and that added by TO itself. When all these coalesce as they are parts of the sum of TO we complete our quest for their accurate and reliable description and proof. Evolution, creation and an ontological proof are examples of the differences that transcend overall forms of varieties.

The following is a summary of the make up TO derived from the proofs of each chapter in the text.

ACTUAL TOTALITY is -

- the gains from non existence to existence, necessary existence, actual pure essences, totalities, the totality, TOT and TO,

- the sum of its unique and typical features and constituents, - the unified, select, qualified, proper, valid, proportional, pure, essences of necessary existence, or actuality,

-highly exclusive of all unqualified and unneeded, inclusive of much qualified externally, retentive of all needed,

- highly complex, massive, predominant and dynamic, different yet contains much that is commonly known,

- not to be thought of or believed in as an undifferentiated, single or indistinct whole. It is definite, distinct, changing and has its own unique, massive, differentiated form and function,

- reversible with TOT in succession by time from integral to differential, worthless to valuable, and unfulfilled to fulfilled,

- arises from existence that is proper/inappropriate, positive /negative, necessary to imperative in potential to actual,

- the actual pure essences in all things whose transformations become the constituent parts and roles of TO.

- not the physical universe, God, philosophical systems or any of the many worlds we experience in life,

- the actual pure essences of combined, universes, deities and added typical pure essences make up actual totality,

- by being constituents of the combined and added typical pure essence universes, gods, systems and worlds actually exist,

-the physical universe and relative universe of best human life and mind are large portions that unite to fill their part and role, - a refutation of lesser and simplistic beliefs and attitudes,

- does not deny the external, interactions, transcendence and other proofs to the degree they are in time increasing with TOT, - of internal relations, different aspects of the same thing, distribution and dynamically balanced,

- from entities to a, any, and the totality, to actual totality,

- founded and formed by the union of its many universals, qualities, modes and continua, each and all in proper balance,

- the five modes of TO are actuality, actual pure essence, objectivity, subjectivity and reality,

- consists of many continua and dualities whose spectra reveal much of its typical and unique form, function and quality,

- its continua and dualities are the basis for much of its basic features, kinds, varieties, mass energy, length, time, dynamics, relations, language, logic, math, proportion, geometry, creation and evolution in life,

- is the select fusion and unity of absolute and relative totality whose divisions and actual pure essences typify the whole,

- is formed by its ten plus generic-specific varieties,

- the sum of the actual pure essences of all its parts, processes, forms, functions and states. produce a typical composite totality, - practically, functionally and by its pure essences relatively permanent in the near-mid term future, treatable as a closed set,- in transition maintained by its form, function and momentum internally and full reversibility and balance on TOT externally,

- exists by its mass energy, spatial and temporal dimensions in three levels that enable an accurate spherical simulation of form,

- founded on its mass energy whose transformations by relativity develop to yield much of its form, function and states,

- permeable, consisting of a continuum of static-dynamic with a spectrum from very static at present, to very dynamic in future, - the potential and actual, action and reaction are guided by more or less inevitable external conditions and highly selected internal powers always imperatively converging on homeostasis,

- by relative totality mass energy levels in TO are subject-object oriented by relativity to provide actual reality to life,

- largely human and personally focused by the powers of the mind in the here and now, advancing with reason and control,

- by RELTO language and logic of ever higher levels of ability, essence and reality produce and prove its unity in totality,

- a whole. neither absolute nor zero, except by its limits, integral to unity, and differentially finite to infinite,

- formed and function as a whole of the infinite and finite united as one in perfect equilibrium of cardinal and fractional,

- accurately proportional with its own precisely typical values,

- by form highly geometric, with physical, chemical and life ordered actual pure essences that increase by relative totality,

- propagated by the accumulated convergence of its design, form, forces, creation and evolution propel and compel life,

- includes increasing attraction or love to create generations and forms by adapting relative and altering absolute totality,

- changes with time, retaining stability, homeostasis, and powers of resistence, reversibility that restores its best existence,

- arises, exists, and will continue by how the relative absolute focuses existence advantageously and is preserved in TOT,

- the far term transitions and transformations will tend to expand toward the absolute relative yet ARTO must balance,

- personal bond is by identity, devotion and TO continuity,

- largely made and powered by its knowledge and proofs,

- assured of existence, survival, success and permanence by the optimal form and function of all its features and constituents,

- proven by disproofs of all that lacks its part or role, reciprocal proofs of all mutually reinforcing, qualified proofs by unaccounted for explanations of exceptions and variances, enigmatic and paradoxical proofs of the reconciling dualities, multiple, combined and total proofs by their convergent sum,

- represented by its form and the positioning and operating of its constituents in various figures and models,

- great happiness and joy can be found by learning, studying, helping improving, realizing and living actual totality,

- one's daily life, work-a-day world and all are manifestations of actual totality we are like its messengers,

- lack of actual totality is all about, and in us, only by learning, knowing, acting by, and continuously upholding it can we be,

- is highly vulnerable, a thread of life in extended time. It has great redundancy and stability in the near term. But continuous lack and loss of quality and substance are a perennial threat that we must reconcile and control,

- can be called by various names, it is only what it is that is right and all, the ultimate unified whole that has, is and will be.

- has given much, and much must be given in return, by natural talents, enhanced capacities and abilities TÓ is sustained,

- time may change, but the zero time focus of TO remains the same, LAZOT, Law of TO always maintains its zero time focus in the here and now, LATOZ. At any time TO is at time zero.

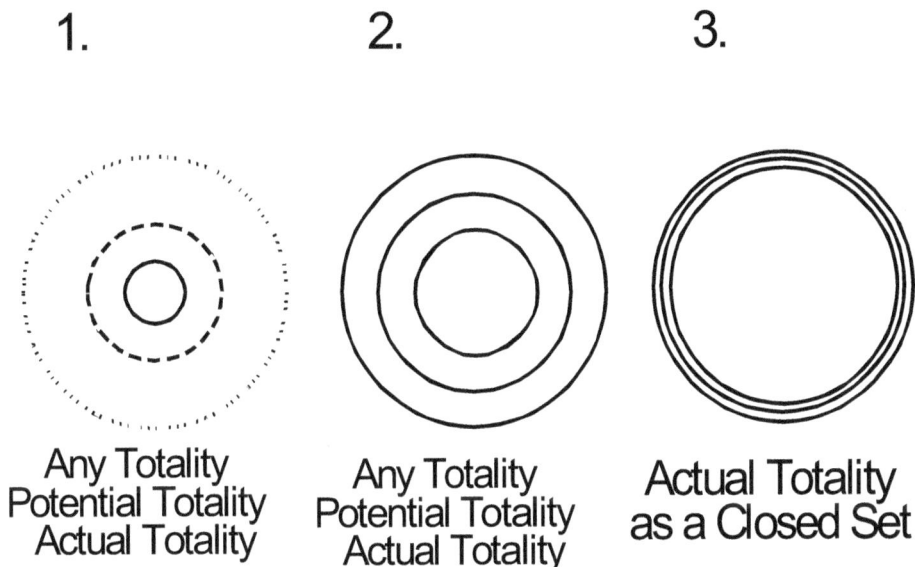

1.

Any Totality
Potential Totality
Actual Totality

2.

Any Totality
Potential Totality
Actual Totality

3.

Actual Totality
as a Closed Set

Fig 15-3 Developing form of TÓ as an exclusive totality

These three different states, from any totality through potential to actual totality as a closed set, show the development of TO, to its exclusive present state. In contrast to Fig. 5-1, the emphasis is on TO and its development, rather than comparatively, indefinite to definite totalities. Actual totality in the present, and especially its concentration, Totality Proper, TP, is shown by its mass, length and volume, ALT, and by the powers of converging presence of relative totality. This is most notable in the relative universe of best human life and its spherical model SPALT. How everything becomes one is by convergence, how all unites in TÓ to one, or The Law of Everything in TO as one, LETO. This is a product of the gains and losses of all as they become, or are less than actual totality. (LOGAT-15) Such concentration by convergence of all in increasing presence is the power that not only defines and shows TO, it is a primary proof. This is needed to contrast with the opposite diverging forces that occur in the existence of TO, as by transition in time. Convergence typifies TO and is signified by the Law of TO Convergence in the Extended Present, LATCEP. Fig 15-3 and its end figure help show and prove TO as a closed set. It shows how TO exists as the exclusive totality, and proves the all powerful convergent forces that bring unity in totality. The intermediary finites of windows of finity, WIF.

By the increasing specificity, sensitivity, relativity and quality of TO and its modes, they become more definite and balanced. This development enables the determination, formulation and totality that make, enable, show, certify and prove the convergence of all by actual totality, TO. Proof by specificity brings to totality, through relativity of its mass energy, the special form of TO. When complete this form of total actuality is largely a result of convergence in actual totality proper, TÄP. It is this feature of actual totality that emphasizes its different, characteristic and exclusive form. Much of TO depends upon this special form and becomes evermore certain as all is collected throughout its presentation. It is shown in the disunity of non totals, and the solid, yet dynamic, powerful and well-directed ambiance of TO and identity in the sum of TES in TO.

Special Form of actual totality, TO, TRU, TIT with TOF is produced, when all fuse interdependently by internal relations in TO, in its special varieties, TRU and TIT, and potential, TOT. It forms a determinative and typical ultimate unified totality. It helps us visualize how TO and TRU can approach absolute certainty with reliable and proven truths. This we accomplish by recognizing what TO-TOV is, can relatively accommodate and exclude as compared to its parts, other universes and the deficits in things. Compounds and complexes have separate properties and inherence in themselves, and in TO. This originates from the different, exclusive and characteristic actual pure essences of TO.

Zero and absolute do not exist in the physical universe, but do exist, by APET in TO. The actual pure essences of TO provide reality to the limit of relativity, the observer observed, and certain proof to the totality of all that enables TO, its varieties, TOV-TRU-TRIR, and adaptation in TOF.

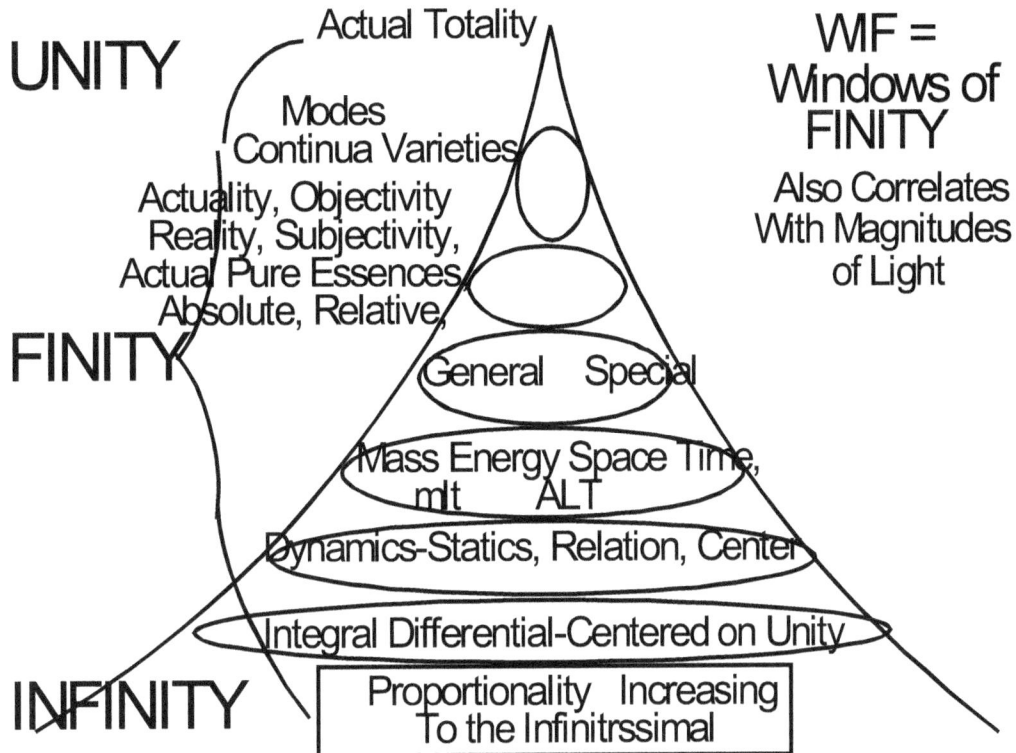

Fig 15-4 Windows of The Finite

Fig. 15-4, provides a good idea of both unity, finite and infinite in TO. How in being a closed set TO can be treated proportionally, mathematically and geometrically with greater accuracy. TO is more or less exclusive, a closed set, acquired from its own actual pure essences and those of its constituents. The greatness or value of the components of TO depend on the degree to which they are larger or smaller portions of the whole. This figure is similar to Fig. 7-2, yet focuses on giving the descending order of the major aspects of TO. The chapters of the first half of this book establish that form from which all proportional parts can be made. The constituents may be viewed in full coexistence in TO or various degrees of separation. To the degree they are separate, they lose form, function and characteristics especially actual pure essences. TIT when considered separately, can be viewed by it's autonomous at once, ATO, state, which with TRIR reversal of TOV to TÓ and TÓ with TOT and TOF is produced. This is crucial to the future, of TO and all, including each person.

Of the many aspects, features, characteristics and properties of TO there are several that are most outstanding providers of proofs. TO is highly diverse, inclusive and demonstrates qualities that represent all things, absolute to relative. TO is manifested by its actual pure essences select and modified from those existing more or less in all entities. Although TO is shown vertically integrated to emphasize unity in totality, it exists equally differentiated in all directions. TO is distinguished by the definite over indefinite, both in its existence, and by representation. Definite existence by relative totality produced from the reality of subjectivity within the limits of absolute totality.

The totality deficit is how much a whole, or TO loses, without some of its modes or parts. By the law of proportionate truth and the unity and balance of TO, especially evident in TRU, there is proof of only one final truth, through TO, and TO with TOT-TOF. What we must remember at all times is how each variety, transformation and stage has and imparts its own characteristics, properties, and actual pure essences on TO. They explain some of the most profound, subtle and critical features and aspects of TO. Without understanding these we rapidly fall into the trap, and wrong beliefs, of lesser and partial knowledge that destroy the coordinated form, and thereby validity of TO. This is most apparent in the seeming incongruence and impossible coexistence of absolute and relative universes, or of the physical and mental. However, these are readily validated by the many proofs and laws of TO, and limits of all extremes in the polarities of continua, kinds and absolute-relative in TO.

In many ways the internal states, conditions, relations and limits of TO provide the evidence needed to show how all, or totality, as TO, can be proven by its existence, actual pure essences and predominance. Best human life or TRU, like the physical universe and evolution, is only a qualified part of TO. It's unique and indispensable inclusion renders its form, function and actual pure essences typical of TO. It is the contribution of these and others that make, save, are, show and prove TO, and provide its guide for all.

As universals, essences, ES, CES, TÍS and TES exist and function together in TRU and TO they reveal internal relations and key built-in mechanisms which make up and show the form, content and proof of TO, TRU, TIT and TOF. Their integration and interdependence produce the added states in totality that gives unity to the whole, TO. The mentality of humanity, and humanity of mentality converge upon unity as it encompasses the whole, TO. They provide additional safeguards and means for survival against extinction. Precisely perfected total comprehension is the absolute idea, or revelation of the whole, TO, and with TOT, as produced interactively by its modes and relative totality, RELTO. This shows the central role of actuality, objectivity, subjectivity, and reality throughout TOV in TO.

Of the special form of the parts of TO and TOV, including TRU, we need to remember and consider their characteristics and properties. Although these are like forms, are also large proportionate components or major parts, their great importance shouldn't be overlooked. It is by its typical modes, kinds, characteristics and properties that TO and its varieties, as TRU, are revealed and proven.

It may be assumed that actual totality cannot exist in more than one way. If TO is changing, even infinitely, or if TO is changing even infinitesimally it is other than itself and cannot be a stable unitary existent. This seems to rule out actual totality as an exclusive and ultimate unified totality. Furthermore it would seem certain that at both zero and absolute time in actual totality do not exist. Yet such interpretation is more by appearance and does not apply the full rational explanation of all that makes up actual totality. For between infinitesimal and infinite change there is a whole body of existence provided by the very aspects, modes, continua, varieties, form, function and characteristics that typify actual totality. Foremost among these are its actual pure essences, by which actual totality exists by its own functional values, practical and relative basics. This is a concept whose state and condition are infinitesimally changing. Until one can grasp the meaning of actual totality as a whole, and the proofs and laws that support it, they never get the complete picture of what actual totality is. This is why we must always remember to supply and apply an adequate and sufficient knowledge and demonstration of the whole of actual totality.

The correct universe and the ultimate relative actual totality consist of a large amount of best human life. This is evident in the near here and now of the special varieties of TO, i.e., TRU, TIT and TRIR. TO is the most fundamental and critical of topics, subject, object, idea and necessarily existent. It has evaded human search for unity, proportion, survival, success and harmony in life. This is because the necessary existence of TO has not been known from lack of actual pure essences, complexity, dynamic changes, and the great detail that are TO. This striking contrast reveals the opposition and alienation from TO that produces our greatest enigma. Many worlds, gods and universes have been and are known that have greatly added to our knowledge, purposes and ways in life. Only their diversity, lack of relation and opposition prove their inadequacy and errors. Only when their selected essentials are transferred from potential to actual in their part and role do they have actual existence in TO, or with TOT.

HIERARCHY to and in ACTUAL TOTALITY

In this progression to and from TO the typical differences are revealed by each transformation. How they are separated, and what they exclude are presented. This provides a highly effective and practical way of showing vertical integrated totality. Each difference and distinction is an important proof that directly or indirectly proves actual totality. Each category and its difference is distinguished by what lost or is lacking in transformation

Existence, a, any and all Entities, everything Lacks or loses presence by absence of all needed for existence

Necessary Existence, Actuality, Actual Pure Essences Lacks or loses presence by absence of all the rest in totalities.

Totalities, any, everything Lacks or loses presence by absence of all potential to TÓ,

Potential to Actual Totality, TOT Lacks or loses presence by absence or less than actual totality

ACTUAL TOTALITY, TO Lacks or loses all not of and for itself

Primary Divisions (Kinds), Lacks or loses all TO less than itself

Absolute Totality ABSTO, Physical Universe, ARTO, Relative Universe RELTO, Relative Totality

Primary Categories, Lacks, loses or less than TO and divisions

Universals, Continua-Spectra, Modes, Varieties, Form, Dynamics and Components

Modes of TO, Less than TO, divisions and categories

Objectivity Actuality Actual Pure Essences Reality Subjectivity

Varieties of TO, TOV (1-10), Less than TO, divisions, categories, modes Extreme Generic Generic-Specific Extreme Specific

Form and Function of TO Loses that less than the above and varieties

Mass Mass-Energy Higher Levels (1-10), S-O, A, SA

Spatial, Length Area, Volume

Temporal Time Change Variation"

Past, Present, Future

Combined, mlt, ALT, ACEXIN

States of Motion of TO Loses that less than above plus form function

Most Static, Static, Static-Dynamic Dynamic, Most Dynamic

Temperature, Zero, Moderate, Extreme, Hot or cold

States of Relation Relativity Non Centered, Centered Here NowLoses most of the above, more or less related

Major Constituents of Divisions, all they effect, only typical of TO Loses most of the above, more less used for

Language and Logic, Recto Loses above, concentrate on essences to Recto

Mathematics of TO of AbstoLess abs to essences for use by those in TO

Proportions of TO of math and Relto Loses most of the above, converges on Relto

Geometry of TO of Absto and spatialLoses most of the above, converges on TO

Life in TO of absto-relto and temporal Loses most of the above, converges on Relto

Animal Life in TO, of Relto and temporal Relto Loses most of the above, converges on Relto

All Human Life in TO of Relto and temporal Loses most of the above, converges on Relto

Transition of TO, of mass energy and temporal Loses some of the above in transition

Completion of TO, TAR, " " Loses absolute totality, converges on TAR

Personal in TO, of all, and " " " Loses much of the above, prime focus of Relto

Knowledge in TO of all, and """ " " " concentrates on essences of Relto

Higher Levels of Knowledge """""" " " " " " " and highest levels of A in TO

Proofs of TO, Disproofs, Qualified Proofs, Reciprocal Proofs ,Less than above, confirms proofs in actual totality

Combined, Multiple, Mixed Proofs, Paradoxes, Total ProofsSurvival/extinction, immortality, """" Less than all not for survival and supports TO

Representation of TO Less than all except showing its object in TO

Applications of, Life in TO,Less than all except best use of object in TO

Increasing differentiation and detail. Less than all except its own part in TO

Major Components and Parts, Less than major constituents, extended parts TO

Classes, Fields and Sciences, extended regions of TO, " " " "

Characteristics of Actual Totality, most typify form of TO, " " " "

Properties of Actual Totality, most typify function of TO, " " " "

<u>Cmponents and Parts, bulk of the quantity of TO Less than all above, except its necessary TO part</u>

TOTAL CONVERGENCE of all on ACTUAL TOTALITYCollective gains of all becoming actual totality

Table 15-1 Hierarchy of TO, Ascending to and Descending from TO, with gains or losses in each level.

This hierarchy is closely related to other figures, e.g., the overall figure of actual totality in the introductory chapter. Their combined use helps to orient and achieve the right perspective. The gains, externally in the top four items, and internally in all the rest are given by a simple reversal of the order. It is the accumulative gain to, and the accumulative lose, or what is lacking from, that determines actual totality, TO. The hierarchy provides the basis for action, acceleration, advance, improvement, planning, prevention, protection, production and all needed for optimal well being of actual relative totality, and best human life in TO. Each different chapter and category of TO should be applied separately, interactively and totally for distinction, and depending on its focus, contribution, value and the problem or need for action.

Such a table as 15-1 is most important, for it reveals and proves the coherence, inclusiveness and plausibility of actual totality. Each step, and its accumulated gain and loss, is the evidence and information necessary for the adaptation, modification and alteration of that occurs in the transitions and transformations, externally to internally, of actual totality. The gains or losses in each step delineate the various levels of categories out of and in actual totality. Such a clear distinction helps to prove actual totality and the category. Each proof can act as a test for their accuracy and value. It shows how absolute totality and relative totality exist only in that way consistent with actual totality. It shows and proves how the constituents of actual totality are formed and function by actual totality, and how actual totality consists of them. This is proof by, and the Law of, Gains and Losses of Actual Totality, LOGAT. This premise of actual totality has added power in stimulating the correct path and guide for all persons' actions in the future.

The figures of actual totality, and its absolute and relative components produce unique and great contributions. Along with that of their many less-known concepts they clarify and enforce TO. We must always remember the absolute and relative as equal polarities that tend to perfect balance in TO, and in the past present and future with TOT. Yet each is, as shown, loses and is less than actual totality. This both justifies the roles of the absolute universe or matter and relative totality or mind in TO, as well as proving and showing their existence only as portions of its form and function, aspect, category or constituent of TO.

The whole form and function of TO are both simple to complex depending on how much detail and effort goes into its determination and demonstration. Simply it is absolute to relative totality more or less together, and in time, as shown by the ten varieties of TO from most generic to specific. Much can, and needs, to be added to this if the actual pure essences and its full existence are to be comprehended and simulated. The content of TO and each TOV consist of a vast array of aspects, features, qualities, characteristics and properties typical of TO and each variety. For instance TO at the first of ten levels, starting from absolute totality, will be much like the physical universe, only adapted by its actual pure essences, and all else to the actual pure essences and that which best exists in all TO. This includes that needed for existence and the many objective qualities TO requires for fulfillment. At succeeding levels of TOV, each level consists of differences and changes from adjacent levels to accommodate the degree to which absolute or relative totalities exist in TO and predominate. Each level is a cross section of TO, and to a large extent is similar to how TO-P existed at each point in development or evolution. In a sum, this enables us to take an average of all and provides a single replica. This is how we can adapt The Relative Universe to TOV-TO, show and prove TO by its combined absolute to relative form, TRIT, and SPALT. To the extent the form and function of TO are known, is the degree to which its proof is complete.

Summative Reinforcements, SRE, are combined powers that exist in TO. They provide the inclusiveness, exclusiveness, solidity, certainty and proof of actual existence and the state in which TO is all, TRIA. This is supported by convergence, coherence, extension, intensity, and the Principle of Ultimate Coincidence or PUC.

Actual totality is different from what it would appear to most people. It is also much different from any totality, the physical universe, God, systems or worlds. This is the beauty, greatness and proof of actual totality. It eliminates all that other totalities deviate from TO, and adds actual pure essences and qualities positive for TO. Actual totality is all from the extreme of absolute to the extreme relative totality, and the best that human life entails. To the degree each variety is positive, supports and enhances is actual totality. To the degree all that is outside of, or less than or detracts from TO, is negative, deters, diminishes and tends to destroy actual totality. Thus actual totality is a great, massive, well founded and dynamic totality, unlike any known heretofore.

If TO is all, to the extent TO is all, or to the extent TO when perfected as the ultimate unified totality is all, all else does not exist, except what it is by TO. This is the Law of TO is All, all by Extent and Perfection, LATEXP. The implications of this law are the fact that nothing exists separately, only by, and as necessary for TO. It confirms DAOST, or Different Aspects of The Same Thing, why and how absolute and relative exist not separately but as continua and dualities fused as, and by, their spectra in TO. It is from LATEXP that we prove and confirm much of the features, modes, continua, absolute-relative, varieties, form and function of all, as they exist in TO. It is by LATEXP that we must not only discover and develop all associated with the transitions and transformation of TO and its consistency, we must work to perfect TO, and TO with TOT in the future.

TO to TRU to TRIR with TOT are highly interactive and interdependent. This proves the actuality, greatness, exclusiveness and stability of TO in the present, and increasing adaptations and reversible stability in the future. It also proves the limits of all and interdependence of absolute and relative. Most significantly it proves the fine variable and somewhat indeterminate line that exists between TO, TRIR and TOT. It also proves the requirements of all to be TO, including the positive imperative and the power of the will and life in TO.

The varieties of TO form a whole that replicates TO, and qualities of the absolute and relative in TO. The hierarchy of TO show how all that is gained and lost results in its typical unity. The specific or relative, and the generic or absolute, actually exist in TO as a duality of the same continuum, two different aspects of the same thing, TO. It is this same and different that unite in ABSTO-RELTO to produce TO that is one of its greatest qualities and proofs. The absolute and relative totalities, and the physical and relative universes, are polar opposites in TO. They do not exist as separate entities, things or senses, but together as extremes in, and most basic aspects of actual totality, TO. The absolute and relative totalities, and the physical and relative universes are modified by their actual pure essences, and these are again modified by the actual pure essences in TO, and its modes and varieties. The same is the case for man, mankind, homo sapiens, and knowledge or the human mental world. They are modified in stages to absolute relative, and actual pure essence in TO. So that in the extreme absolute pole of TO, all that exists of the relative is like its initial and most necessary existence in TO. This is confirmed by relativity. Likewise in the extreme relative pole of TO, all that exists of the absolute is its final and most necessary existence in TO. Separately they must not be thought of as complete nor an end, they are only what this initial and final gradient, as it is its part and role in TO.

By the hierarchy of TO and to and through its varieties, that of the lower ones often do not partake of the higher. What is of the higher exists in the lower in different, and more generic, way. What is in the higher as ABSTO and RELTO is much of the basic form and frame for the existence of all succeeding varieties and transformations. The identity is formative, not a replica. Internally all of each preceding class includes all of each succeeding class. Externally, each preceding class exists in actual totality only as each succeeding class gains existence in actual totality. These processes of the formation of actual totality are a method that is both very demanding and clarifying in demonstrating the make up of TO.

The figures in this, and other chapters, help to derive, understand, show and use the major themes and topics of the text to give the necessary identity of all with TO. These include,

1. Always remember that our immediate awareness and instantaneous attention are highly limited.
2. Recognize the functional value and practical use of all as these are the basis for actual pure essences.
3. Realize how and how much actual pure essences must be separated from all that is less, and how they are brought to their right part and role in TO.
4. Recognize the absolute-relative continuum and different varieties, and their spectra and gradients in TO.
5. Realize that the absolute-relative and generic-specific continua are neither the only continua nor gradients, nor exclude all the rest of that exists within TO.
6. Continuously move about in the hierarchies and form of TÓ to all its constituent parts and processes, from its unity to either infinitesimal or infinite, and finite levels, by returning the given and encountered to their full identity with the functional values and practical actual pure essences in and of TO.
7. Operate the finite parts and processes, using their positions to enhance our work, activities and lives to increasingly bring all to their most valid, actual, proportional and total contribution in TO.

The greatness and power of actual totality and contribution of relative totality in actual totality, are not largely nor fundamentally altered, and thereby practically stable. For roughly the past two and a half thousand years all the basics of actual totality, both physical and potentially mental have not been markedly different from what they are now. The difference has been more quantitative than qualitative. It has been the quantitative expansion of knowledge and all its benefits. Because we have advanced these quantitatively, or do not use them well today, do not disprove, only help to prove the dominance and degree of stability of actual totality. If we are wise enough to use actual totality well, we should be good for another two and a half thousand years. Through that time with the benefit of brilliant minds, great discoveries, and the guidance of actual totality we should clear the way for many more thousands of years, and accelerate at a rate that assures overwhelming sufficiency and permanence. This stability enables us to treat actual totality as a closed set, helps to prove much of its doubts and reveal the greatest hope from its future potential. It is a stability that proves the far reaching presence of actual totality, and disproves much that may seem to be its disproof.

The totality of less than or greater than a species is proven by the relativity or mass energy, observer and observed, in all that enables TO. Also there can be a totality of TO, and another in the totality of its models that are more or less separate, or less than TO. The a-priori stages and the convergent unity of reality in TO, propel the growth, development, form, function and power of TO. Life, by actual pure essence, is externally one indivisible whole, and by actual pure essences of TO internally a divisible whole. By the law of conformity the same and relative produce conformity to become the whole, totality of TO. This unity in totality with its own set dimensions is what enables accurate representation and models, e.g., SPALT.   In the principle of organic unities, the whole is more than a sum of its modes or parts. This reveals the form and function, actual pure essences, superiority and importance of both the type of form unique to the whole itself, and the new and specially activated combined modes and constituents in TO. Together such a sum provides the law of validity in totality. This is shown by the coexistence of all that becomes and exists in TO and each level of variety, including the natural order of combined dimensional form, mass, length, time or mlt, ALT, ACEXIN and REM.

TO is a work in progress, largely ordered by ABSTO and RELTO. It is often caught in opposition between an external physical universe versus human life. In this way the hope and optimism of discovery and a new world held by the ancient Greek Philosophers has not been met by mankind, or other world views. Only by the increasing development in time of TOT-TO-TOF of a well formed and functioning ABSTO with the same of RELTO will TO become well ordered objectively and subjectively, in actuality and reality, physically and mentally. Such an order is simple to allude to, yet most complex in the full actuality and reality of TO. There is a great equilibrium in the massive whole of TO, wherein all not only function optimally, but the many polarized dualities vary in perfect harmony. These are guided by the actual pure essences of ABSTO and RELTO and the perfected powers of the advancing rational mind of humans. Such an optimistic goal is the only actual one, of the great propelling and compelling forces that both sustain, and drives, actual totality ever forward.

The degree to which all that makes up human existence, evident in the massive greatness of RELTO, ARTO and TO, enables us to better the environment, alter and adapt to the physical universe, ABSTO, and TOT. This is the degree TO by its APET and all, maintain, sustain, advance and assure its exclusive, and all-inclusive existence. This is a corollary of LORAMT, and is signified by, the Law of RELTO Alteration and Adaptation Assures EXclusive and Inclusive EXistence, LARIEX. This law is important for more than one reason. It proves the sustaining power of actual pure essences and actual totality. It proves the reversible transition of TO and TOT. It proves the combined existence form and function of the relative and physical universes in TO. It disproves belief in the exclusiveness of a physical universe by showing what it lacks, and loses as it exists only in actual totality.

That actual totality hasn't been recognized in the entire course of human knowledge can be explained. Since humans first attempts to understand the world, and since the Preocratic Philosophers from Anaximander to Anaxagoras held cosmos and nous as the two ultimate totalities, humans have been left with a polarized perspective of all. Is it the physical universe or the universe of philosophy, mind and God? The two great steps that remained are to realize that neither one is complete. Physical universes are less than absolute totality, and the universe of philosophy, mind and God is less than relative totality, and both are less than TO.

The second step that has never been recognized is the fact they are not separate. They only exist as a continuum or duality, whose convergence, fusion and unity are the form and function of actual totality. Only when these two steps are correctly discovered, described and demonstrated can actual totality be well known and proven, Only then can all the diverse systems, purposes, priorities, and actions lose their lack of foundation and gain the benefit of a highly effective participation in actual totality. This is the Law of the two step identity of Absolute and Relative totality, and their Union in actual totality, LARUTO.

CHAPTER 16 - MASS ENERGY, SPACE, TIME, to mass energy in TO

Forms of length, area, volume, time, duration, variation, and mass energy, mass = matter times volume, other combinations of dimensions, separate and united as mlt to ALT in TO

The spatial, temporal and mass energy dimensions provide much of the form and function in the physical world, and objective totality for TO. Although only three of many continua of TO, when altered, adapted and combined, they greatly add to its proof, representation and use. They exist in TO in different forms and at different levels, including those of the varieties of TO, higher levels of mass energy transitions, length, area, volume, duration, change and variation. The various combinations of mass energy, space and time like matter equaling mass and volume and the other combinations help to complete the panoply of form, and proof of, TO. The many different kinds of mass, length and time permeate the whole of TO. They progressively become more specific for TO by RELTO to reflect what exists in each variety and the whole.

Dimensions are fundamental continua which divide and measure the physical form of an entity, object, system or totality. They are accurate means of representing its form and the relationship between its physical or absolute, and mental or relative basis in TO. There are set relationships between the form and function of basic dimensions or continua. There is a reversibility between the dimensions as well as between the poles of each, the mass energy, spatial and temporal TO relationship and orientation of its continua or dimensions. This is the law of reversal and proof of TO by a result of relativity, and variance of an observer and observed. The dimensional form of mass energy, length, and time forms TO, as TO forms them. This is proven by the fact that mass and length are highly static, energy and time highly dynamic. For their higher states, whose total existence forms TO, is a major proof and helps to show how TO exists as a unified, and largely closed, set.

As absolute totality and relative totality fuse in TO and absolute totality is transformed by relativity, relative totality is transformed by the absolute and developing TO, new forms of mass energy, length and time are created. This is like the transformation between internal and external of the big bang. Relative totality induces changes in absolute totality and TO, forming a new kind of mass energy and matter in time, consisting of S-O, with concepts whose actual pure essences exist as a new kind of matter in time, in the burgeoning actual totality. They, as TO matures, become like the matter we have today, the major kind of elements in the prime mass energy spectrum of TO. This is the mass energy spectrum in TO, whose common form is called its Á dimension.

Many dialectic, opposite, and contrary divisions of TO are less than primary dimensions. Yet they can be interpreted in terms of mlt and used as lesser continua, parameters or systems. This is evident in three ways, one by adapting altering and showing the nature of ALT the mass length time dimensional form that exists in TO, how polarities can be positioned in TO, and how these divisions, contraries, opposites, polarities and dialectical dynamic aspects combine to make up, form and prove TO. It shows how the mass or Á dimension of TRU-TO is more than any S-O relationship, but is set by the predominant form, frequency and function of TO.

What is length, time and mass energy? Why do they seem so intangible? Why is it so hard to achieve any order and total relation between them? To do so we must look at what they denote; what their basis is. Length is the relationship between the large and the small. This includes, area and volume, as well as the medium and all in between, when carried to its fullest meaning. For without the medium, the means and the extremes we have no focus, length, set area or volume and thereby no order. Time is the relationship between the past and the future. This, of course, and especially, includes the present and all in between. Without the present or some set point of reference, as the mean, we have no focus or order in time. These apply similarly to higher states of time as duration and variation. Mass-energy is the state and dynamic relationship of what is predominant for an entity. It is between that which has the greatest weight, force, power, and effect and what has the least. This especially includes the world in which we live and how it relates to us, or, MIDST = Mass is in different states in entities, and especially formed in

TO. This is signified by LESEM, the Law of Specific Energies and Masses. Without relation to us and our world mass energy has a different focus or order.

The importance, and proof, of the general to special nature and form of ABSTO-RELTO and ARTO in actual totality, TO, is to be found in the measures of length, time and mass energy. It is impossible to accurately know and show length, time and mass without a point of reference, a center on which it is based. This is provided by here and now, or present of RELTO and special varieties of TO, e.g., TRU. Since ABSTO is not centered in here and now, only in fusion with RELTO does TO acquire its focus, stability and the closed set form from here, now, and mass energy. For time this now, and commonly determined cycles, regular, less regular and non regular events set the form upon which RELTO and TO are based. The pivotal measure is the center of TO which is associated with equality of large small, past future and mass energy, with equilibrium for each and all in the whole, TO. These typify TO, as other characteristics and properties that collectively prove TO, TOV and TRU.

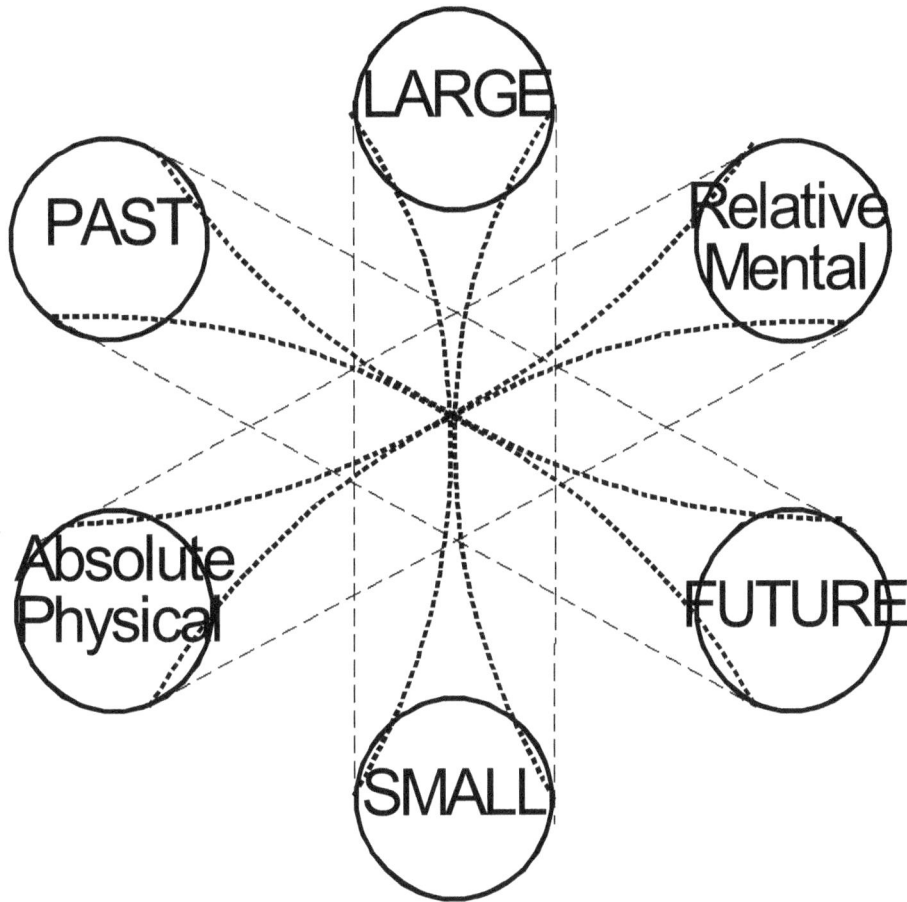

Fig 16-1 Mass Length Time Extremes, Basis for Form and Function, ACEXIN, Equilibrium, Standards and Perspective

There are different kinds of mass energy in space time. The relativity equation of mass energy doesn't tell us what mass, energy, and mass energy are? All it tells us is that they are related by light, length and time in a certain fixed equation and relation. Even this in astronomical measures may not be absolute. In astronomical measure and physics generally mass and energy are major forms and functions of matter in time, which like continua are static to dynamic, extremely cold and solid to extremely hot and gaseous. Mass and energy are also of a continuum whose spectrum is known as the electromagnetic spectrum or ems. The numerous interactions and states of different levels of this spectrum display the inner workings of mass energy physically, and by actual pure essences of relative totality, mentally in actual totality.

All entities and objects must have symmetry, which can be of various kinds. It will depend upon what type of entity or object it is, e.g., subatomic particle, atom, molecule, macromolecule, RU, TRU, ROT, MOT and TO. Symmetry is not only a result of the form of the entity itself, it is also the result of the form imposed both from outside and by the observer, subjectivity, as an ENIT becomes ESIN and ES, ESIT or the ES in and of itself or aspect of TO. The symmetry of TO is the basis for much of its representation, modulation and presentation. The basic, most simple and unifying material is mirrored across the horizon toward the most complex and infinitesimal. This mirroring, although of great advantage in many ways in constructing a concrete vertically integrated approach, is not expected to provide definitive solutions. It is more a frame, enabled by the symmetry of the object and the symmetry provided by the subject of the object, TO. The mirroring does provide a cross check and proof of the balance symmetry provides. It provides a check, more proportional and complete demonstration of the object and its presentation of TO, (POT).

To understand the polarities of the dimensions of TO, is to gain greater appreciation for its form, frame, function, and existence. Some of their characteristics and properties include, plasticity or being pliable, interactive and flexible, reversibility, highly transposing and retracting dynamics, redundancy or ability to compensate, retain form and existence, power of tolerance, relativity or how all both form and function interdependently, different aspects of the same thing, and adaptability or the continual direct and indirect readjustment of its form function and constituents. They help to maximally sustain being and actuality of TO. TO exists by having nothing that is superfluous, non existent, intangible or invalid. Also included are, unity or existence as a single and exclusive whole, with proof by its integrative processes. TO also has diversity or existence in finite diffusion to the infinite and infinitesimal, extension with formative strength sufficient to maintain and sustain itself in space, and intensity or functional power sufficient to maintain and sustain itself in time.

TO has sphericity or optimal form to assure inherence, proportion and efficiency. Accompanying these characteristics and properties are those of the modes, other continua, kinds, varieties and constituents of TO. More are noted through the other chapters, with whose sum provides the definitive kind of identity needed for differentiation and representation typical of TO, TOV and TOT.

TO consists of many forms or states of mass energy. The highest order and power of the mass energy dimension, duality and polarity in TO is signified by that of relative totality and its specific varieties, mass energy by Á. By being named, this most important form of mass energy in the more specific varieties of TO, and in TO itself, can be better understood and proved. It constitutes the unique, typical and special predominant aspect and form of TO. The typical mass energy dimension, Á in TO is not all TO, nor all of its form, nor even a 1:1 figure of its frame, for TO extends infinitely and its dimensions are altered, warped and not so simply delineated because of the curvature and distribution of space, time and mass energy. Yet the primary Á dimension does, when corrected for the Laws of Limits of Representing TO, ROT, and objectivity, provides a good idea of the frame, structure, and function of TO by RELTO and it's specific varieties. The fact that forms of entities and objects are fundamentally different and change or undergo alteration is supported by fractals and entasis. Entasis is the curvilinear, and often almost imperceptible, alterations in the form of structures that adds to their support.

Dimensional Level or Power Mass EnergySpatial Temporal

Primary base linear 1 (Mass Energy)11 Length 1 Time

Secondary squaredplanar 2 (Mass Energy)2 2 Area 2 Duration

Tertiary cubed spherical 3 (Mass Energy)3 3Volume 3 Variation

List 16-2 The physical classification of prime dimensions and their higher powers

The classification of the form of the world is similar for the spatial and temporal dimensions. It is less similar for the mass energy dimension, yet being a physical dimension it also has three powers of existence, which completes the total form. In the relative and specific varieties of TO mass energy increasingly focuses by relativity on the observer-observed and subject-object continua, adapted and altered by their functional, practical and proportional existence in actual totality, TO. For mass energy in TRU, it is dissimilar and unique both by quality and quantity, the higher the level or power, the more it follows actual pure essences special for TO and TRU. This is more than gravity, electromagnetic, strong and weak forces of physics. It is mass energy adapted and altered by increasing relativity of the predominant and typical quality of relative totality, or subject-object, which in its highest three dimensional forms approaches full reality. This is produced by the focus of all modes, especially the practical of actual pure essences and other properties that are most self sustaining. There is a confluence of matter and mind in which object and subject of ABSTO-RELTO compose the mass energy continua whose gradients recapitulate human evolution, adapted and altered through time, transitions and transformations. (qv. Fig. 24-1 MESOA)

When mlt, and their combined higher levels Á, L, T exist in TO, they are signified by ALT. ALT will vary with each variety of TO, yet is stable for each and for TO. In TO it is a sum of each variety, and provides an excellent basis for the form, TRIT, of TO, and its representation, presentation and applications. When correctly defined, designed and represented ALT is a simple and practical way of looking at, showing and proving, TO.

Time, by RELTO, tends to be centered on the present, or time zero. Since the present is an extension of zero time, and since it, rather than time zero is the relative and practical concentration of existence of TO it provides a better focus and fulcrum for time in TO. Both exist in a paradoxical relationship in TO, by the contrast between time in absolute and relative totality. The Paradox of Time Present, PATI, in TO, ATO, TIT and SPALT is evident from, 1. TO is not an exclusive absolute, only an absolute-relative one. 2. TO, has internal relevance and powers that exclude all that is less and external, 3. All that is reversible is compatible with TO, only to the degree of actual pure essences, quality and quantity that is positive for TO. 4. The past, present and future exist in different forms and ways externally and internally of TO. 5. The present in TO and TIT is not like any simple object, it is a sum of many of their features and factors whose resultant is of a special form in the present. 6. It is the intension of the temporal continuum whose actual pure essences are from past through present to future, none of which exist separately but are averaged aspects of the sum of TO. Paradox of Time Zero, PATIMO. Zero time in TO, TOV and TIT is the vanishing point of integral time past and future to the present toward zero time. It becomes relatively, seemingly and actually zero, by essence and for practical purposes. This depends on, and is in effect partly by our representations and attention, as their ATO is different and moves in transition. PATI is fully existent by the actual pure essence extension of PATIMO in TO. Both are very subtle but important concepts to master if we are to realize TO, and its exclusive difference from all else, including our own beliefs, the physical universe, God, etc. Zero time is the opposite of absolute time, as are zero and absolute. Time is a continuum as duration, which like so many others with an absence of its zero and absolute ends, except when interpreted by the actual pure essences of TO. PATI and PATIMO, present and zero time are gradients in a continuum, with present being the relative average for TO, and zero time being a vanishing point. Such a vanishing point is without existence except as a means of centering time in TO. In this way present is simply an extension or duration, of zero time, as all focuses by RELTO and its total distribution in TO. The enigma of time is resolved, and TO is proven, when time is properly ordered from the mlt or absolute totality to ALT of TO.

Much of the form, structure, arrangement, and order of TO-TOV are provided by the spectra of it's combined dimensions. How their parts are gradients and ordered by these combined dimensions, is the proof and benefit of the ordered form of TO, byTOV-TRU-TIT-TRIR-TOT. A coordinate frame renders parts and processes much more amenable to position in a spherical geometric model. The design of such a model with dimensional polarity can identify parts and processes, mechanisms and dynamics that enable and prove necessary existence, by TO-TOV-TRU. (DEPENT).

Unless the form and structure of actual totality, TO, can be well shown, we can never achieve successful understanding. TO is shown by its 3D spherical mass energy, length and time construct, SPALT. This provides the optimal representation and model for use. Many other illustrations and representations are possible and exist for TRU, TOV, TO, TOT and their constituent parts. This, when spherical, model replicates the dimensional form of TO, yet the reader must remember that it is not exactly the same as TO. It tends to be an average sum of TO and its varieties, ABSTO-RELTO or ARTO.

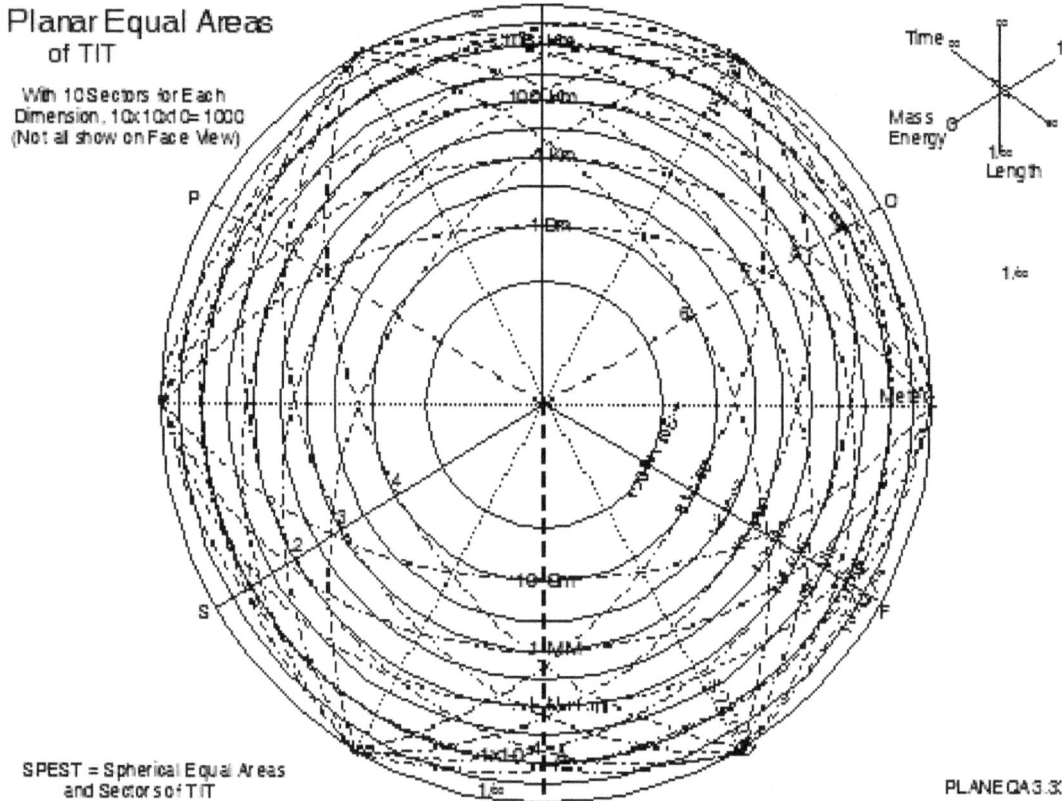

Fig.16-3 Mass Energy Spatiotemporal Form of TO, Based on Average Sum of all TOV, From RELTO and TRU, with Spherical Form on a planar field, of equal volumes

When SPALT is viewed with figures SC-1, 14-3 and 38-1 the importance of time in the proof of TO is clarified. With the absolute - centered - relative of ABSTO and RELTO in ARTO, there exists a dynamic of reversible expansion and contraction, externally and internally in TO. The human world centering of the relative tends to contraction internally and expansion externally. The physical world's lack of centering tends to expansion internally, and contraction externally. These two opposing powers, tend to coalesce by convergence and union, in the power of actual totality. These most fundamental fusion processes are most unique and fundamental for actual totality and its proof. They are the basis for many characteristics and properties of TO. They are at the core of its dynamics, both as near static, a closed set, a dynamic open existent, and a highly interactive largely reversible variant externally with TOT-TOF. This form and function of TO is the product of the spatial or length, temporal or time, and mass energy or mass roles in and out of TO, The various aspect and views when compared with SPALT provide much evidence for, proof of, and clarity for TO.

TO, as the sum of TOV, consists of many variations of dimensional coordinate form. TO is not only centered as shown in Fig. 16-2, this is an average of all, TO and TOV. Each variety of TO has its own scale representative of the mass, length and time of their part and roles in ABSTO-RELTO of TO. When TOV is divided into 10 varieties, most generic to most specific, each variety has a spectrum that is typical of its contribution in TO. This is the law of the correspondence, correlation and correction of the Varieties of TO with their models, SPALT 0-10, or LACVAS. For dimensions this becomes a subclass of LACVAS, or the law of Equivalence of TO, its Varieties and Dimensional form with their Representation or model,

SPALT. (LETORD). ABSTO or the most generic TOV will be largely of the objective, large, past, and RELTO, the most specific, TIT and TRIS will be largely of the subjective, small, future.

The set state of TO by actual pure essences is largely a result of the mechanics, momentum, redundancies, inertia, form and function in and of ALT. ACEXIN in TO helps indicate, show and prove the make up of the mass energy, Á continuum, completed by the spaciotemporal frame of TO. Together they consolidate and establish the structure and form of TO, TRIT, that is available for representation and modulation.

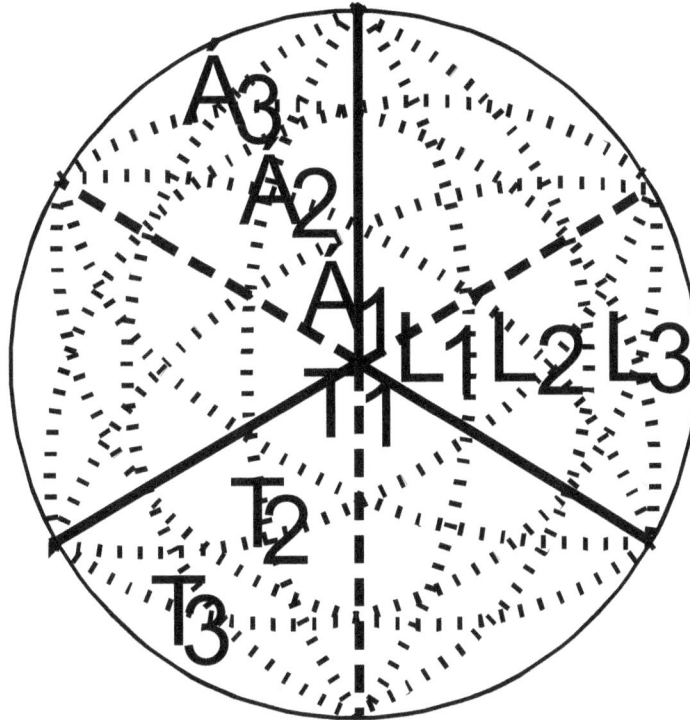

Fig 16-4 Á, L and T Corporeal Forms, States, Levels and their Higher Dimensions in TO

To emphasize the importance of a spherical rendering of TO, the distributions of its three dimensions on a planar field are given. In thus figure the higher levels, (ALTÄ), and distribution of the Á or mass energy, L or length, area and volume, and T or time, duration and variation dimensions reveal the full embodiment or corporeality of TO. These dimensions may be shown, as existing in TO not so much as simple lines, but as spectrums, continua, and sectors in its overall form and makeup.

The actuation, extension and intension or higher powers of actuality, space and time of TO, ACEXIN is the combined powers and state of the more specific varieties of TO. It indicates, shows and proves the make up, and levels, of the mass energy, Á continuum, when completed by the spaciotemporal frame of TO. Together they fix and set the structure of TO and TOV that is available for representation. ACEXIN also helps to show and prove how TO exists as an exclusive, or closed set. ACEXIN is preeminent, although like much in TO, it permeates the whole, only in quantities that correspond to ABSTO-RELTO. ACEXIN signifies to emphasize the state and quality of TO, in which its actuation, extension and intension, ALT, fully and proportionately encompasses the whole. The total of their combined levels reveals the unique form and function, characteristic of TO-TRU-TOT. Further proof of combined and unified continua is evident when we position many of the greatest existing and new concepts in the model SPALT and envision them through ACEXIN. It is by these means we find the most direct, clearest, most accurate and best perspective, thereby proof of TO, TOV, TRU, TOT and all. Actuation, Extension and Intension, ACEXIN signifies the basic integrated form of the unity of TO, TOV and the power of their fusion in TOF. Its higher powers permeate all within, and whose form determines TO-TOV. It is for TRU, and more generally for present TO and TOT, the highest states of mass energy, volume and variation. ACEXIN most characterizes and proves TO and TO-TOF by revealing the precise sum of their form. Without its understanding TRU, TO and TOT are less known. ACEXIN and the enclosure of dimensional formulation establish and prove representation and models of TO, such as SPALT. With TOT they are produced by coordinate transformations of mass energy, space and time that, have, are and will continue to occur in the progression of ARTO by RELTO, human life and mind. This differentiation when well formed leads to the coordinated essential reversal of relativity in subject-object, primary A dimension of mass energy in the best of worlds, TRU in TO.

Reality and ACEXIN differ. ACEXIN is the third and highest state or level of mass energy, size and time in TO. Reality is, the union of objectivity and subjectivity, and is largely limited to, yet includes, all levels of the mass energy dimension, especially its higher levels of the subject-object relationship in TO. By this difference and contrast reality and ACEXIN are coexistent subclasses of the mode and form of TO. Yet because of the concentration of reality in RELTO and the more specific varieties of TO, and TO with TOF, reality portends much more in the present and future. Together they provide concrete evidence and proof of the explicit and exact proportional totality required of TO and TO-TOF, reality by its combined levels and power of increasing Á in TO, and ACEXIN by being an accurate, proportionate and all inclusive form of TO-TIT.

Structure-Function Theory has given support to the fact, and role of combined form and function in the proof of an object, totality and TO. It describes the set order of TO in which the complexity of dynamics occurs. Since TO is a compound of absolute-relative totalities and varieties the presence of structure and function varies with each level of ARTO, or variety in TO, TOV.   Change occurs not only singularly or by increments, but in different multiples, and often massive and by precipitous quanta, ways and degrees. Duration, change and variation are ubiquitous, universal and occur continually in and by all levels of the spatial, temporal and mass energy continua from infinitesimal to a build-up of finites, their constructs, systems to all in TO. Complex change is evident in both the infinite nature of TO and the infinite nature of the interchange between TO and all that is external. Differential change reveals the expanding and contracting existence in, and proof of TO and TOV-TRU.

These two tables (16-5) provide, in the large to medium finite range, a fairly accurate and reliable measure of the scales of dimensions, that form TO, ALT and SPALT. The second table recapitulates the first. Its purpose is to show the position of centers in each sector, and to emphasize the range and limits in all dimensions. These provide, at least in the large to medium finite range, a fairly accurate and reliable measure for position. With use, future development and refinement they can be extended, improved or perfected. For even more refinement put in key limits or clines at different levels. Key limits or clines at different levels can produce even more refinement. By the Law of the Cosmotic Absolute, (LACOSM), if we have a sufficiently actual and distinct totality and if we have a sufficiently distinct 1:1 relation of an observer and observed in its representation we can have a concrete representation that provides a valid, proportional and balanced self inclusive system in which accurate and increasingly complex detail can be provided. This is the great means and method human life so greatly needs, to provide vertical integration of a closed set. Yet we must always remember that mass length and time do not exist in any standardized formal way, nor in actual totality like they do in absolute totality, and the physical universe. They are neither linear nor exponential, but a hybrid of these two plus the kind of alteration, or warping, noted in perspective.

Perspective in art has gradually developed through history and became a major feature of renaissance painting. What is close to the observer is most prominent and as we get farther away it loses prominence, yet still retains the right proportion at least to the viewer. In this way it exists in the average sum of actual totality and SPALT. The Ideal positive objective approach is where there is 'a place for everything, and everything has its place'. This is not only demanded of TO, it is a clear proof of TO, to the extent the vertical integration is right and detail is optimal. This is largely, increasingly and practically provided by TO, its form, proofs and representation. The subjective cannot exist without the objective, as the mental of human life without the physical. Equally, without the subjective and mental, the objective and physical lack the necessary existence, actuality and reality that objectivity and the physical alone cannot provide. It is not any mlt, only mlt existing as ALT and provided by the actual pure essences of mass energy, extension and intension that exist in TO. The mlt dimensions, and ALT exist in TO as a continuum, from the most generic to the most specific seen in each successive variety.

ALT signifies the average sum of the mlt of TO, as well as its successive, and specific stages.

It is not so much that humans make knowledge, mentality, subjectivity and reality so great. It is reality, subjectivity, mentality and knowledge that make humans so great, and together making and proving actual totality. Neither one is so great without the mlt to ALT that forms and best functions in their existence, with simultaneous integration and beautiful intricate detail of differentiation that is correct, most accurately, efficiently and practically applied. The Law of Peripheral Reduction Physical Universe by TO is signified by LAPRAUT. There is a relative reduction of importance of the extremes of space and time by the here and now of RELTO, TRU and TO. To expect the subatomic Higgs Particle to be the "God particle" is largely illusion. Yet with more knowledge of the roles of ABSTO and RELTO some modifications of fundamental relations in TO may be forthcoming. Humans are largely only the custodians and vehicles of knowledge, mentality, subjectivity, reality and the form and function of all in TO. It is a large part of their duty to the life they have been given to correctly fulfill and complete this duty. This is why by being their actual pure essences and achieving their right transformations they must exist in TO.

#	Mass-Energy	Length	Time	#	Mass-Energy	Length - Size of	Time
0	Infinite Objective	Infinitely Large	Infinite Past	0-1 0.5	Completely Physical, objective	Astronomical to more than Earth	Before Earth
1.0	Highly Objective	1 Million Km	10M Yrs Past	1-2 1.5	Highly Physical to early Involuntary	Earth to county >10 B -100,000	Earth to 1 Millennia P
2.0	Mostly Objective	66.7 Km	1,000 Yrs Past	2-3 2.5	Physical/Real Inv-oluntary to reflex	County to block 100,000-300 Pop.	1 Millennia to 10 yrs
3.0	Real / Objective	100 Meters	10 Years past	3-4 3.5	Real/Physical Reflex to voluntary	Block to house 300-20 people	10 Years to 15 Minutes
4.0	Mostly Real/ Little Objective	6.67 Meter - 10K Kg	15 Min Past	4-5 4.5	Mostly Real/ Little Objective, Words	House to a person 20 - one person	15 Minutes to Present
5.0	Perfectly Real Evenly O and S	1 Meter @ 100 Kg	Exact Present	5-6 5.5	Mostly Real/Little Subjective - mental	Person to an organ Brain Size (1 Kg+)	Present to 15 Minutes
6.0	Mostly Real/ Little Subjective	0.33 Meter @ 1 Kg	15 Min Future	6-7 6.5	Largely Real to Conversation	Brain to size of tooth or one gram	15 Minutes to 10 Yrs F
7.0	Real/Subjective	1 Cm @ 1 gram	10 years Future	7-8 7.5	Mental to Thought to Cognitive	Tooth to Size of Average Cell	10-1000 Yrs Future
8.0	Subjective/Real	33 Microns	1,000 Yrs. Future	8-9 8.5	Largely Mental to rational & logical	Ave. Cells to Size Atom, Proteins>	1000-10M Yrs Future
9.0	Highly Subjective	1 Angstrom	10 M Yrs Future	9-10 9.5	Highly Mental to all comprehensive	Atom size to what is Infinitely Small	10 M - In fix nite Future
10.0	Infinite Subjective	Infinitely Small	Infinite Future				

Table 16-5 Scales and Limits of Dimensions in Actual Totality

Orientation and Ordering of Parameters - By investigating different parameters and identifying them with TO, their relations and the form and function in TO can be better known. It is crucial that we bring this together to see how they are oriented and can be ordered. This is what will enable their use for a positive objective approach that can convincingly prove TO.

Does empty space exist? Is a perfect vacuum possible? If zero and absolute are non existent then what exists in seemingly empty space? If the law of more or less, and never perfectly absolute or relative holds, there is always something in space, if only infinitesimally small and diffuse. Just like a zero point in length or time doesn't exist, except theoretically, the same would hold for empty space and a vacuum. Is there a continuum between infinitely concentrated space and infinitely diffuse space? What are its characteristics and the content of infinitely concentrated and diffuse space? If nature abhors a vacuum how are these facts reconciled? What regions of space are we considering? Is it astronomical space, space between subatomic particles, the space we encounter in the world around us, or the space of ALT in TO? These are problems involving the length dimension in its highest form, volume. It reveals how physical dimensions do not exist separately. They exist only in the larger context of the system, world or totality they are a part. This is evident from the contraction and expansion of the farthest known extremes of astrophysics to the closest extremes of length and size about us. To the extent they are valid, and must have necessary existence, they will hold for TO. It is here their explanations must be found as shown in SPALT.   Just as the great energy produced by the big bang altered mass, length and time, so is the great energy of burgeoning human life and mentality altering mass, length and time. Not only does mass energy become that characterized by the Á dimension, space and time become ALT, as shown in SPALT. TO, being a composite of many varieties, TOV, retains some of the m l t form of each variety, but is characterized by their combined balance, the sum of their averages plus its own actual pure essences form, i.e., ALT.

There is a huge difference between the physics of absolute totality, the physical universe and that which exists in actual totality. One of these is that between the lack of relativity in absolute totality, relativity of the physical universe and its role in the relative universe and relative totality. Another is the fact that the astrophysics that borders on absolute totality, although in essences and ways a part of actual totality, is of the extremes of space and time relative to all the rest in actual totality. These extremes contain a myriad of subatomic particles and their variants or characteristics, and a myriad of stellar objects, such as suns and galaxies, etc. Because our knowledge of these extremes is incomplete, and because discoveries of their questions, problems, paradoxes and enigmas should or do not change the essence nor the overall content of absolute and relative totalities, ARTO, in TO, the classical and relativistic interpretation holds in the near, mid, and even much of the far term of TO with TOT. They should hold for the practical existence of TO, and into the future. It is for this reason we can utilize mass length and time, and their modification by actual totality, as a practical and largely accurate means of determining the form, function and content of actual totality.

CHAPTER 17 - MASS ENERGY Form in TO to Dynamics-Statics

Kinds and Levels of, Matter and Energy, Inertial Energy, Transformations, Progression, Adaptations, Separate to United and S-O in prime Á Dimension, and Form in TO

Mass Energy and its forms, states and levels are key to an ultimate understanding of the actual universe, TO, and TOV-TRIR-TOF. Through integration of its mass energy gradients from an object through real to subject, spatial and temporal dimensions produce an accurate 3D form of TO. Mass, the static polarity of the mass energy duality with many universals, continua and qualities combine and determine the form and proof of TO and its varieties. With expanding transformations these change, to maintain optimal existence in time, TOF of TOT with TO.

In Pre Socratic times, especially of Anaximander and Anaxagoras, the belief in typical form was established. This conclusion arose from the fact that mountains, seas, and human bodies were formed from their rocks, water and organs. Various names were given to the form of the cosmos, or rational principle. A cosmos that is the actual totality is not quite so simple, for it contains a mass energy spectrum from each of its varieties, from those of the physical universe to relative totality, especially typified by that of its highest levels, best human life, and it's massive, intensive power of thought and reason. This is its primary parameter, or Á dimension, a major proof and law of TO. Cosmos and nous are not two separate deities, they exist, as all, only as they are their part and role in TO. Proof of TRU is from the propelling and compelling forces that inevitably combine its mechanisms and powers. Many of these forces represent the inertial energy of TO, energy that produces the momentum that carries TO ever forward. They tend to assure eternal continuity. Combined with other kinds of energy and mechanisms, and added steps in the development of the whole course of mass energy, TO is established.

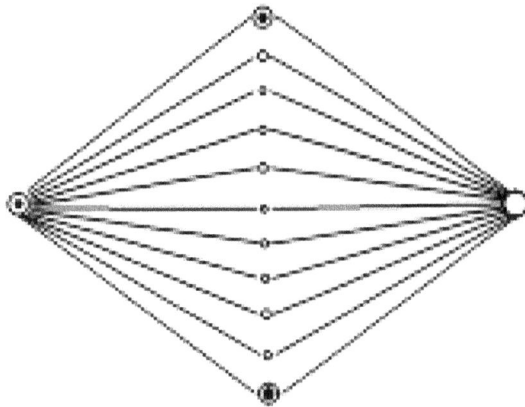

Fig 17-1 Frequency Becomes Mass, or Matter x volume

That frequency, as the amount of matter, becomes mass is a proven physical principle. In totality, frequency of its predominant constituents, is the determinant of the form and function of mass energy. It is how frequency becomes mass that life and humans are formed and function. Both qualitatively and quantitatively what is predominant, most characteristic and typical for TO, becomes its kind of mass-energy, its Á dimension or continuum. This great primary form and function render's humans able to distinguish, make discrete, select, decide and develop all the physical and mental essentials needed. These are all levels of mass energy, especially that are special for predominance, as senses, ideas, language, meaning, concepts, principles, laws, reason, intellect, comprehension, wisdom and right action. It is what occurs when the most frequent or predominant relates and combines with other parts and processes to make a new entity. It provides the proof of necessary existence or actuality that certifies the actual pure essences of ABSTO-RELTO and those of TO that give it its typical and unique form. This is like the transformation between internal and external of the big bang. More energy was produced by the big bang than any other known source. Next to that are the long gamma ray explosions of many supernovae. The third and others were from versions of the several kinds of supernova, and black holes, detected in the last ten or so years. The continuous build up of frequency in mass of relative totality gives TO its typical and unique mass energy, form and function. With space and time, the mass-energy, Á dimension, forms the actuation, extension and intension, or 3D ALT basis of TO-TRU. By RELTO ALT provides that form necessary for representing TO.

All mass energies, like matter and even more than length and time, are not the same, especially with the development in TO, and the progression from general to specific varieties. The continuum from mass energy in absolute totality to its most advanced form in the most specific varieties of TO forms a spectrum in which mass and energy fuse and become the Á parameter and dimension of TO. Each variety of TO does not have the same Á form of mass energy. Since this is the predominant form in TO, by the force of relative totality it is the Á form we largely use in presentation. Yet the limits of each form, throughout TO, need to be remembered and applied.

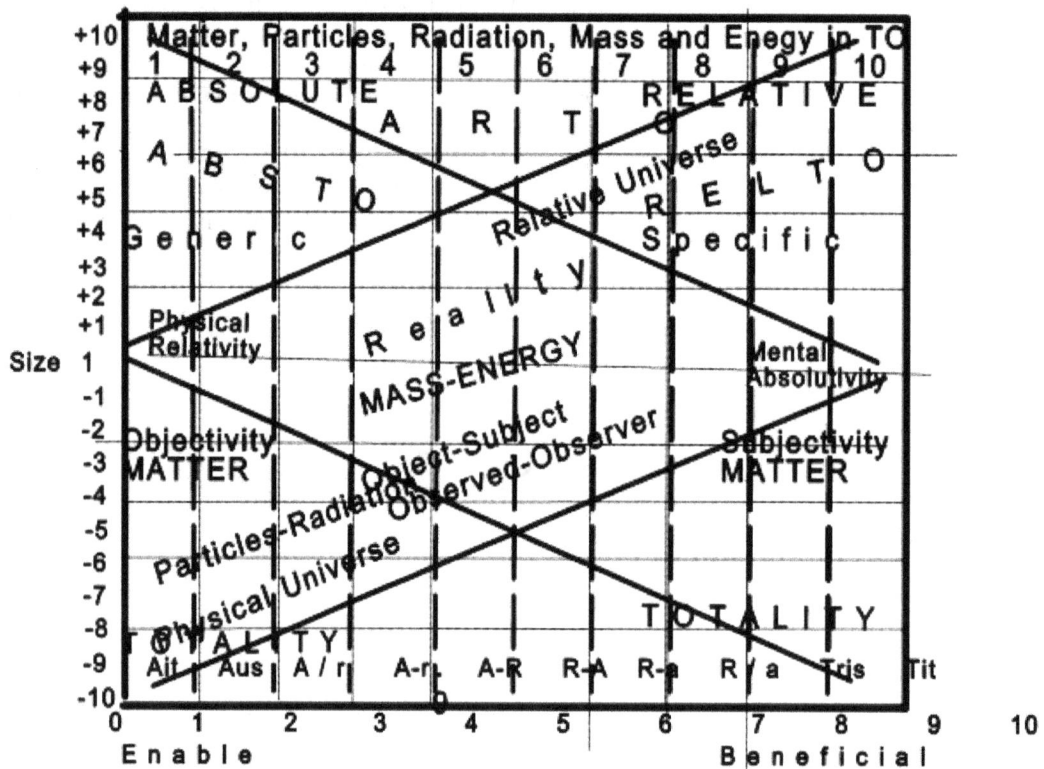

Fig 17-2 Matter, particles, radiation, mass and energy in actual totality

A good beginning for explaining the kinds and forms of matter is to start with absolute totality and the physical universe. This is with the different subatomic particles and radiation of the electromagnetic spectrum. They are followed by the atoms and molecules of inorganic and organic chemicals. Each variety of actual totality consists of its own typical kind or form of predominant matter. So that as we pass from ABSTO to RELTO in ARTO of TO, matter - and its combination in space or volume, mass-energy becomes characterized by the predominant form, function and power of observation, consciousness, attention and higher levels of thinking and reason. Thus all together in a sum, TO consists of the entire spectrum of matter, or mass energy that make it predominant, and make it most successful and self sustaining.

Since matter is the fundamental stuff or "material' all is composed it is to matter, and matter in relation to the other primary categories, features, modes, continua and varieties of TO we must direct our attention if we are to ever reduce all to their least common denominators and order in TO. Matter like so many great terms we use, are indefinite, even for physical science. To make matter more definite it is imperative, if we are to learn, know and show TO, we must give it's most definite form, function and relation in actual totality. To make matter more definite and to explain how the most basic material is not only static matter, but the whole continuum between the most static particles and their greatest energy and activity, e.g., the radiation of particles. Since like "substance," matter or particles and radiation are what is predominant in any world, universe, deity, system or totality, we need to describe and show both the various totalities that makes up actual totality, as its varieties, and the kinds of matter that is predominant in each. How matter, mass energy, space and time are altered by transitions to, from and within TO depend on the current form and state of TO. How the changes that occur, increasingly with time, in TO with TOT, and the many factors and forces are involved. How matter, mass energy, space and time exist, or were and will be altered by transitions before, within and after the big bang, are poorly known. They are reversible, with forms that help to explain transitions, which have, are and will occur in TO.

Matter corrected for volume equals mass. This enables matter to assume its role in the mass energy dimension of the physical and relative, ABSTO-RELTO inTO.Matter and radiation, like particles and radiation, or light and its waves, are their static and dynamic forms in the physical universe. Dark matter and dark energy are the largest components - more than 95% - of the astrophysical universe. Of the total matter in the universe dark matter is estimated to be 84.5%. Although dark energy is about 68.7 % of the mass energy, dark matter is about 26.8%, which means the relative distribution of mass energy is a factor of about 2.5x in favor of mass energy. This is confirmed by the fact that dark matter occurs more by attraction, in clumps, and is not universally distributed, whereas dark energy is globally and evenly distributed or with varying density in space and time depending on the form it takes. There are many optional theories or variants including standard and cyclical models, effects of pressure and temperature, and changes relative to the observer. These latter changes bring dark energy into the absolute and relative totality as it exists in actual totality. It depends on the effect of relative totality and how its predominance is much more than any observer, but the whole realm of observed-observed of the absolute and physical by the knowledge, technology, activity and life of relative totality in actual totality.

Mass energy interactions, creation and expansion of TO, with and by its specific varieties - When matter is conveyed into energy, as in the linear accelerator, a little mass is equivalent to a lot of energy. This is a fundamental relationship with immense applications to technology and TO. It not only reveals the availability and potential power of energy. It reveals the transformation of energy in TOV, and both the positive and negative potential and actual roles of humans in TO-TRU-TOF. It explains the greatness of mass energy, their identity, the power they generate, and comparative proof in TO and TRU. Like the big bang, a tremendous amount of energy was stored in the tiniest mass of matter. So is it of the most specific varieties of TO, an increasingly accelerative tremendous amount of energy is stored in such a relatively small world and volume of subjectivity, by the actual pure essences of RELTO in the human brain, cells and synapses, compared to the physical universe. It is with such potential and actual power and energy of the human brain that the dynamics of TO are created and expand. This causes and effects much of the subsequent parts, processes and whole, of TO.

It is the unknown of dark energy, like antimatter, such as composition, variation of distribution and role with the relative in mass energy that can potentially, resolve many of the problems of the physical universe and help better show and prove its role in actual totality. Like many unknowns, when known, become so obvious. Yet without the evidence, facts or key mechanism they remain unknown. How dark matter and energy, consist, like APET, of a continuum, especially those qualities that are not associated with physical particles or chemical elements alone. The riddle is ameliorated by the low probability of astrophysical factors affecting the near, or even mid, term of TO. However some factors are so close to TÓ, and its discovery and knowledge that they can less directly be beneficial. Like how TO, and God, exist within each individual, so do dark matter and energies consist of a continuum of states of mass and energy. Thereby it is how each exists in each variety, and all that is TO that will help to resolve this question, and give further proof of TO.

Mass-energy, matter-antimatter, absolute-relative exists in totality and in transition providing proof of TO, TOV, TRU. The two poles of mass-energy in the physical universe are exemplified by dark matter and dark energy. Dark matter is associated with expansion, dark energy contraction. The fact that expansion appears to be overwhelming, and contraction less so suggests that there are more forces and more to it than an absolute physical universe alone can account for. There are many basic changes in absolute and relative totality, TOT, TO, TOV and with TOT in time, TOF and TAR, more than a simple non existence before, and unity of everything now, and an indefinite disunity and diversity after. This makes the transition of mass-energy in the absolute to relative makeup of TO, and its varieties in transformation, more probable and evident. With 96% of the physical universe consisting of dark matter and energy the questions are what did they come from, what kind of matter do they consist of, and how are they distributed? They are probably not evenly distributed, nor of even consistency, and may have, at least in part, existed before the big bang. Most likely they were, are, and will be continually emitted from stars, galaxies, and a variety of explosions. They are also likely to be of absolute totality with a limited relation to either indefinite or definite relative totality. The significance of this is the separation and exclusive existence of TO. This means that what helps TO is either, 1. externally, needed potential to actual, 2. Internally, already built into its existence, or, 3. Internally, up to what already exists that assures the continuity of TO. The latter are the positive imperative of TO. Each and all its parts, especially each and all persons, must be their part and fulfill the needs of TO, as it does for them. By the sense of reality in objective-subjective, S-R-O, that such transformations have, are, and will occur in the past, present and future, and the making of actual totality.

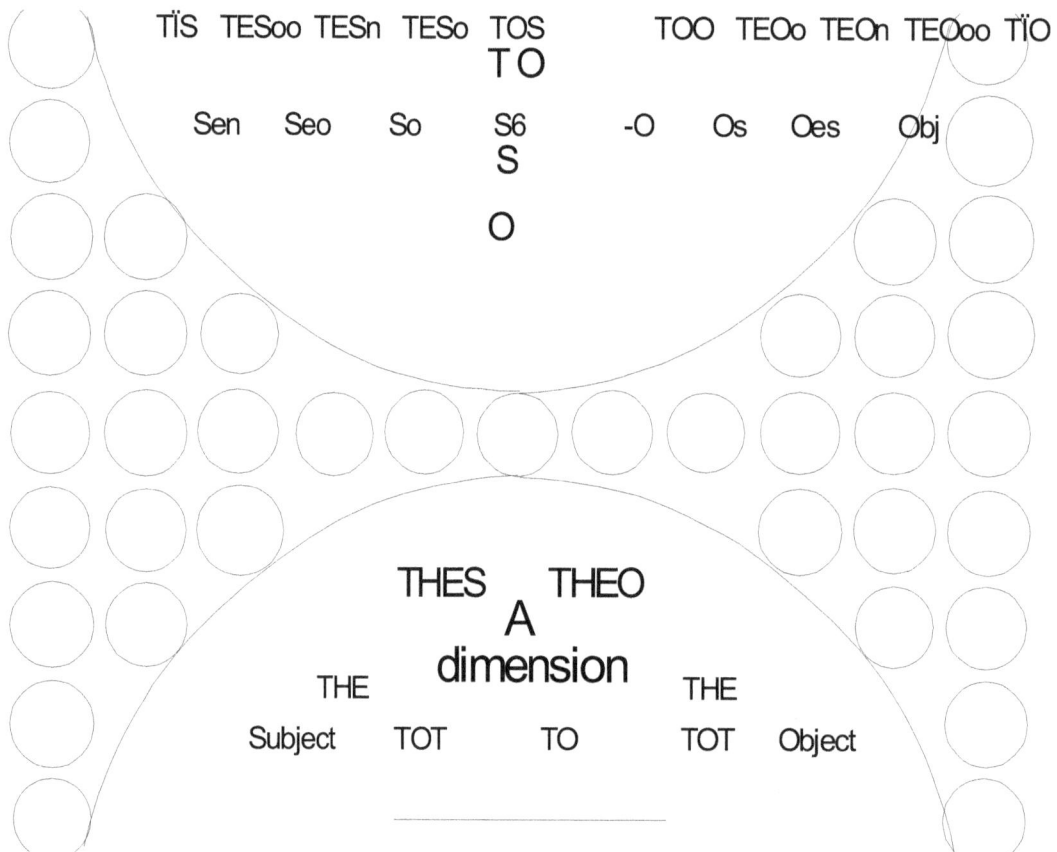

Fig 17-3 Convergent subjectivity-objectivity, S-O, proofs of observer-observed in mass energy and the Á dimension in TO, TOV and TOF

This is a standardized perspective of the observed-observer or Objective-Subjective. It is one of the greatest of vistas, for if one can grasp the gist of TO and feel the revelation and inspiration that come from understanding the actuality of S-O and Á one can realize much of the form and proof of TO. It is fortified by S-R-O or the great expanding reality of actual totality relativity and observer observed provide. This is because it not only gives a good account of the parts of entities - senses and things - as they become TO. It gives us the basis for the prime, S-O or Á dimension of TRU that is the essence of the fusion of absolute and relative totality in TO. It results from the extreme lability of our position as observer, instantaneous attention, IA, and the observed, as entities and objects. This dependence is the essence of special relativity that reveals how the physical universe is a step away from absolute totality, and a qualified part of TO. The exclusive ability of meanings by SO or A for TO is the Exclusive Ability Principle, EXAB, the hallmark of reality and proof of RELTO, TRU and TO. Each and all of us must not only have senses, perceptions, apperception, feelings, thoughts, reason, etc. we must be able to bring these together and sort them out. We must have some kind of overall world view and perspective. This is impossible without form, order and standardization that with perspective of the objective-subjective fulfill their purpose in TO. It takes knowledge combined with learning, training, and intellect. It is a great part of one's maturation in life. With it a person can better cope in life. Without it one flounders, is quickly lost, confused and defeated. With its cure and end the absolute of objectivity and relative of subjectivity are incorporated and formally modeled to best bring representation and presentation for totality proper, the actual totality, TO.

The speed of light is constant for the physical universe and all within the big bang. All not within the big bang has basic alterations of mass, space and time, wherein the speed of light can be more or less, reversed. Matter exists as particles or waves, photons and their movement limited by the speed of light, and other spectra like the electromagnetic and color spectra by their actual pure essences they are part and contribute in TO. TO is confirmed by LANERA, the law of non existence of absolutely absolute or relative totality. It is also confirmed by the non existence of independent form, function, mass, energy, observer, observed, subject and objects, (LASOT). This is the basic fact of relative totality that permeates all TO, and is the foundation and proof of the mass energy continuum, Á or subject object dimension of TO. It holds that nothing is absolutely only subject or object, only physical or mental in actual totality, TO.

By being predominant and highly autonomous, the special modes and properties of TO, driven by the human mind and all it compromises provide the essential mass energy whose operations we seek to show. The potential and actual energy and power of the human brain arises from the continua of actuality, or ever higher levels of necessary existence that has, is and will transform the extreme specificity of potential TOT to actual TÓ to ever higher and greater levels of capacity and ability through TRIR in future TÓ, with TOF of TOT.

Doubt is a feature of subjectivity-objectivity reality in TO, and thereby of its mass energy. Although uncertainty is opposite to knowledge, correct knowledge and TO, doubt by stimulating the higher powers of mass energy in the more specific varieties of TO, becomes highly positive and a great propelling force for TO. This shows the fundamental relationship and interdependence of progressively greater relative totality in TO, and the role of subject-object and reality. This is the way and mechanism by which the highest powers of mental, cognitive, rational, total mass energy and reality encompass and help to unite, consolidate and expand TO-TOV and by TRIR with TOF. The existence in other mammals, birds and animals of lesser levels of mass energy is associated with uncertainty, doubt and their stimulus to response by resolution. This helps to describe and prove the most important gradients and distribution of the continua of mass energy in the varieties, of TOV, of TO and their course of evolution in TOT.

With TOT, TO develops new entities or varieties of special kinds of mass-energy, form, state and dimension. Each has their own qualified, typical mass-energy states and dimensions. For TO the typical predominant frequency, relatively, have added states in totality which we call its Á dimension, state and form. This is manifested in a spectrum, by levels of Á. It is this dimension and spectrum, SÁ, that typifies the form, function, state, conditions, limits and relations of TO-TRU-TOT. It characterizes TO, explains and proves its existence and distinction from all else. The gradients of this spectrum, like frequency in mass, show the substantive makeup of TO and TOV.

Mass energy undergoes many transformations in the transition from its state in the physical universe through the combined ABSTO-RELTO in TO and into TOF. From the most absolute to most relative, generic to specific totality in the varieties of actual totality, TO, mass energy exists in its most actual form and function in TO, that of its primary, or Á dimension. It is the mass energy continuum or Á dimension that signifies much of the form and function of TO, and TOV. The O-S spectrum provides positioning in the form of TO, TRIT, in SPALT and other models. The levels and expansion of mass energy in TO and TOV are part of its basic form and function. The transformations and progression tend toward fulfillment or completion by the powers of TO-TOT. The form and function of mass energy include characteristics of their role in the physical world, adapted by their actual pure essences in TO. The actuality of mass energy in TO and TOV-TRU-TIT-TRIR-TOF is most unique. The great convergent power of RELTO combines, unites and typifies TO and TOV-TRU-TIT-TRIR-TOF, by the average sum of all the varieties in TO, and the effect of its own actual pure essences.

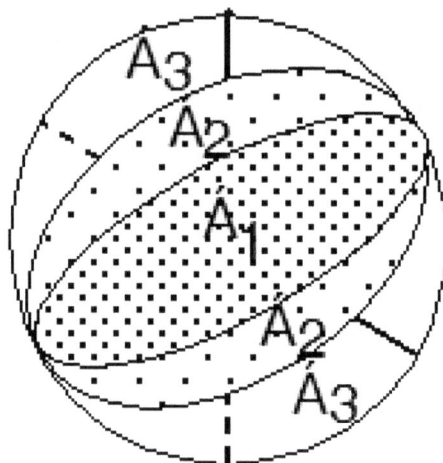

Fig 17-4 Higher Levels, Orders and Powers of Á

It is by the higher dimensions of Á in TO that all can be better understood and proven. It constitutes the unique, typical and special predominant aspect and form of TO. It is a key feature of TO and its form. When corrected by the limits of TO other features of form and representation, they provide a good idea of the frame, structure, and function in TO. The fact that forms of entities and objects are fundamentally different and change or undergo alteration is supported by certain determinants and principles of form, e.g., entasis, the curvilinear alterations in structural form. Such determinants and principles of form can be used in more than one way. This is by more accurate representation, to increase the effect on the observer of a quality or of the structure, or most important to show the actual alterations that exist in the object itself. This reveals and proves the role of form in developing mass energy, and the development of life forms. This separates the absolute and relative totalities, as well as shows how for either to exist they are combined, as ARTO, in TO. For the form of TO, this is called: LASTIL Law of the Suspended State of Structural Limits. It does so by not only showing a more actual form, but by showing how the observation and reproduction of this form produce altered, different, and added states in its structure. By the mass energy dimension, Á, of TO and TOV-TRU-TIT-TRIS we can better discover, learn, understand and prove all, all TO. Á is the major constituent of the unique, typical and special predominant aspect and forms of TO and TOV, TRU and TIT. It is not TIT, nor even a 1:1 figure of its frame, for TIT extends infinitely and its dimensions are not so simply delineated in its extreme. Yet it does, when corrected by the limits, LOLS, of representation, ROT, and objectivity, provide a good idea of the frame, structure, and function in TIT and TO.

As we proceed through the text, we continually add these superimposed states and structures to supplement the form of TO. Yet we must continually return to the exclusive form that exists in TO, what this alteration or warping of the form of TO necessitates when made an object or modeled. This is the Absolute Necessity, Correction for Short Warping in objects and TO. It is conditioned by, Law of TO Eternal. No matter if TOP, TOT or TOF are 10, 100, 1,000, 1 million or 1 billion years it is still TO eternal. For within this time frame there are concentrated the whole infinite and eternal being of this special entity, TO by TOV-TRU. Correction of short warping is how TO eternal can be within a set time. If the actual world of man, TO, should be terminated or become extinct in a relatively short time, then it would still behave and function as the totality, maintaining its ALT configuration as a solid mass, or as by SPALT, a sphere. This is related to Zeno's Paradoxes. As function is abbreviated so are other dimensions balanced and abbreviated by concentration and reapportioning. This is, Law of Proportional and Abbreviated Reaction Equilibrium, Equal Abbreviated Reaction, (LEQUAR), and vividly illustrates the forces of proportion and internal equilibrium that occur through time. These are some of the adaptive powers of relative totality in TO. The maintenance of optimal shape is signified by: TASPA, the form, shape and structure of TO are not necessarily spherical, but will vary with kind, and showing, of TO. It will be increasingly solid and spherical when viewed from the distance, when the limits of its dimensions are set, all its FAF is incorporated, and its representation and modeling best replicate the dimensional form of TO.

Since light is a kind of matter and mass energy, and since mass and energy are two poles of the same continuum they are aspects of the same thing. This is the problem with a physical or objective universe only, its math applies largely to separate entities, not their essential existence when combined in the ABSTO-RELTO ARTO of TO. From relativity there are the following relations,

Energy = Mass x $1 \sqrt{lt}^{2}$ The speed of light or, Mass = 1/ Energy $\sqrt{lt}$ x ^{2} The speed of light or,

Mass Energy = $1/l/t^{2}$ The speed of light or,

mass energy = kinds, and speed of, matter or,

mass energy = matter x T

Light = a kind and form of mass energy, since

Light = a kind and form of observer-observed and,

Mass energy = a kind of object-subject and,

Reality = a kind of subject-object, S-R-O, then

Mass Energy into = reality of TO and,

Reality = Light/ $\sqrt{lt}$

These equalities or correlations show the bond between relativity in the physical universe and the basic relation between the mass energy of subject-object in the relative universe of best human life, RECTO, ART and TO. They provide a reciprocal proof of the equation, and thereby how reality contributes to the proof of TO. Since mass is a collective of matter, material and exists with and depends on motion, practically they are of mass energy.

Light as photons and waves are of mass and energy.

By the above, matter, and photons, is reality in TO.

Matter = ME, (bosons or Higgs Particles).

Matter $\cong$ O-S, TO = O-S balanced with S-O

TOV = direction by specific TO = power of O-S

This tends to establish the identity of light, mass energy, subjectivity-objectivity and reality in TO. Thereby the absolute and relative combined basis for and proof of TO. Since mass energy is a function of light, and since light is a component of the relation between observers and observed, or subjectivity-objectivity, SO, it is a kind of mass energy. Since reality is a compound of SO, it is a kind of mass energy and a kind of matter. Matter, mass energy, light, relativity of SO, and reality thereby is different aspect of the same thing. This is how they exist together in TO, yet manifested in different ways at different levels of TOV, the varieties of TO. The effect and stimuli of light, along with other forms of matter and stages of mass energy, were major factors in the origin and development of early life. In many ways light led to the development of relativity of observer-observed, the nervous system and the subjective-objective form of mass energy whose highest stages are revealed in reality and the power of language, conceptual thinking and logic. The central role of light in TO is evident in the brilliance of the big picture of one world, the unity revealed by all that is great in actual totality.

Since reality, and SO, are prime continua or basis of TO-TOV-TOT they help to show and prove equality or proportionality, and the role of relativity in TO. This is how matter, mass, objects and aspects of reality are largely static, and energy, subjects and other aspects of reality are dynamic. Each cannot exist, in TO, without the other. As both light and SO are a kind of matter and mass energy, so are they more or less the same. As SO is a kind of matter, so is reality. As all are great by their actual pure essences in TO, so together are they more greatly. As both particles and waves are mass energy so are objects and subjects, and the highest complement reality a kind of mass energy. Since all are different aspects of the same thing in TO, so is light and reality, as all in TO, more or less the same. As reality is a kind of SO, so is reality a kind of matter and mass energy. TO cannot exist without all the kinds of mechanics in dimensional formulation whose actual pure essences combined in SO and ALT forms make up much of the detailed form of TO.

The law of specific masses and energies, LESEM, signifies stages in transformations of TO and TOV. Basically for Á, this law is the special case of the highest levels of mass energy in the most special varieties of TO. TO consists of all levels of mass energy transformed from most generic to specific. Á is the total focus of all common, relative, and actual that makes up the mass energy that becomes the special form and function of all that enables best human life, TRU, in TO. Entities, objects and universes are determined by their specific or special and predominant energies and masses. For RELTO and the special varieties of TO, e.g.TRU and TIT, this is their mass energy continuum, or Á dimension. It is the fundamental mode and division that selects, identifies, assures and makes exclusive and permanent their essential existence. More generally, extended from APET, Á and LESEM, the law of specific masses and energies, are the total focus of all common, relative, and actual that makes up their mass energy, spatial and temporal form and function to become the special form of all that enables best human life, RELTO and TO. It is the total of all their parts and contributions to the whole as they exist together in the special relative set of TO-TOV and TOT-F. It is a large determinant of special independent necessary existence, holds for, and is, a proof of TO. For parts of things or anything taken independently of TO, this law does not apply, for they will lose form as an object in itself and will more or less adapt to the characteristics, energies and masses of the totality to which they belong. The law of specific masses and energies occurs in and helps define constituents, parts and processes external and within TO. Further proof of combined and unified continua, their gradients and parts are evident when we position many of the greatest existing and new concepts in models, as SPALT, and the combined total proofs, of TO. It is the total of all their parts and contributions to the whole as they exist together in the special relative set of TOT-TO-TOV-TRU-TIT-TRIR-TOT. This concept provides further proof of combined and unified continua evident when we position many of the greatest existing and new concepts in SPALT or models of TO. When combined with the mass energy, mlt, and especially time form and function of absolute totality, the total form and function of TO are determined, cf. Table 16-1.

Mass energy transformations, gravity and adaptations - The role of gravity in mass energy and dynamics has been not fully known. Although one of the weaker forces in the physical universe its part and role in both absolute totality and TO is more than the dynamics of mass energy, it is part of the great transformations that occur between the spatial, temporal and mass energy. This is evident in the fact that as we approach the speed of light we become heavier. It helps to prove the interdependence and interchange of mass, length and time in totality, and how this differs between the physical universe, as ABSTO, and TO as ABSTO- RELTO. The importance of gravity is commonly seen in astronomy as well as our existence here on earth. It is how gravity, other forces, mass energy and its higher states exist in TO. They present to us the limits we must adapt to, or alter, if we are to manage, control and survive, now and in all time.

For TO the mass energy of absolute totality or the physical universe exists as a continuum of expanded higher levels, called the Á, or prime of mass energy that exists in a state of expansion, explosion and acceleration to completion, ACOMEX. This explains the transformations and transition of TO into TOF, continued presence of their modes and the direction of TOF, especially by the speed and acceleration of higher cognitive powers of special TOV and TRU. These can be of various modes, qualities and rates, as well as various types of completion, as relative, absolute, toward completion, or through completion reversibly through maintaining optimal existing states and limits. ACOMEX is a great concept that both reveals and shows how all comes together to show the continuity and prove TO-TRU-TRIR by their power to reversibly adapt and alter TOF. ACOMEX may also be compared with expansion of the physical universe, and the limits associated with each. If ACOMEX far exceeds the expansion of the universe, actual totality will be much more successful in it's inter actions with TOT-TOF, especially in the long term.

If objectivity is the mass energy of absolute totality, and if subjectivity is the mass energy of relative totality, then reality is the mass energy of the average sum of actual totality. This assumes reality is the intermediary of objectivity and subjectivity. It also assumes objectivity and subjectivity that represent the physical and mental, physical and relative universes. In this way the various kinds of mass energy, as physical forces, ems and all higher states of mass energy in actual totality, whose average sum is the Á dimension, unite to form the whole spectrum of mass energy in TO. It proves Á, the mass energy make up of TO, and TO itself. Subject Object Standardization is the method of formalizing the state of subject-object in TO and TOV to optimize representation. It includes steps like dividing the spectrum of S-O, SA, into ten equal levels. It also includes the various vantages by which TO can be given perspective to optimize our view and comprehension of TO. Mind, mentality, subjective and spirit in TO-TIT - The mental portion of mass energy subject object spectrum in relative totality, and effect on all actual totality, is the salient of human evolution and basis for and proof of the mass energy continua in TO and TIT. The mind, mentality, subjective and spirit of humans are the great, although not only, feature of TO-TIT that best human life along with much of the rest of RELTO and TO have developed. The higher the form of mass energy they provide, the more critically essential they are necessarily existent in TO.

CHAPTER 18 - DYNAMICS, in TÓ, statics-dynamics, mechanics to Relations and Centering

Functions, Forces, Action - Reaction, Processes, Mechanisms, Interactions, Integrative Processes

Dynamics, the dynamic-static spectrum, positive/ negative, in and out dynamics and optimal function provide the best and greatest growth, improvement and permanence of TO. Dynamics is the active side of the static to a dynamic spectrum. It includes all that has or is in motion, functions, processes, forces, actions and many time related items in dimensional formulation that are the mechanisms that have a large role in TO. The continuum of static to dynamic is one of the most important. By discovering and understanding the part and role static to dynamic play in the proofs of TO and its progression, a much more accurate and fuller knowledge can be provided. Like the absolutes of all continua perfectly static and perfectly dynamic do not exist, no more than a zero or absolute temperature exists. Yet relatively and practically static, and especially the dynamic, are most important as well as common features of existence, in and out of TO. Much in this book, of proofs of TO, is directly or indirectly related to their dynamics, provided by the positioning in static form of TO-TRU and representation.

Most interesting of continua and relations is that among all that are equal or similar to statics and dynamics. There are form and function, objectivity and subjectivity, mass and energy, matter and energy, particles and waves, darkness and light, density and non density, the states of matter and anti-matter, solid and gas, and the gradients of their spectra. The degree and state of dynamic displayed in all of these are the doors to understanding and proving actual totality. When the degree and state of dynamic are viewed in the transition from mass energy in absolute totality to its highest forms of reality in actual totality we can begin to see the light that shines so bright on the most valid, proportional, actual and total existence, TO and all that it consists.

A most basic continuum is that between nothing and all, zero and the absolute, darkness and light, which are equivalent to length, as static and dynamic are to time. The facts that with increasing temperature solids become liquids and then gas is typical of absolute totality and the inorganic of the physical universe. The fact that with increasing temperature, as when an egg is boiled, liquids become solids is typical of relative totality and the organic world. This demonstrates and proves the reversibility of mass energy and certain physical characteristics from absolute to relative totality, and the inadequacy of separate physical universes. It is the combination, altered, adapted and reversible states, and their actual pure essences that justify and prove TO. This is by the Law of Alteration, Adaptation and Reversibility of all in TO. LARIT. This is a primary proof of TO. By the Law of Matter and Anti-Matter, LAMA, anti-matter and matter exist as a most basic dimension. Relativity to be correct should adapt its math and physics of mass energy accordingly. Since mass and energy are opposite sectors in DIMFOR that parallel LAMA, this produces roughly the same effect as antimatter and matter when a primary dimension with space and time.

Mechanisms are modes, dynamic functions, forms, states and conditions, derived from mechanics that are highly applicable to TÓ and all its varieties, TOV. They explain the way an entity, object, system, universe or TO exist and function. It is evident in evolution and teleology, as well as the key dynamic mechanism of the human body. By learning and using the correct mechanisms of TO, their most actual, extensive and intensive form and dynamics can be shown, proven and applied. It is by mechanisms that the exact, relevant, proportional and pragmatic form and functions of TO explain and prove the greatness and predominance of its existence. They contrast with the unknowns and flaws in the physical universe, God and other worlds or systems by revealing the valid, actual, proportional and total detailed form, function and processes of TO that can be shown, especially when by a closed set. Many mechanisms can help manage and control TO, and all actions that require reaction and guides for best human life in TO.

Processes based on stimulus response and subjective objective dependence signify that dimensions are not only static parameters, they exist in TO as continua in highly active forms and states. Their effects are major mechanisms in the dynamics of TO. The Compulsive Dimensions of TO-TRU, CODIT, denote why a person, as their part, TES, must be, think, act, and do as they are in accord with TO and TOV-TRU. It also expresses why TO and TOV-TRU must be and function as it is. TO is proven by its Integrative Processes, PIP. How processes are related, interdependent and integrated into a whole, the whole of all that is ABSTO-ARTO and ARTO of the actual pure essences of TO. Their relation and inter dependence is a large contributor to focal convergence in the center, here and now of the reality in TO. This is how TO becomes united in its oneness and exclusive totality. Totality Proper, TP, is largely the center of ALT. Totality Proper also equals the core of Reality. Since reality is the center of combined S-O, or S-R-O, it converges and concentrates in here and now. This is as one might expect from experiences in life, reality being the S-O identity and interactions by immediate awareness, attention and close relations that need response or resolution. It is the power of these convergent and integrative processes in here and now of reality and totality proper that is a large and determining portion of the dynamics and proof of actual totality. This is signified by, LIPRIT, Law of the integrative processes in present reality of totality proper. This is a most crucial law and proof of TO. Without knowing the most important laws and proofs, especially of its complex dynamics and mechanics, actual totality cannot be adequately known, (LAKIMPAT). Without being known TP and TO can never be lived by. Without being lived by, each and all individuals, and all that constitutes, as well as actual totality can't survive.

Law of the Absolute Necessity of Integrative modes, mechanisms, forces and processes of TO, LATANIP. This, concept and law, along with proof of integrative processes, PIP, is highly evident in all about us, as in the structure of the human body. The various structures have optimal functional use, as its special senses, bones, and rim of the acetabulum of the hip. Formatively, functionally and pragmatically all fits together to provide the best possible dynamic, actual totality needs. This has been the long term effect of natural selection, whose integrative processes working in mass have continuously adapted form to necessary existence in TOP-TO-TOF. It has been, is and continues to be in response to the environment and all conditions interacting differentially and integrally, individually, collectively, by varieties, kinds, continua and modes to produce optimal actual existence of TO. This is a major proof of TO, for it not only proves the integration of absolute totality and relative totality, it proves the exclusive unified totality that is TO.

Actual totality is highly dynamic, internally and externally. It is a great mass that includes all that is typical of itself. It is externally in a state of intense, and somewhat variable, interaction with all that is not of actual totality. This is the great reservoir of non existence, fragments, existence, necessary existence, actual pure essences of various totalities that lack, or are less than, the actuality of TO. It is all outside of TO shown in Fig. 1-Intd. Unless a person can keep a panoramic view of the massiveness and dynamics of TO, and sufficient non TO in mind, they will fail to know and interpret all in life. Much we experience in life, as we ourselves, is not of TO, but of its great reservoir. We are caught up in the many and simultaneous actions and reactions continually occurring between the two. The positive and negative, helpful and harmful, identity and non identity are the determinants of the forces that are received or rejected by TO. TO is alive, all else is either dead, less than alive, potential or in transition. To know these two great totalities, their form and function, interactions and courses of action are to know all, by actual totality. TOT or what is potential to TÓ is in the middle ground, not all non existent, nor of TO, only more or less potential. It is more important in transition in time of TO, wherein all totality's interactions are increasingly altered, adapted or reversed. By revealing the great masses, form, function and how actual totality and other lesser totalities and existence interacts, are known and shown, proof of TO is achieved.

It is by specificity and sensitivity of the student, reader, observer, each and all persons that the focus of their part and contribution become the great unifying feature of the whole, TO. These plus others are the interface and site of interaction, accepting and rejecting processes between what is other, external, potential, TOT and TO. They are evidence that proves and shows how totalities are permeable, highly dynamic and transitive both externally and internally. They do not disprove, only prove the right existence and correct state and form of each.

The mass energy continuum in TO consists of three stages, physical, of relativity S-O, and their highest meanings in reality. It may also be viewed from the outside to inside, objective to subjective, or from the inside to outside subjective to objective. All that forms TO can be viewed reversibly looking from outside in, or inside out, OITIO. So that it becomes subjectivity of subjectivity plus all S-O or the form and function of the prime mass energy dimension. The validity, actuality, proportionality, and totality of both this dimension and TO-TOV-TRU-TIT-TRIS must be maintained if we are to know and show it. It is most interesting, and important, to realize how such a view is much the same as what MI, the moment of inhibition, yields in the instantaneous attention, IA, TINI, and TIR of TRIS and TIT in TRU-TOV-TO.

In the typical form, function and actual pure essences of each TOV much of the dynamics of TO exists. The effects of varieties are more or less in opposition and synergism, or repulsion and attraction to each other in and in and out of TO. This explains how the varieties function by propelling and compelling forces to maintain their part and role in TO. It explains the absolute and relative functions in each variety and in TO, how they attract and repel each other. This is most critical for the most extreme varieties, especially most specific and relative varieties of TOV, like best human life, TRU, TIT, TRIR. Gravity, subatomic forces, geology and weather are both helpful and harmful in generic varieties. How viruses, bacteria, organisms, animals, are helpful and harmful in middle varieties, and humans, their groups, cultures and mentality are helpful and harmful to each other in the latest and most specific varieties of TO. It is closely related to the contingency of solutions, cf., Fig. 35-1.

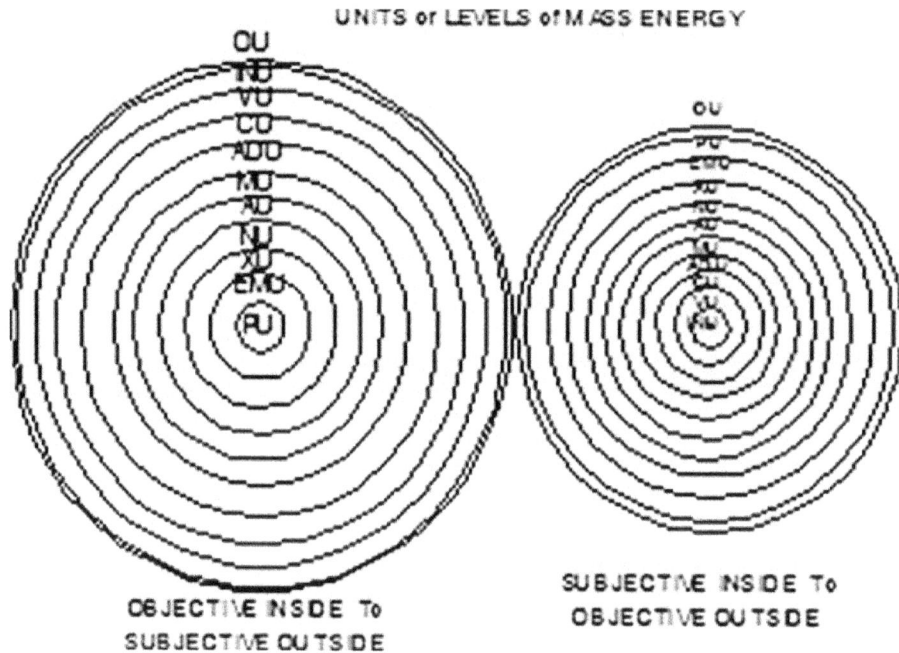

Fig 18-1 Energy Systems, S-O and Reality Levels in TO-TOV-TRU-TIT-TRIS and Representation

To signify a most fundamental fact of the relative in actual totality is the Law of Subjective TO Objective Dependence, (LOSTO). The subjective and objective occur together, are interdependent and the basis for observer-observed relativity, reality and proof of TO. Their relation, interdependence and fusion exist as a continuum and spectrum which as the mass energy of TO, TOV and TRU are its primary mass energy dimensions. This continuum, spectrum or dimension is the preeminent basic form and function, other than the spatial and temporal of TO and TRU. It largely determines and guides the convergence and action-reaction processes in TO. All that is complementary and positive for life, in TO, and complementary proportion in the constituents, components, and dynamic functions of TRU is activated by and through the continuous functions of the observer-observed, subject-object, or S-O continuum, SA, and Á dimension. Their differential elements and steps are amenable to positioning and operation in the models, SPALT and SPAT, to reveal the valid, ordered detail that a unified coordinated

TO can provide.

Special form of participants in TRU-TO-TOF - Human responses participate in a vast array of direct and indirect, simple to total interactions that reveal the great mass of positive to negative actions and reactions in TO, and worlds on the fringe of the actuality and limits of TO, TOV and their form and function. There is nothing without at least a tiny bit of significance, whether readily conceived as TO, or non TO, when carried to its infinitesimal participation in the massive human responses by TOV in TO. The stimulus-response and action reaction patterns in life are developing forms, functions and processes of the objective and subjective in TO. Even negative and harmful stimuli and actions in response and reaction become positive through compensatory mechanisms of TO and RELTO. Herein is the proof of TRU and TO. They are the answers to those critics who disclaim TO and TRU to be to narrow, or fail to allow TO and its modes. They are objective totality on the one hand, and relative totality, like God and the more special varieties, as TRU on the other, as shown by MESOA, ALT, SPALT, SPAT, and other figures by convergence on TO. We must conceive of SPALT and TO not as separate static objects. They are objective to subjective, static to dynamic and potential to actual modeled by the varieties in TO. We can better comprehend this when we expand our vision to encompass the massive intricacy of all the form and function that minimally to totally have their role in TO and TOV-TRIR-TOF.

Of the selective forces of determination proportional to TÓ, and adaptation of TOV-TRU, there are continual fusion and fission, union and separation, in the combined mix of the totality of TO. This is, and explains, the problem to determine, completely and continuously fill out and adjust TÓ to conform to its actual totality. There is common ground in universals and universes whose superimposed location collectively reveals and proves the whole, the totality of TO. The principle of viability, and selective forces of determination are only proportional to TÓ. Such selection is exemplified by the fact that a person can't do right in one department of life while occupied in doing wrong in another. Life, as TOV in TO, is one largely indivisible whole, its parts are like the notes in a great symphony. Transformation and transposition of mlt to ALT, to (ALT)3, the coincidence of actuation, extension and intension, ACEXIN, is the specification and adaptation law of TO and TOV. The totalities of TOT, TO, TOV, TRU, TIT, TRIR and TOF together and separated, when compared, reveal the validity of each, and the adaptations and alterations needed for the practical ultimate unified totality of the extended present, that is TO.

Concord is convergence or agreement of peoples or parts, often opposite, in which an ideal resolution or settlement is needed and reached. All types of concord represent common and related dynamic and convergent ideals by which TO, TOV and especially TRU is enabled, improved, enhanced, perfected, succeeds and survives. It is by the equilibrium of basic universals, continua, modes, and other major components, whose convergence produces the coordination, harmony and concord that reals and reality show and proves TO. When combined with proportions, values and other continua and actual pure essences the evidence is reinforced, and proof is more complete. Much of actual totality, many of its problems, and most of the experiences we must endure have existed for thousands of years. This is a lesson, most people in shortest present don't know, or fail to heed. Oppositely that of the distant past and future fails to credit existence in the here and now. This is why proof, a good understanding of, and life by TO is critical for the individual and TO. The complexity of the dynamics of TO is mastered only in unified totality.

This diagram (18-2) is devised to show relations between and among operations of processes of entities or TO. In this case it also shows how these apply to the relations and operations between subjectivity and objectivity, the observer observed, or relativity. The element's C, D, E, F, G, H, and I, show how SO is composed of many varieties. They include anything that acts as the means or vehicle of relativity, the interaction of subject and object. The words we use in language are excellent examples of these elements. Identifications of other elements are,

External Parameters Internal Constituents

S - Subject A - Subjective Sphere

O = Object B - Objective Sphere

Subject ObjectA' - S to OSO Subject Object B' - O to S

The arrows represent vectors of the processes and show the vast, important, often rapid and elusive, interchange of processes and the relativity that make up much of the mass energy, or Á continuum FAF of human existence. When extra elements were added, as D, the alteration in A or S and TO provide the evidence and proof needed to show how necessary existence and proof of integrative processes proceed totally. This and the basic absolute relative or objective subjective dependence is key to proofs of TO, and Á, PRUT and PRÁ.

TO is much more than life, even human mentality when best, alone or combined. With TRU internally, TOT externally, and TO, the actual world, TRU is much of its relative aspect. And TOT is the past present and future potential and approach to TÓ of the extended present. Like a perfected God, most glorious and predominant, never to be denied or defied, TO is always for immediate attention, devotion, highest exaltation and target in life. The united centripetal forces of divine-like powers prove the reversal of continual centrifugal forces in TO and determine much of its form, function and modes. Without TO humans regress backward slowly or more rapidly into less and less actuality, objectivity, subjectivity and reality. Without relativity and specificity it is for relative totality a world of physically chaotic obliteration, alien to much TO has become. The positive imperative of life in TO is propelled and compelled by the absolute relative in ever better and greater 3D coordination, of mlt-ALT in TO. This enables equilibrium, harmony and concord whose existence will be driven by the limits of an absolute relative, TAR, to make it relatively and dynamically permanent.

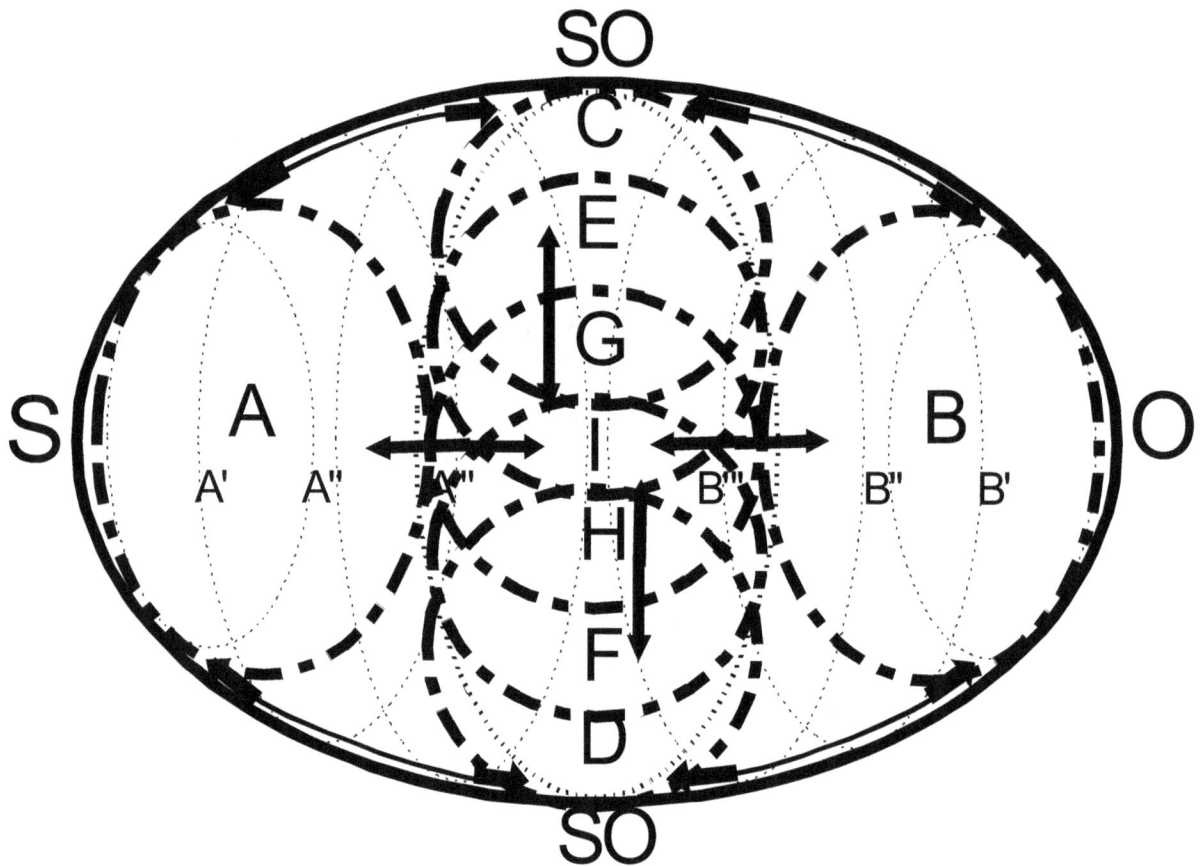

PROCOPE1.074

Fig 18-2 Processes and Operations of S-O in TO

From mechanics in physics dimensional divisions provide the formal distribution and distinctions needed for the static to dynamic basis of all in totality. This becomes more accurate and complete when, the length and time, l and t, dimensions are joined by mass energy, m or ME. They become more specific and actual when, by their pure essences, are formed in TO, A, L. and T. Together, this forms a field or cube of many sectors. Their differences, positions and characteristics provide a comprehensive basis for showing the key elements in mechanics and the form and function of TO-TOV-TOT.

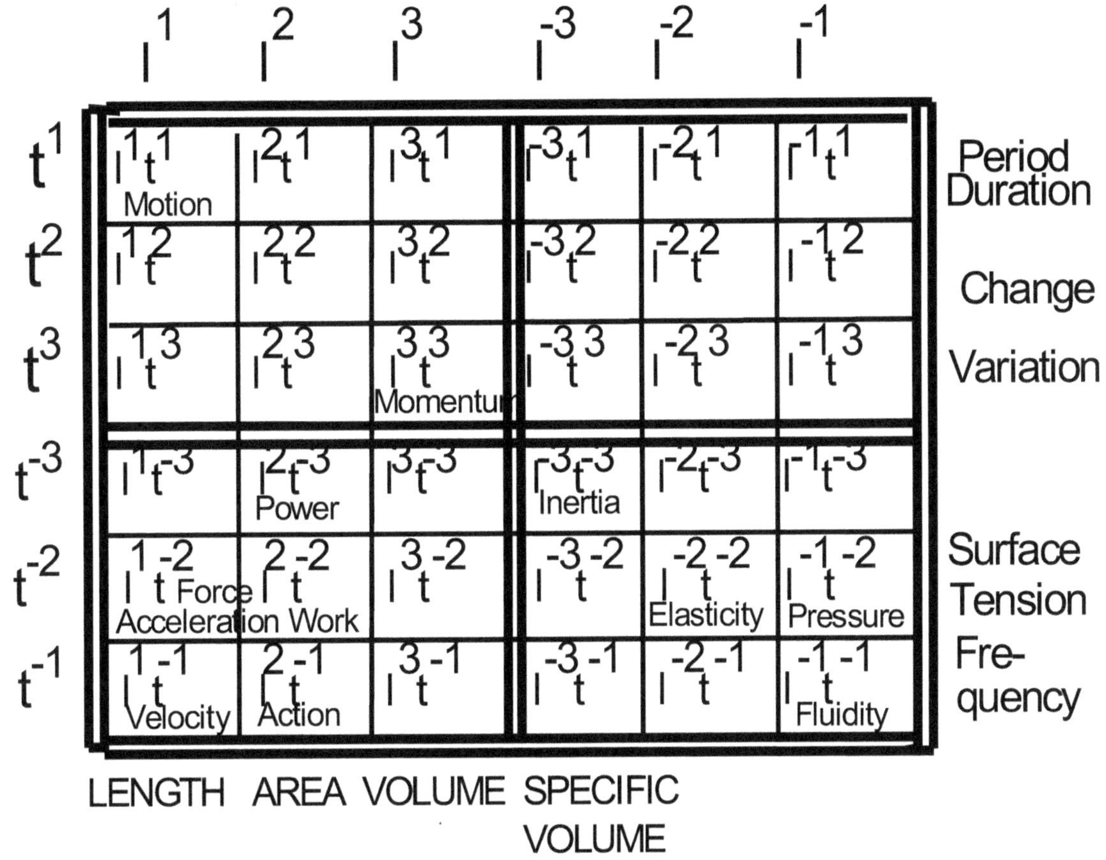

	l^1	l^2	l^3	l^{-3}	l^{-2}	l^{-1}	
t^1	l^1t^1 Motion	l^2t^1	l^3t^1	$l^{-3}t^1$	$l^{-2}t^1$	$l^{-1}t^1$	Period Duration
t^2	l^1t^2	l^2t^2	l^3t^2	$l^{-3}t^2$	$l^{-2}t^2$	$l^{-1}t^2$	Change
t^3	l^1t^3	l^2t^3	l^3t^3 Momentum	$l^{-3}t^3$	$l^{-2}t^3$	$l^{-1}t^3$	Variation
t^{-3}	l^1t^{-3}	l^2t^{-3} Power	l^3t^{-3}	$l^{-3}t^{-3}$ Inertia	$l^{-2}t^{-3}$	$l^{-1}t^{-3}$	
t^{-2}	l^1t^{-2} Force Acceleration	l^2t^{-2} Work	l^3t^{-2}	$l^{-3}t^{-2}$	$l^{-2}t^{-2}$ Elasticity	$l^{-1}t^{-2}$ Pressure	Surface Tension
t^{-1}	l^1t^{-1} Velocity	l^2t^{-1} Action	l^3t^{-1}	$l^{-3}t^{-1}$	$l^{-2}t^{-1}$	$l^{-1}t^{-1}$ Fluidity	Frequency

LENGTH   AREA   VOLUME   SPECIFIC VOLUME

Fig 18-3 Dimensional formulation of length and time, (DIMFOR), mass in the sagittal dimension is not shown.

Dimensional Formulation is a formal representation of a solid frame that contains the constituents or elements of mechanics in which mass length time, mlt, and ALT, are the three dimensions. It provides both a positive and negative approach that yields 27 in 2D, or 81 in 3D squares or sectors. Each square or sector is characterized by a specific kind of mechanism of absolute to relative totality in which all are, when best balanced functioning more or less, highly beneficial parts and processes in TO. This figure provides a highly effective means of showing the processes, combinations and summation in TO. When viewed from different perspectives, the best approach can be selected to prove the detail of TO. The sectors can be correlated with the SPALT model to produce an accurate spreadsheet like simulation of much of the beautiful intricate detail of the sectors, foci and vectors of components that contribute to TÓ. All sectors and groups of sectors need to be shown if a full and balanced showing is to be provided, e.g., ACEXIN and parts of the total form, LT, AL and AT. Interpretation of the mechanisms with pressure, volume and temperature, light, and the electromotive series expand and explain more of the basic detail of absolute relative states and dynamics in TO. Many mechanisms are not known or named, much of those known and named do not cover future needed work. The accompanying figure presents an enumeration of many of these mechanisms and some major interpretations. They go far, especially when reduced to their actual pure essences, to show and prove the detailed dynamics of the form and function of TO. Such a representation of mlt and ACEXIN is supported by, and can be used to interpret the volume, temperature and pressure features in physics, and in actual totality. For volume is equivalent to extension, temperature to intensity and pressure dependent on the energy component of actuality. Their form and function in actual totality are not the same as in physics, only typical of actual totality. Thus we must continually adapt, alter and apply actual pure essences to show their parts and roles in TO.

Fig. 18-3 is a good example, and proof, of the dynamics of actual totality. The interrelationships and combined effects are highly significant processes, e.g., temperature, volume and pressure in the gas laws. When these and many other sectors of dimensional formulation are combined and expanded to show their various roles in actual totality many of the unknowns of TO are discovered. Dimensional formulation and the various mechanisms in actual totality help to reveal how dynamics is largely a function of temperature, being greatest with high, and least with lowest heat. These are related to the form and function of mass and energy in, and proof of, actual totality. An example of this is the human condition on earth. With increasing population and other forces producing increased density, human life on earth is under increasing pressure in shrinking volume. This is also affected by increasing temperature, cf., Chapter 32.

Proofs of TO and TOV, by the teleological argument -

The teleological argument is used to prove the existence of God which when correctly qualified also increases the proof of TO, TOV and especially TRU, TRIR-TO with TOF. The evidence for form, design and purpose in nature and the world suggests more than a material, random or coincidental process is occurring. Not only does an object, TO or TRU, retain much of its form and motion, the self- sustaining momentum of this motion tends to qualitative and quantitative expansion. TO, propelled by its relative varieties, is expanding toward self perfection, completion and fulfillment. For the extreme specific variety, TIT, it is an autonomous, at once, or ATO state, derived, developed and determined by concentrated powers of relation and the associated propelling and compelling forces. This is also the similar to the core of TO, actual totality proper, TP. The powers and forces continually operate to sustain, maintain and assure the existence of TIT, TP and TO whereby their forms, designs and purposes are fulfilled.

To produce the greaest ends, action itself is largely its final test and proof in TO, ACIT. This is largely a function of dependence on their actual pure essences. Meaning and truths are in the action with greatest right identity, actuality, response, operation and will to fulfill their part and role in TO, and TRU-TOF. Right action is the ratio of benefit over harm that results in TO, TRU itself, TIT, and the proof of TRU, and TO, (PRUT). The greatness of right action is determined by a full accounting of the forces and factors involved, the better it will satisfy its need and that of TO.

CHAPTER 19 - RELATIONS, CENTERING and TOTALITY, to

Language and Logic

Convergence, Integration and Unity in TO, Relativity, Focus, Subjectivity, Here and Now in TO, Constant Present, Time Zero, action and effect of ARTO in TO.

Relation and union of absolute and relative totality in TO and TOV are proven by their interdependence, integrative processes and use of SPALT. That, and how, the physical universe exists in TO and TOV are proven by relativity and quantum mechanics, their actual pure essences and those of TO. By both special relativity, observer-observed, geometry in general relativity and relation in quantum mechanics actual totality requires a point of reference or center. By relation and reference TO, its varieties TOV, TRU, and TIT are largely proven, served and saved. How they interact, forms the basis for an ultimate unified totality and this book. Relation activates essences, potential to actual pure essences, as they become the actual existence of TO. By quantum theory the time and place of matter, entities, objects and their gradients and intervals can't be determined without relation, a point of reference or observer. By special relativity of the observer-observed, and general relativity of geometry, a point of reference is necessary. These form the basis for transformations from absolutely absolute totality, its extreme limit, in ABSTO-RELTO to each successive variety in TO. They also form the basis for increasing convergence in TO, and by averaged sum, and the centering of TO.

Form, dimensional form, point of reference, center, roles of absolute-relative and DAOST in TO -

If the physical universe has a center how does it differ from TO? What is its center of mass, length and time in TO? Is the center, here and now, the present, some other, variant, or none? This problem is critical in revealing differences between the physical universe, other totalities, TOT, TO, its varieties, TOV, and the proof of TO. With infinite universes, or absolute totality such as modern astrophysics holds, there is no center, no here and now. There is no union of expansion and contraction. Dimensions are in flux and changeable. This reveals the absolute of absolute totality, and the necessity for ABSTO-RELTO in TO in the right form, way and proportions. It helps to show how TO is DAOST. All in TO are different aspects of the same thing. This is most evident in the balance of differences in TO. In TO, differences largely arise from the absolute, sameness from the relative. Since absolute and relative are only poles of the same continua, neither are absolutely absolute nor absolutely relative, by LANERA. Its solution is the Law of More or Less, LAMOL and DAOST, or different aspects of the same thing. Ultimately this same thing must be actual totality. This is signified by, LUSITO, Law of ultimately the same thing is actual totality. The opposite of LUSITO is the Law of ultimate different things is actual totality, LUDATO. To complete the same and different laws there is, the Law of the Spectrum of Same and Different in TO, LASDIT, and the Ultimate Sum and Average of Same and Different in TO or LUSDATO.

These laws of same and different are more or less obvious, yet of tremendous help in distinguishing the content of TO. They show the static to dynamic spectrum, and add to the accuracy of making the distinction of TO, TOT and all that is less than TO-TOT. Relativity is ordinarily thought of by astrophysicists and scientists as its state of existence relative to mass and energy, e=mc2. In the actual universe, and most generally, relativity is much more than this limited relation between mass and energy. It is the relation between mass, length and time, the role of light, and their adaptation, alteration, transition and transformation in TOT and ABSTO-RELTO to TÓ with TRU, TIT, TRIR and TOF. Light is one form of mass energy, photons and waves. It is what light does, its actual pure essence in the subjective-objective reality mode of TO that light becomes increasingly higher states of S-O in TO. As ABSTO-RELTO develops and grows so are the stages and levels of mass energy, light and S-O transformed to become the reality that permeates all TO. It is how proofs of MLT, mass energy, dynamics, relation, realities in language, mathematics, solid geometry, proportions and the state of matter in life, prove TO. It is how TO does not exist as either static or dynamic, particle and motion, or only one kind of mass energy but all together typical of itself, evident in ALT and ACEXIN.

By light, its relativity in mass energy space time, and our S-O or observation by way of light as a beacon, we are enabled to see, know, think, represent, and live by TO, and TOV with TOF. Such observation of the observed takes many forms. One of these is how we watch, and are being watched, in so many ways, even from the distant future on the present, (WABEW). Such an ubiquitous and dynamic interaction in ARTO of TO, helps to explain how neither absolute nor physical alone, or relative and mental alone, but only the actual totality, TO, shown by its modes, continua, kinds, varieties and components, fulfill necessary existence in ultimate unified totality.

Relative totality is based on the present, of medium size and objective-subjective orientation. Absolute totality is based on the limits of time, space and mass energy orientation, e.g., the limits of infinitesimal and infinite, or zero and absolute. This is the clearest distinction and proof of TO, and enables a solid functional basis for a closed set with the fusion and union of absolute and relative. When TO is in transition as TOT, each set time becomes a new time and its own optimal unified totality. For practical purposes, however to the extent the gradient of change is infinitesimal, small, or practically unified, the totality functionally and essentially becomes the optimally unified one, actual totality, TO, or a variant of TOT, TOT-TO proper. The most characteristic feature of TO, is its center in the present. This distinguishes TO from TOT and absolute totality, proves TO and disproves the others. When its essences, by relativity, are altered to their role in TO, the physical universe acquires altered and new actual pure essences or coexistence, essential existence with, and its part in TO. It is in this way we can see, make more discretely, realize and prove TO. Relativity initiates reality by expanding and creating new and higher forms of mass energy. These depend on what is necessary and essential for life. Actual pure essences are rapidly and selectively learned, qualified and used by immediate awareness and instantaneous attention in the observer-observed and all its higher levels of reality in TIT-TRU-TOV-TO.

The figure (19-1) typifies existence in the core of life, especially in ever higher forms of subject-object reality, S-R-O of all that enables best human life, TRU and TIT. The stages of relativity to reality in mass energy are necessary for the existence of TRU-TO. They also help to establish the actuality and proof of all in TO-TRU. There is no Object without Subject, and no Subject without Object, (LASAO). Like observer and observed this is the ever-present identity and bond that exist between them as they are aspects and modes of the same thing, the mass energy Á continuum in actual totality, TO. This is confirmed by the fact that without subject or object TO is only a polarity or the opposite absolute, which cannot exist unless outside of the realm of all that enables TO-TRU. It is by TRIT, or the coordinate form of mass energy, spatial and temporal that enables representation and presentation of TRU, A form that can be comprehended from the inside or outside, by variance in subject-object relations.

Zero time present is the universal constant of here and now, of present position in TO. TO is absolutely stable only at time zero, the vanishing point and fleeting present between past and future. It is highly and practically stable in the near term, increasingly less in the mid to far term. This is corrected by the form, forces and mechanisms of TO, including inertia, momentum, and redundancies. It varies with TOV, being largely a function of relative totality or the specific varieties of TO. That the dynamic, change and variation are much the same is evident in zero time. At zero time, time is motionless, without dynamics and change. Totality at time zero in its state of here and now is static, lacks motion, even internally, and exists as a perfectly closed set. This fact, combined with how TO exists as an exclusively unified totality of a typical form of mass energy in space time, is how it is formed and functions, and explains many of its unique characteristics and properties. This is what provides many of the basic proofs that are key to knowledge of TO.

A major proof of TO-TOV-TRU-TIT is whether or not they have a center and what this center is. To be an independent set and unified totality requires a center. *The center of the universe is not any physical object or region, it is the mean sum product of all that is and enables the optimal union of absolute and relative totality, TO, by its modes, continua, kinds, and TOV, or special varieties, LACTSAR.* By the focus of its continua this is most concentrated in the specific varieties of TO and converges on the center of its mass energy, spatial and temporal dimensions, or reality of here and now. The center can be no less than the degree to which TO is a unified totality with autonomy, at once, ATO, as produced by all that enables best human life in itself, TIT. The 'here and now', center and relative focus of TO is the medial confluence of its mass energy or objectivity-subjectivity S-O, spatial-temporal dimensions. The center or focus is the medium of each dimension, the real in the 'here and now', or center of subject-object in mass energy, medium in size, and present in time. It is the relative and total distribution of these dimensions that show and prove the special form of TRU and total form of TO. This is a major and deciding characteristic and proof of TO. It is the broad inclusions of all in TO, and the power of its special varieties and RELTO that produce and prove TO. Especially from qualifications by actual pure essences we are enabled to achieve the correct super vision of all, with TO. This is how TO depends on unity in totality which arises from the centering of specificity, here and now of TO and TOV in an average sum. The continuum of relation or reference, is a reverse of the continuum of absolute totality to relative totality, and of an infinite concentration to infinite diffusion, Chapter 11. When most focused or concentrated it is our mass energy in here and now. When most diffuse it is all that exists at the periphery of RELTO, or all that exists at or beyond the periphery of all TO. Neither diffusion nor concentrations are all, nor exist alone, they are modes, limits, part and processes in all that enables TO and the universe of best human life, TRU.

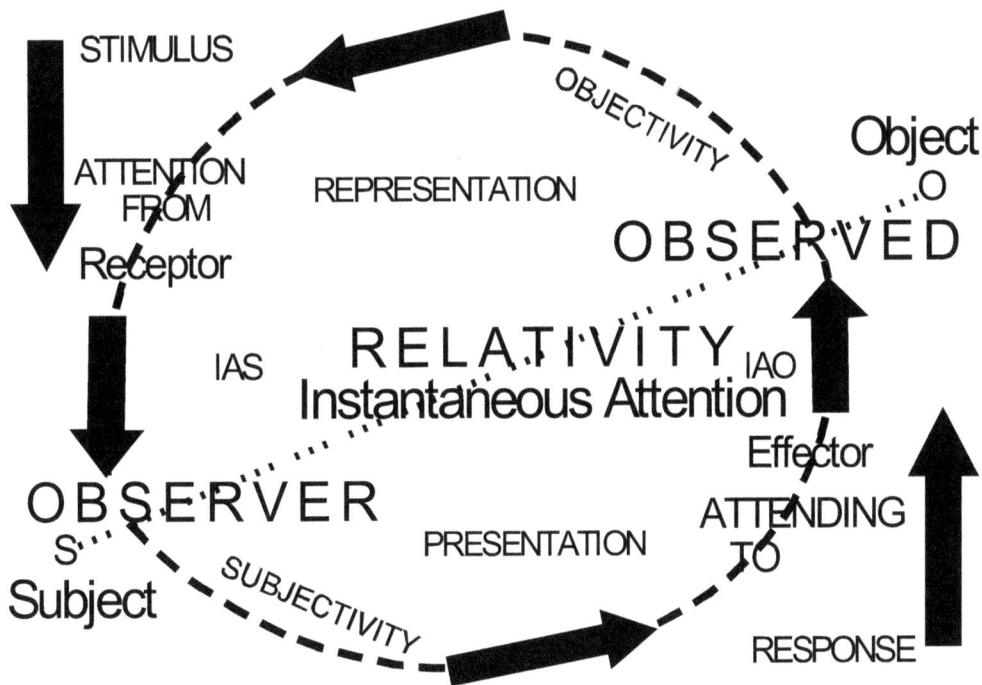

Fig. 19-1 The Relativity of Observer-Observed, and subject-object is the basis for the mass energy dimension

Totalities and universes are constructed from zero and infinitesimal through fractions to a unit and through increasing finites to the infinite and absolute. The importance of this diagram is to show how numerically form converges on one or the unity in totality of TO, how the combined ABSTO-RELTO form TOV in TO, and how poorly they have been understood. The integration differentiation construct shown in the figure helps to show the form of ultimate unified totality and its representation. The integral differential exists throughout any and all totalities, including TO, only in changed forms. It is the effect of the more special and general varieties on each other in, and by TO that enables its convergence on one in the averaged sum of all levels of mass, length and time, here and now. The more refined relative varieties of TO, e.g., centering of the ALT dimensions in TIT, TRU and TO, gives it the form that enables treatment as a closed set with the validity, accuracy and order an effective representation and understandings require.

The continuum of relation or reference is central to the actual world, TRU, in actual totality, TO. Its form is provided for us by its basic dualities, continua, parameters or dimensions as they together make up the whole. This is by the ultimate focus of the specific variety of TO, whose combination of universals, of mass energy, space and time, although not the only ones, are those that primarily determine much of its form. It is their combination that makes up the absolute-relative general to special varieties and form of TO. When combined in their higher levels, stages, or quanta, as by ACEXIN, they provide the most beautiful and glorious image of our existence. When we discover, learn, know, remember and live by the actual pure essences of these varieties, universes, universals and partials, occurrences and continua that form TO we can finally reach an excellent supervised way of life that achieves and fulfills its meaning and purpose.

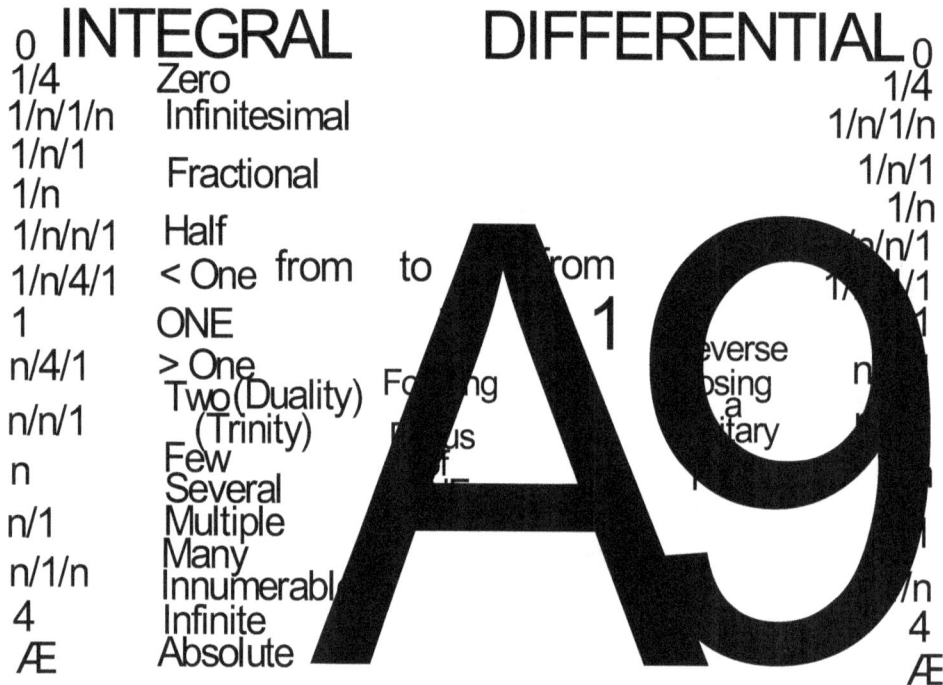

Fig. 19-2 Unity in Reverse Integral-Differential Divisions, Focus, Construct and Interpretation in TO and TOV

Time, space and mass energy change externally and internally in TO, TOV, TRU, TIT and TRIS. TO, and all internal and external to it are in continuity in which an absolute zero and fixed present does not exist, NETIC, except subjectively and functionally by actual pure essence. It is the diminishing and starting point of a zero center. This is quite different externally than internally. Externally it is by all that is less or other than TO. Internally it is largely by all within that changes both as TO changes in transition, and by the interaction of its constituents and components. By ABSTO, now or, the present is only an interval of time, between the nearest past to the nearest future, unless relative and APET. It has no fixed absolute center. TO is a continuum of spatial, temporal and the mass energy or Á dimension, ÁCOM that reflects the progression of completion of the continua of all in TO. On the other hand now for RELTO is opposite in the duality by concentration on the most set and stable fixed centers, shown in ALT. It is the balance of these two opposing modes of actual absolute-relative existence whose duality produces the continuity of TO, in which the fixed existence of relative totality is centered on here and now.

The many continua of TO, and their different actual pure essences, transitions, interactions, and effects are crucial to the concepts of TO and its understanding if we are to correctly recognize the greatest proofs of their actual existence. This is the Law of the Existence of an Absolute Center in actual totality, TO, LEXÂC. Since this is only approached in the extreme special varieties, TIT, TRIS, TINI, there is an absolute center, only relatively, subjectively and practically. This is a product of RELTO, TOV, TIT-TRIS and their APET, the fact that all exists not as things or objects in themselves, but by the actual pure essences in TO. An absolute center by actual pure essences, is of great formative, functional and practical identity, presence and value, and thereby signifies that LEXAC exists. The critical difference lies in the continuum between no center and an absolute center. It is the difference between the extreme generic and extreme specific in TO by ABSTO-RELTO. It is by the extreme specific varieties TIT, TRIS, TINI that give the absolute center justification that proves TO and its representation, presentation and perspective. How it can be practically and reversibly fixed as a closed set for all TO. Like TO, this closed set is not absolute, only relatively so by their actual pure essences.

Relation, and its opposite, non relation or the absolute, form the primary distinguishing feature of actual totality. As absolute totality is material, so is relative totality mental. Yet both are much more. Mind, mental and mentality consist of many variants, such as consciousness, subjectivity, attention, thought, reason and higher cognitive powers. It is what these variants produce, affect, and bring into application that has, is and will form the vast and dynamic state of relative totality and its great part and contribution to actual totality. The interaction of absolute and relative is similar to that of the positive and negative, the compelling and repelling forces by both are altered and adapt to each other in actual totality. It is this continual dynamic state of positive and negative, gains, losses and interactions that characterize actual totality. They set the stage and initiate the progressive growth and development of life, animals, language, humans or relative totality in actual totality. It is their combined mechanisms and product that provide much of the proof of actual totality.

Viability, the great product of relative totality, is actual reality by relation becoming specific form and proof in TO and the developing and expanding generic to specific varieties and TRU in TO. The centering of TO on here and now is not only a product of the consciousness, subjectivity, attention of individuals, nor all humanity. Although they are large, driving and predominant components they are not all. The centering is a sum of all the varieties and constituents of TO, including absolute totality, as evident in its interdependence with relativity, in TO. This fact is similar to the fact that the proofs of TO are complete not by one or two proofs, but by all together and reflect the correct proportions of viability and all in TO.   The acuteness of relation is to be found in the Law of Absolute Distinction in Relativity, LARD. This is the ultimate state of centripetal powers and focus of the observer-observed in the combined moment of here and now. Like an immense IA, or instantaneous attention, when combined with the autonomous at once state of the most specific varieties of TO, TIT and TRIR it reveals many of the most profound aspects of TO. This includes the existence, role and power of extreme RELTO through all TO. LARD is what proves and assures TO, by confirming the bond of observer-observed, subjectivity-objectivity and RELTO-ABSTO in TO.

By Man Point Time, as past increases and future decreases, we move in the direction of extinction. If the past decreases and future increases we move in the direction of continuity and eternal existence. There is a set equilibrium which we are in here and now. This does not mean they are at all times equal. Yet by the laws of symmetry this is a natural, form, state, tendency, or inclination. These are shown in the accompanying figure in which: A. Time changes as man approaches extinction, B. Time changes as man approaches the absolute, C. Time changes with human life's variability and indefiniteness. As humans or TO approach extinction they converge on the present. As we approach the absolute, as in TÂ, all, N and now, MTP and MTF expand toward the absolute. As we are variable and indefinite these changes are variable and indefinite so that it depends upon the many factors, forces, processes in, or potentially in, operation. This is an example of, Reversible Proofs of TO, (PRIT), cf. This is shown by how logic proves ontology and ontology proves logic. It is shown by how ethics proves epistemology and epistemology proves ethics. It helps to show how philosophy proves religion and religion proves philosophy.

00   1/00 1/00   00

A.   MPTP Ⓝ MPTF

B. MPTP   Ⓝ   MPTF

C. M P T P Ⓝ M P T F

PAST   **FUTURE**

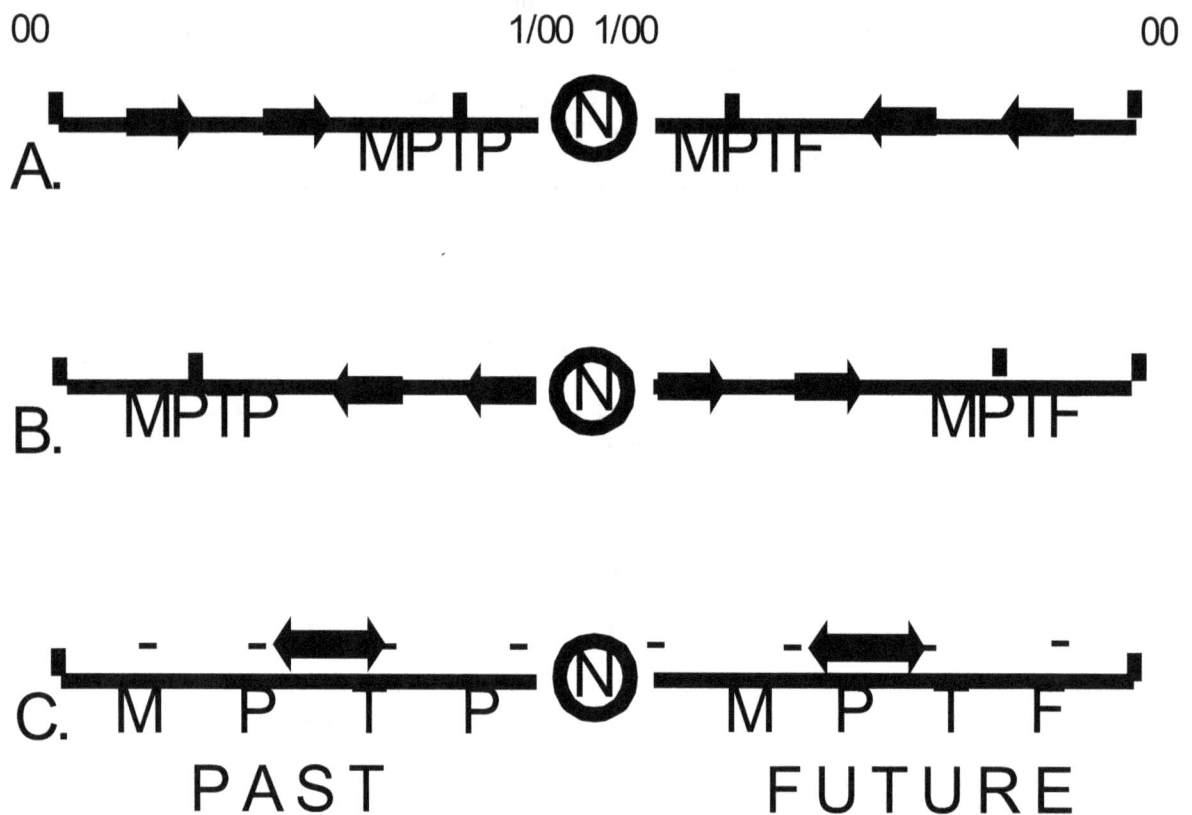

Fig 19-3 Man-Point Time, Focus and Implications

It is how all great categories, topics, subjects, and aspects of TO tend to prove each other and prove TO, as they are properly held to their SCARLS and interactions with each other in or of TO. In order to survive, it can't be left to chance. It can't be in disorder. It can't be in confusion or chaos. It must be ordered and controlled, by us, as TES-TIT-TO is in command. We must be capable, able, predominant, execute all indicated, act, and achieve TO success. We must be able to prevent all possible harm. The basic requirement, key nature of the S-O relationship, is that it functions, as our CNS, coordinately, and massively, for optimal success. The necessary amount of knowledge, reason, learning, logic and comprehension must already be in place to give proper direction, VAPO and SCARLS to this command, management and control. We know we exist! We also know we have existed in the past. We must know that, and how to exist, improve and exist with permanence in the future.

Knowledge of TO, TOV and all revealed by what they are not, leave out, how detailed and shown. To better understand what TO, RELTO, TOV and different universes are we must consider what each is not. We need to discover what each loses that its predecessor had, and in what ways this that is not had partaken in what was. It is obvious from the physical universe and the development of life that only a special, relative, and realistic explanation can prove the existence of an exclusive centering of an autonomous totality with some kind of averaged sum that yields a center. It should also be obvious to all that there is more than a material like diffusion of entities, senses and things. That separate absolute and relative, ABSTO to REITO, are not enough. We need only to consider the positive over the negative, the beneficial over the harmful, the rational over the irrational, and the actual world over all that has no importance, divinity, proportion, order or harmony. In this way human wisdom, and especially monotheism, have come to realize that even in diversity there is unity, and that it can be named. Likewise this can only come from centering that comes from consciousness and instantaneous attention that account for relation. These are the most relative, near and dear to all human existence. They are its universals, qualities, modes, continua, kinds, varieties, mutual equilibria, tolerances, values, loves, virtues, successes, and all that produce predominance, well being and survival. The center must be in perfect proportion and equilibrium with all else that makes up the form and function of TO. This is what each variety of TO supplies to the ABSTO-RELTO continuum, as they form and exist in TO.

Quantity, math and measures of relation are based on the integral-differential, on the one, here and now, and its limits, statistics, probability and correlations. Yet we are not dealing with absolutes in actual totality, only half and half, absolutes and relatives. Thus we can provide correlations, probability and a high degree of statistical accuracy as long as the basic tenets are correct. So also is each of the forty proofs of TO incomplete, only to be fulfilled in the whole of TO. In this way relations, here and now, and meanings progress to higher language, as do all proofs, to form an integrated whole in ultimate unified totality of TO.

From characteristics and properties the typical form and function of actual totality, and its proof, are clearly shown. In the following table some of these are listed. Singularly many are not conclusive of TO. Together in a sum they are, reveal the typical form and function of TO, and produce the most convincing proof and certainty.

Table 19-1 is only a sample of some of the greatest characteristics and properties of TO. A large amount of interchange and overlapping exists between characteristics and properties, as well as interchanging uses in parts of speech. They are typical of the form and function of TO, present in all chapters, and are some of its greatest proofs, cf. 15, 20. Characteristics are primarily of the form, and properties of the function of TO. This depends on whether characteristics are treated as nouns or adjectives and properties as verbs or adverbs. Often their suffixes help to determine this form or function. As a state and limit they are more apt to be a characteristic, and as relational or conditional a property. They are often either characteristic or property. This is seen in suffixes, cf. Ch. 20. Whether terms are characteristics and properties, nouns, verbs, adjectives and adverbs, or other aspects, modes, continua, kinds, varieties of actual totality depends largely on how they are used. The main determinant is whether or not they are and how much they exist in actual totality. By relation, and the order it provides, we better understand TO, its form, and how to represent and present all.

Here and now, or zero time focus, of actual totality is not only a most important characteristic and typical feature. It is the focus of here and now, or zero time that exists throughout time. Time may change, but here and now, or the zero time focus of TO, remains the same. It is critical that this form and state of TO be learned and remembered. The Law of the Zero Time focus, here and now in TO is signified by LAZOT. Within TO there is accommodation for, and existence of time, time change, and time zero, law of Existence and accommodation of time, time change and time zero within TO, LETOZ. TÓ always maintains its zero time focus, at any time in the past or future there has been and will be, at least infinitesimally, a vestige of TO by LATOZ. TO may change in some to many ways, but not it's here and now, zero time focus. The time zero focus imparts on TO of TOP and TOF forces that cause alteration or adaptation, less in the mid to near term, more in the long to far term future. These are largely reversible, depending on the capacity and power of TO. The existence of time, time change and time zero in TO, are increasingly dependent on RELTO, TRU, and especially success of the higher powers of knowledge, mentality, applications and production to maintain TO. This is the Law of Time, time change, and zero time on Dependence on the success of higher powers of knowledge and achievement, LATSAK. The capacities of TO for the laws of time zero, and other features of time in TO, arise from different causes and states in TO.

You cannot have totality without unity, you can't have unity without centering. Absolute totality and the physical universe are largely without a center, thereby lack full actual totality. Thus, they are not the ultimate totality. This is why relation, centering and unity are prerequisite for all, if it is to be a completely a well-centered totality, TO. The fact that TO is in transition, and changes with time, does not alter this completion, only modifies it. It is increasingly modified to include more and different relations, ones that include the relation and centering by lesser and other kinds of matter whose complement fulfills the totality. One reason why Einstein could never find or prove a theory of everything, as easy as it seemed after General Relativity, is exactly that, the difference between Special and General Relativity in absolute totality of the physical universe. This is because there is anywhere from one to many totalities in absolute totality. TO conforms to the relativity provided by special relativity and its observer observed basis. This is what brings totality to TÓ, the Law of special relativity proof of TO, and disproof of a totality or "theory of everything," LASREPT. This accompanies the facts that there were are and will be multiple physical universes, as before the bag bang, and these most probably had markedly different form and function.

Characteristics	Properties
Integration, differentiation	Uniting, diversifying
Space, length, volume	time, change, variation
Mass, matter - dark, color	anti-matter, dark energy
Objectivity, observed	Subjectivity, observing
Position, in out, inclusion	Size, Shape, input output
Centered on here and now, or time zero, Present,	Transition, Half Life Transformation
Proportionality, value	Disproportion, disvalue
Equilibrium,	convergence, divergence
Conserve, save, durability	Liberal, spend, expend,
Practicality, Selectivity,	Practicing, Selecting,
Sensitivity, Specificity	Feeling, Pain, Misery, Joy
Supply, Demand,	Needing, Want, absorbing
Thought, Memory	Thinking, Remembering
Occupation, a job, skill	Working, skillful, doing,
Capacity, Selectivity,	Predominance, Fulfilling
Reversibility, Permanence	Reversible, Adapt, Thrive,

Table 19-1 Characteristics and Contrasting Properties, Actual Totality

IV Proofs of Altered Form and Function by Relative in Actual Totality

CHAPTER 20 - LANGUAGE and LOGIC in TO to Mathematics

Communication, Dialectics, Essence, Expression of S-O Reality, Formal Logic

Of all the products relative totality contributed to actual totality none have been greater carriers or vehicles of mass energy than language. From its earliest incipient utterances through common conversation to the most advanced stages and forms of subjectivity-reality-objectivity, conceptual thinking and logical use in the beneficial activities of life by language has been the great driving media of TO. To mean what we say, and say what we mean, to mean what is TO, and have TO with a meaning that is as perfect as possible and practical by actual pure essences is the basis of actual totality. Language and logic are the carriers of objectivity-subjectivity and reality with increasing meaning for precise and proper representation of TO. TO can't be shown when based on little and wrong language, false beliefs and narrow mindedness. It can only be based on language that is qualitatively and quantitatively precise, accurate and complete of all that is and proves TO.

The abbreviation or acronym for actual totality is TÓ, pronounced with the long O, as in toe or tow. If this is not sufficiently identified, actual totality can be supplemented and signified by is proper status, or Totality Proper, TP. Likewise it can be supported by its major determinants, such as the combined absolute and relative, ARTO, ABSTO-RELTO, equivalents in common usage as the physical universe or God, if as TREQ, equivalent to TÓ. Like any new field or topic, TO requires its own language. This is to avoid lack of identity in existing terms, and the advantage of being a root upon which further new concepts can be both based, well meaning and ordered by TO.

Vocal sounds, words, symbols, languages, gradients of meaning, learning, and representation in the mass energy dimension, Á, and spectrum SÁ of human life are the medium, carriers and closely related major determinants of all in TO. Through the ages for lack of identity and will it has been exceedingly effective to elevate objects, emblems, figures or words to prominence by unifying people, to give focus and purpose for achievement, as to win any and all in contention and conflict. Similarly language is a common means humans have to show and make their world more meaningful. It is the great strength and power of many words, as those of the thousand greatest components of TO, by SPALT, that reveal the role of language in the mass energy spectrum, proof of TO-TOV-TIT-TRIS-TRIR and their form and function. Language is continually changing, if better we predominate and survive, if worse we are oppressed and become extinct. It is one of the major vehicles of mass energy, especially in humans by its role as mediator in relativity between observers and observed. This is why the quality, and improved form and functions of language, are so important.

Just as the great energy produced by the big bang altered mass, length and time, so is the great energy of burgeoning human life and mentality altering mass, length and time. Not only does mass energy become that characterized by the Á dimension, space and time become ALT, as shown in SPALT. TO, being a composite of many varieties, TOV, retains some of the m l t form of each variety, But is characterized by their combined balance, the sum of their averages plus its own actual pure essences form, i.e., ALT. It is from the creation of these new forms of mass energy space and time that new forms of matter that exists in TO have been produced. These are the conceptual elements of TO, the actual pure essence's definitions of words that make all most meaningful. LOCET, the Law of Logical Conceptual Elements of Language in TO, whose parts, ES, is the basic material of actual pure essences, and proof of the necessary existence of TO.

Language and Logic are central to the form and proof of TO. They provide the material and mass energy by which the actual pure essences of TO are realized. It is this subjective-objective realization that makes TO separate - and different - than any other totality. It makes TO definite. By distinguishing the conceptual, linguistic and logical elements of TO, we can practically advance life to the level human mentality exists in TO. It is a salient proof of TO. It is the logical use of mass energy, S-O and these elemental parts, ES, that make mentality, humans and TO what they are. This is the Law of Logical Conceptual Elements of TO, LOCET, a central and indispensable proof of TO. If we are to accept any totality, absolute, relative, physical universe or deity as the actual totality they must account for LOCET, the notations, factors or elements of great change in the material form and function of actual totality. This is evident in the creative forces of existence of mass energy, spacial and temporal forms outside, before, during and after the big bang of the astrological universe.

A simple and very pertinent test and proof of TO are questions posed to another person, or themselves, looking at an object, "What is, or why is that?" Answered by a name or description and, "What is its functional value or essence?" Why is it the way it is? It is by such dialectics that meaning and actual pure essence of what is, and why all is the way it is, reveals the subjective and objective, relative and absolute continua. Their combined make up of form and function is the key test and proof of TO. By question and answer, yes and no, the continua, dualities and differential dialectics have been, is and will be the basic way humans have of communicating their world. This was never more well presented than by Plato describing Socrates, especially when guided by a search for the best and greatest meanings, as for what is Good or equivalent, TO.

The only exception to a state of being a kind of continua is actual totality itself in some frames of reference. This is also apparent in words, ideas, concepts and meanings that are equivalent or close to actual totality. Such are normal and abnormal, where normal is what is or tends to TÓ, and abnormal what is not, less than, or tends to depart from TO. It depends on how the terms, entities and objects are used and the meaning or manner of this use. It is the flexibility in use of language that both give it strength and weakens it, depending on each occurrence. The strength comes from freedom and broadening in usage. Its weakness comes from the multiplicity of meaning, making duplication, superfluous and differentiation more complicated and often less than actual totality. The Law of expanding and exacting meaning, is signified by LAMEX. This law like that initiated by actuality or necessary existence and actual pure essences, is a product of advancing knowledge, communication and writing. It is a natural outgrowth of humans basic needs to make sense of their world, and their success in it. This is evident in the explosive refinement of language, the written word, and knowledge in Ancient Greece 600-400BC. It was championed by Lycophron, and expanded upon by many of the great names of philosophy. It was the beginning of many major sciences, whose accumulated knowledge is the basis and proof of much of actual totality.

Languages are media and higher powers by which the subjective and objective interact to produce reality. Language does this by fusing matter and mind, the absolute and the relative, so that things make sense. In this way language is a great proof of actual totality. It is a major tool in the arsenal of human life interacting with and ordering the physical and relative universes to provide higher unity in the right, actual totality. With knowledge, the highest powers of analytical and comprehensive thinking, reason, logic, judgement and action characterize the salient powers of best human life in its expansion and enhancement of actual totality.

Language is the great bond that gives life, human life and actual pure essences to actual totality. It is the core of relative totality, RELTO and TO in which life in here and now and powers of relation connect the objective world, absolute totality with the subjective world, relative totality to converge and fuse in actual totality, TO. Languages are the great media of observer-observed, subject and object, S-O or O-S. This is the continuum is manifested in the great bond that gives reality to mass energy and human life. This is how the higher to highest powers of mass energy, in their transitions and transformations in the nervous systems by stimulus and response join objects to subjects, matter to mind to produce, by actual pure essences, what is practical, functional and essential for life. In this way it is language that we must realize its own reality as the bond that forges object and subject of mass energy to produce knowledge and all its applications. These applications have, are and will give humans prominence and predominance in cultures, civilizations in the relative universe of best human life, TRU and TO. This is why language in its central role, 20, of these 40 proofs is the center of combined proofs from modes to varieties, mass length time, mass energy, relation with mathematics, proportions, models, life, humans, knowledge and all from the beginning to end of the proofs of TO. It is in the quality and quantity of language as it adapts to best human life and TO that has developed, grown and become the central power of TO, how words, language and the communication of knowledge bring the physical and mental worlds together as one, the current ultimate unified totality. TO.

The interchangeability, as well as identity, of subject and object, shows and proves the convergence of mass energy, modes, kinds, varieties and all in TO. In the structure of language a sentence's subject may be physical and its object, mental, even though its subjects are subjective and objects are objective. Such a reversal does not disprove either, it merely illustrates the dual and reversible form and functions of TO. It also illustrates the flexibility of actual pure essences in, and out, of TO. It is evidence of the external representative nature and effect of language in TO, and how it can be misleading if the subject and object are not held in the reality of their reversibility in TO.

It is largely by pragmatics that language and knowledge are communicated in quantity, and achieve greater quality. It is the functional use and practical value of language, whose essential qualities make it most effective and efficient as they contribute to, and enable best human life, TRU in TO. They most simply reveal what is, the proof of TO. Yet there are limits, as with all, by the degree language and pragmatism can improve to equal the meaning and actual pure essences of all that is and enables TO. It is by the power of what is practical or essential that actuality becomes the mode that guides us life, a life that is melded by, and melds actual totality. Actualities, as actual pure essences, provide the functional values and dynamics of our existence. This is a natural product of the mutual coalescence and proof by pragmatism, functional value and actual pure essences of TO.

Of the many languages, and language groups, that exist today, and have developed over many thousands of years the words used have tended to become more numerous and suited to human needs. The structure of grammar has adapted to human expression, with increasing complexity. Many aspects of grammar have added to the power of meaning. Of these the use of prefixes and suffixes reveal ways grammar adds power. They especially identify various kinds of characteristics and properties of TO.

Characteristic suffixes - tion, ation, ar, ous, ment, ity, ery, ence, ance, tor, er, ist,

Properties as suffixes, ate, size, en, once, fy, ly, ing, sity and able.

Some of the most basic categories and features of actual totality can be and act as characteristics and properties, e.g., existence, being, exists, to be, form, state, condition, function, relation, limit, part, parts, mechanism, process and processes

Each and all persons exist and participate, more or less, in TO through TOV, RELTO, TRU and best human life. The language, we so often take for granted, is central to the reality of our existence. The words we use are not commonly directed toward actual totality, nor are their actual pure essences realized. They are primarily directed toward senses and things in themselves. As a result we have had languages, words and terminology that fail to give us the perspective of what we actually are. To correct this will take its own terminology, in addition to what language is in common use. To this end the following concepts are proposed: Learning language and its best use is a large part of human maturation. Language groups cultures and civilizations have come and gone, replacing those that are less beneficial. If we are to have the best in a unified totality we must adapt our language to it. This is why in this book we initiate and emphasize those ideas and terms of actual totality that are needed to make it successful.

The relations, integration and differentiation, identity, language, figures, and representations of TO enable us to produce ever greater formal distinctions of all that is TO. It enhances and accelerates our processes of subject object operations, realization, and mass energy form and functions, Á FAF, whose continuum, SÁ, gives us the predominance so dear to life. Such a language of TO, LANT, Notation is absolutely necessary, LATAN. A language and terminology for all great sciences are indispensable tools for signifying and communicating its ideas, concepts, principles, laws and mechanisms. Since TRU is both a unit and a totality that can be treated as a closed set, a logical and symbolic system for demonstrating it provides an ever greater means of discovery, learning, knowledge and use. This enables a reduction to optimal positive objective approach and how to apply logical theory and methods to the appraisal of argument, reason and all thought. It is how to best fulfill their role in providing the needed perfected identity and replication of TO.

We call the New Concepts of TO, NEC in contrast to OC for Old Concepts, EC for Essences of OC and Existing Concepts, and ECT or Existing Concepts with TO meanings. NECS of the highest order are called TOPNEC. This separation is done because most of our concepts and language are less than EC, much less than ECT, and few of adequate NEC equivalence. This lack, or their continual variance in and out of TO, OITIO, causes endless problems when not upgraded to show their actual existence in TO. It is one thing to mean something by realizing an idea, concept, or principle, as in TO. It is quite another to develop a good word and definition, especially one that will best communicate to others and get others to know and use their meaning in TO. This is STÖP or the Sterile State Of Principles, Concepts, Laws and

Language. The reader will find it much easier if they pause for a moment to analyze, recognize and remember key terms.

If: T = The, Totality	Ė = A Finite or Part	TÂ = TO-TRU Absolute
and: TO = Actual Totality	Î = The Finite, Part, Person, I	TE = Finite, or Part in General
and: R = Relative, U = Universe,	Î = Infinite and Infinitesimal	TÏ = Finite, or Part, Special
Then: TRU = The Relative Universe	Ó = Total	TI = Infinite or Infinitesimal
And if, Ä = Form	U = Unity (also)	TÓ = Actual Totality
À = Mass Energy	Then: TÄ = TO-TRU Form	TU = TO Unity
Â = Absolute	TÀ = Mass Energy of TO-TRU	TOT = TO Potential

List 20-1 Root letters for Concepts of Actual Totality,

All other letters, as consonants, or syllables, with separate useful phonetics, can be given denotations that best represent their form and function in TO. Thereby: D = Differentiation, F = Fallacy, L = Law, Ô = Object, S = Subject. P = Presentation, R = Representation, ES = Parts, and CES = Processes. Also, s when added to a word as in common usage signifies plural, as do m, l, and t stand for mass, length and time. Certain combinations of letters, or acronyms, are supplied to best signify new concepts. These include some that have been in use, especially in scientific contexts. However they use some of the same letters so limited to the topic under discussion and are so noted, including: EN(S) = Entity(ies) R = Response, and S = Stimulus. Examples of Combinations include, S/R = Stimulus Response and CNS = Central Nervous System.

The whole of knowledge, indeed, the whole of human existence in actual totality, depend on clear, distinct and appropriate concepts, words, definitions and usage. They also depend on the step-by-step reasoning provided by these concepts and definitions. This is the great power of language, logic and philosophy by existence in and contribution to actual totality. In a summation they cannot only produce a good showing of actual totality, by their presence in it, they are the gifts of the mass energy of subjectivity, objectivity and reality of relative totality to actual totality.   The power of the most highly developed existing languages and the growth in knowledge and human achievement, accompany and strengthen each other. There is more or less continual trend to upgrade and improve meanings and languages by communication, vocabularies, dictionaries, reference works and most literature. To the degree meanings are supported by a highly advanced literate population that properly uses language and its actual pure essence definitions of words, meanings are basic to the strength and power of a society, culture and TO, including TOV, TRU-TIT-ATO, ACEXIN, and their proof. The absolute necessity of developing and using language by TO-TOV-TRU, including emphasis on its actual pure essences, has, is, and will become an ever more certain purpose and way in best human life, TRU, and RELTO, ARTO in TO.

The answers to questions, solutions to problems and resolution of paradoxes that only language and its improvements provide are what sets humans apart and prove the great value and power of language and the exclusiveness and essential existence of TO, and TO-TOV-TRU with TOF.

As simple and effective, as logos, seems, it can be no better than it is right to the degree the one is the only one. For this reason, among others, knowledge has been, and continues to be, driven by identity with a perfected and certain unity in totality, TO. It is with this background logic developed to guide habits of thinking in the most rational and orderly ways. It relies on the certainty that results from step wise deductions in which no step is incorrect or in doubt. In this way logic demonstrates the essences of math. This is the Law of Logic as the Essence of Math, (LOGEM). Math is reduced to its part and role in TO by logical notations, formulae and operations. It is seen in neutrality, set theory and formal logic. This is by the role of logic of propositions in language, and representation in general. By logic TO is provided with the major components and contributors to the increasingly higher levels of mass energy, S-O and reality, whose dimension, Á, in TIT-TO completes the 3D structure that demonstrates and resolves all by formal modes, as SPALT and SPAT. This is the Law of Logic's Á TO Resolution of All, LATRO. Herein is a Proof of Á, PRÁ, a large proof of TO, and reversibly of the roles of logic, math and representation.

TUS	TOS	TRIS	TÂS
TUS = Kinds of TOT-TO, Those less determinant sets and approaches to TOT-TO-TRU-TIT-TOFTOT = TO TOT = Potential, past, present, future OITIO = Out-in-TO in-out TO = Actual Totality, All Inclusive, The Topic and all within, TOG = TO in General Overall Aspects TOCH = TO Change, general TOV = TO Varieties, Variability TOD = TO Dynamics, in general TOP = TO Past, overall TOPRE = TO Present, Now, overall TOF = TO Future, overall TROQ = TO Equivalents ABSTO - Absolute Totality in TO	TOS = States of TO, Unitary, Finite, Infinite, Total and Absolute TU = TO as a Unit, The Unitary TO TE = The Finite TO, Parts of, smallest to largest TI = The Infinite of TO TINI = The Infinitesimal TO TO = The Total - Actual Total TÂ = TO Absolute TITO = TO Total as a set by TIT RELTO = Relative Totality TO	TRIS = Special Kinds of TO, In and of itself, TO Relative and TO Proper, TP TRU = The Relative Universe of RECTO TIT = TRU In and of Itself, TRINI = TRU extended Infinitesimally TRIN = TRU extended only Infinitely TRIPRE = The Present in TIT TRIF = The Future in TIT and TRIS TRIP = The Past in TIT and TRIS TRIR = The Relative TRIS TRIT = Form, Structure, and Frame of the object TIT focused and available for Standardized Representation	TÂS = Relative Absolute, Preliminary and Formative TIA = TO is all, exclusive TAIA = TOT Absolute is All (TÂO) TÂEF = TO Absolute is Eternal Future TÂIT = TO Absolute is this static now TÂF = TO Âbsolute Future TÂS = TO Absolute Plural TÂ = Extending Indefinitely into the Future

Table 20-1 TO Classification by systems, TUS, TOS, TRIS and TÂS

Table 20-1 presents and signifies some of the kinds and transformation of TO, TRUS, and mathematical forms, TOS. Closely related are those of RECTO, TRU, TIT, TRIS, and the absolute forms of TO, TAS. The Differential of TRIS, is that variety TIT, whose forms and states converge on TO. To the degree TIT is practical, relative and not limited to zero time, it must be of some time duration that indicates change, alteration and variation. So that when we speak of TIT we aim at its most perfected set form, centered on the infinitesimal present and encompassing extant TRU in the most concrete and stable way. We must remember that this is never absolutely absolute, only relatively absolute. It is relatively temporal, spatial, functional and practical, which is its transitory nature in TRIS. Likewise TRIS is of a transitory nature in TRU and TUS. By the mathematical kinds of TO-TOV-TRU-TIT, TROS, TO varies less quantitatively as its varieties become more specific. With increasing relation and specificity all will converge on a unified and fixed center and form a closed set in TIT-TOV. To understand TÓ-TOV-TRU-TIT most precisely and specifically we need to increasingly and exclusively focus on TRU in and itself at time zero, TIT, and how it exists presently TRIP, relatively TRIR, and of other kinds, TRIS.

Language tends to be relative, differential and diverse, which is somewhat compensated by logical comprehension, its overall approach and methods that make language more precise and orderly. Yet this is not enough to give language integration and representation of TO This is provided by mathematics and associated methods, including proportions, wherein language can be both represented and valued to give systematic unity to TÓ.

It is from mixed definitions of words, multiplicity of changing languages, altered and lost accuracy of essential meanings that their consistency and benefits suffer. Even worse, their rough, loose usage, and incompatibilities with each other compound their detrimental effect on human life and TO. This is most evident in the levels of definitions, from being vague, non discrete, lacking in actuality, actual pure essence, reality and identity with any, the, or the actual, totality, TO. The remedy is massive, quantitatively also qualitatively by depreciation. It requires the combined endeavor of each separate concept or word by each person, in their total interactive meaningful lives in TO, as evident in the totality of proofs diagram, SPALT, other models, their operations and applications in life.

Logic is of more than one kind, including, the language, and its use, that determine and guide mathematics, and formal methods derived from reasoning with syllogisms, This was A way of thinking, by propositions and of essences that enable valid knowledge and understanding. The latter are typified by, and largely the derived from Socrates. It was by what has since been called the Socratic Method that human thinking made a giant step toward valid knowledge. By Socrates the method was simple in design, profound in effect. Topics, or subjects, favored in public conversation, are questioned and analyzed, to determine the deeper meaning and essentials of the good, beauty, love, right, knowledge, gods, politics, justice and belief, etc. By repeated questioning examining, analyzing and narrowing its implications, by what its fundamental essences are, the best answer is produced. The teacher brings out of the student the chain of thought that is most productive. Mature and logical thinking provides greater understanding of the right meanings in all. Such knowledge, via a love of wisdom or philosophy, leads to relative convergence, and identity with actual totality. Logic arises from, and relates to, the highest form and power of language. The imperative uses of rational thought are guided by the most basic forms, proofs and laws of actual totality. Since before Heraclitus logos has been known. The Word was the beginning of God. It is the power and rules that emanate from the 'one', and govern all. Those are the rules and laws that order all, and all must follow. This is the Law or doctrine of the Logos, Imperative Order from the Unity of Totality, LIOUT

CHAPTER 21 - MATHEMATICS in Actual Totality, to Proportions

Kinds of, number theory, integral-differential, division, quantity, gradients of continua, same and other or different, unity, integral, the finite, parts, interval, measure, equality, distribution, position, field and cube, operations, statistics, accuracy and perfection, quantitative-qualitative, focus on role and use in TO

Mathematics in TO is somewhat different than its customary usage. In TO, it is highly qualified, by the actual pure essences of TO, and the form of its modes, continua and adapted spectra. TO exists by its basic mathematical form, of unity, totality, and plurality. This is seen when approaching a closed set, as in specific TOV. It allows the special varieties as TRU and TIT to provide a more static form devoid of change and variability, (TOS). Yet TO is never without change and both external and internal dynamic. Some of these make mathematics conditional, relative and limited. This is why great discoveries in mathematics and physics that have produced such profound understanding of the physical universe are not so applicable to, and must be modified and altered to fit their role in TO. God never subscribed to, nor was highly limited by mathematical reduction. Mathematics can be no better than the degree to which its logic it is based on accurate and correctly applied knowledge, of absolute and relative totality in TO. Where mathematics applies well in absolute totality and the physical universe, we lose valuable proofs of TO if it is not modified by RELTO and how it exists in TO. Mathematics and numbers are the exclusive realms of neither absolute nor relative totality. In fact in their own ways they permeate all and are typical of each variety of TO. It is seen in the focus supplied by relation, the role of relativity, the ordering of mathematics by logic and many others. This is the Law of mathematics by Absolute and Relative Totalities of Varieties and Logic in TO. LARVIT.

The many operations of mathematics reveal their uses in TO. From algebra to differential equations to the most complex functions math offers an almost unlimited means of proving and showing TO. Statistics is one of the means needed that show how to measure and interpret large collections of data. Many applications of statistics provide statistical proofs for TO. This is seen in probabilities. That math can't make TO or its representation perfectly accurate isn't disqualifying for either. For it is in making perfect TO-TOV-TOT, rather than being perfect in itself, that logic, math and representation are in their highest powers and glory. Examples of this are zero and absolute which in math are valid concepts, although not natural numbers and of TO poorly known. In TOV and TO them have practical existence by their actual pure essences meanings which act as limits. It is the perfection of these and related concepts that qualify and give TO and TRU better proof.

The limits of language, logic and mathematics are inherent in actual totality from relative totality. This is a product of LANERA and LAMOL, the Laws of never absolutely absolute nor relative, and More or Less. This does not mean they are less useful, it only means that they must be held to their parts and roles in TO. They can only more or less be taken with the strict objectivity of an absolute mathematical form and sequence. It can be used, but must be modified or altered to fit the form and function of TO. This is seen in the design of SPALT, as well as the design, math and geometry that are associated with the origin and development of life. Least Common Denominators, LCD, are ideal for vertical integration, reduction of differential to integral in a class. They help produce a closed set, for TO and solve the problems of the complexities of differentiation and proportions. This is signified by Law of More or Less, Reduction to Formal Simplicity, LARS, the complex and enigmatic difficulties of vertical integration, cf. LAMOL, This is how TO is developed by vertical integration, the reduction of all to its minimal state in totality, TO. As mass is a function of frequency so is TO and TOV set by its finite quantity. As TIT in TOV of TO approaches zero, its infinitesimal state loses finitely. The contraction of TIT, in the extreme ABSTO-RELTO of TO, features the origins of the relative construct of the form of TO. The finite is set in each stage of TIT, a quantified mode of extension and measure. By quantity the finite and gradients of the spectrum of continua and dimensions can be provided. The distinctions these terms and means provide illustrate the ontological and theological proofs of GOD as TREQ, TRU-TO Equivalents. This is by the actual pure essences all are adequately limited and qualified in TO.

Division and specific form, measures, position, operation and proof of TO - Division is critical, and especially beneficial if the form is correct and carried to a level equal to an observer's capacity to understand TO. It is the finite separation and division with measures, between extremes and means, of dimensions that provide the most accurate proportions and order to show TO. Division typifies and helps to prove Totality Proper, TO, and it is by division we can separate the general from the specific, or TOV. By specifying the separate or divided components we can develop relation and order. When TO and its divisions and continua are correct and sufficiently known, position can be given, analyzed and operated. The varieties of TO form a continuum or spectrum from the more generic to the most specific and return to combined totality. It is a continuum and spectrum from the absolute to relative in totality. There are many gradientsand divisions in this continuum. They contrast and reveal changes between separate existence and that in TO and TOV, and the form and function each gradient produces in TO. The change from, a or an, to any or some generic form, and to 'the', or the most specific, expresses and provides their increasingly different finite and mathematical distinctions. It is same and different part and role of mathematics in TO and each of its varieties that helps to give accuracy and proof to the whole.

Same and different, and integral and differential give TO-TOV accuracy and proof. The integral and differential do not exist as either extreme, but as a continuum, duality, spectrum, parameter and dimension of the object, system or totality they are in. Since the body of increments in this continuum is its major determinant, it is what the major parts are that largely characterizes that needed to best learn, know and show it. Among these major determinants, modes and parts are validity, actuality, proportionality and totality. These determinants carry special and crucial significance for the development and proof of TO and TOV-TRU-TIT-TRIR-TOT.

Quantity expressed as exponential finite increments from the most integral to differential provides the measures of actual totality, its constituents and representation. They extend the meaning of TO, what makes, saves, forms, and show it, its proportions and why they are like divinity. It is by the changes and development of what is sacred, proportional, integral and differential and their mathematical and geometric representation that we can begin to recognize and prove what actual totality, TO is, how it is formed, ordered, exists and functions as an exclusive whole. Reduction to measure, precision and proof by essential convergence on the whole of TO and TOV. It is by mathematics that all must be reduced for ultimate precision and proof, as long as this math is quality selected and directed by the logic of an actual and ordered totality. This is shown by the corrections often needed, to provide a most practical interpretation of data. Two of these are, Determinants by Exponential Analogy, (DEXA), and the exponential spectra of basic dimensions.

Figure 21-1 gives one coordinate field of length and time centered on unity. The physical universe and much of nature are exponentially formed. This is an adaptation of the meter and time zero, of CMS, the centimeter and second standard for physical representation. Along with a combined convergence of limits, three dimensions and the speed of light by relativity show how to provide the form whose ordered shape and structure represents TO and TOV. Since all, as TO, is balanced in unity between infinite and infinitesimal it is necessary to reach their best - and happy - medium. From James Gregory in solving Zeno's Paradox, of Achilles and the Tortoise, an infinite number of numbers can add up to a finite number. This is a convergent series, which, with quantum theory, helps prove the form of unity in TO-TOV. This is seen in the center of a subject-object and integral-differential diagram, SOID, Fig 3-23. In the combined unitary and infinite of actual worlds, the vast and massive finite is of the highest form of mass energy, or reality, in space time, ACEXIN. How much, then, is each to all persons, capable and able to encounter, manage, and control increasing and decreasing differentiation and probability? This largely answers a question in quantum theory, "is absolute certainty ever probable?" Because of the personal limits of all humans and all in and of TO, it approaches certainty and can come close to being all. This holds with the provision that all TO is included and exceptions made, i.e., TOT, TRIP and TO. This solution to the paradox of infinite and finite in TO, is like the solution for many continua and dualities, including infinitesimal, zero and absolute. It is confirmed by, and confirms the fact that, in TO, infinite and finite are not opposite, they are parts of different aspects of the same thing. It is this same thing that is TO, by actual pure essences and all, that enables its necessary existence, unity in totality, exclusiveness, extended presence and continuity. Law of Integral and Differential Equality. This law may seem simple, yet it is needed to confirm a most basic fact, correct a tendency to prevent its failure, and prove TO.

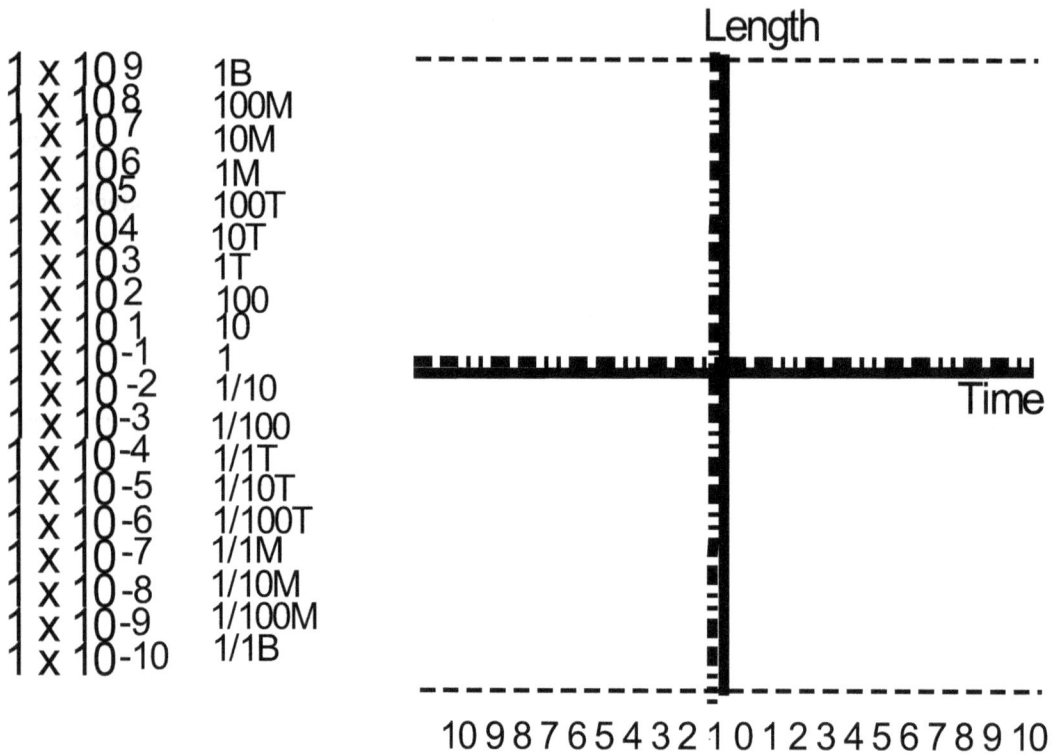

1 X 10⁹	1B	

Fig 2I-1 Exponential progression and proof of the form of an object, TRU or TO.

All kinds of numbers, from the simplest digits through rational and irrational, to the most complex numbers are representative of some aspect, form, or function of TO and TOV. Each has their significance that is often highly indicative of certain fundamental categories, aspects, modes, forms, structures, spectra, continua, or mechanisms of TO. Their discovery and showing not only resolve many of the greatest problems and paradoxes of existence, they are essential to an accurate and successful representation, comprehension, and application of TO, TOV, TRU, TOT, and associated categories.   E Pluribus Unum, one for all and all for one, is a common motto, that are both right and less than right for TO. It balances one and many, yet does not mean that all are equal in TO. Equality in TO is not any equality, only equality of proportional value in TO. The unity, parts, infinite limits and totality that converge, make up and combine in TO provide set form, a closed set whose fulfilling form can be well known, represented and shown.

Many of the most basic doubts, problems, paradoxes and enigmas in knowledge arise from differences in the use and interpretation of mathematics. Is it to be used and interpreted directly, as in simple division, or more indirectly, logically, theoretically or relatively? This is at the core of actual totality, the differences and sameness of absolute and relative totality, as they exist separately, or together in actual totality. The doubts and problems with division were first brought to a head by the atomists, e.g., Democrats and expanded on by Anaxagoras to account for the paradoxes of Zeno, of Elea, in the 5th century BC. Is there an infinite division, as of atoms, or what are its limits? Aristotle took up the problem and helped to clarify, but not resolve it. The solution is to be found in actual totality. The role of both theoretical and logical division is a product of RELTO, relativity of space time and mass energy, combined with physical or absolute totality. By the limits imposed by relation and actual totality, division and mathematics, retain much of its direct and physical use and interpretation, yet must be modified by the limits and division of relative and actual totality. Since the paradoxes arose from the primitive forms in all, what was the form of this totality? Did it lack unity, or diversified? Was it concentrated and convergent on unity? Or was it some intermediary? Although actual totality is intermediary by transition in time, it is practically unified. It is by this fact that all must be ultimately determined and proven, how they exist in the absolute-relative unity of TO and its varieties.

Many TO relations and functions can be expressed mathematically. Many more will be developed as TO becomes better known, e.g. TO = ABSTO x RELTO, and absolutivity,RELTO::ABSTO:Relativity, etc. In this book the emphasis is on proofs, whose grammatical expression better suits their origins than mathematically. The qualitative and quantitative, with degrees or values that can be used as measures, are largely the work of determining proportions, this we see in the following chapter.

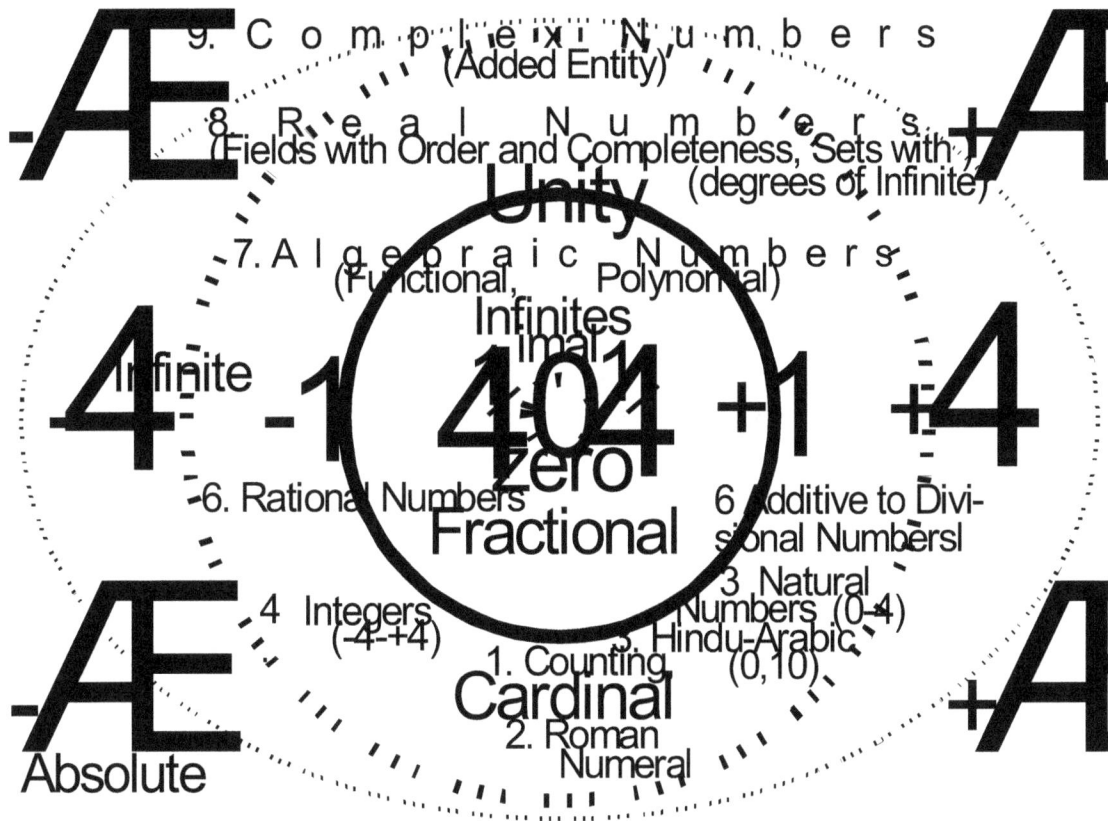

Fig 21-2 Number Systems, math, relative totality, TO, TOV and their representation

CHAPTER 22 - PROPORTIONS in TO, to Geometry

Worth, Values, Importance, Greatness, Relative Magnitude-REM, Ordering, Quantum Mechanics, Fractals, TO, TOV, TRU, their Proportions and Values -

The emphasis, interpretation and use of values come from the simple proportions of all in TO. It is called axiology and investigates, measures and treats, worth, importance and what is most to least, and best to worst. Axiology is a quantitative measure of quality. It is closely related to, and largely determines ethics, what is right and wrong, and the best conduct in a culture. Proportions, values and importance and their descending and ascending levels are critical to knowledge of, and the proof of TO. It is by proportions, values and worth that the mass energy continuum of the observer-observed relation in TO, Á, makes choices and selections. Hereby the superiority of proportional totality can be seen and their proof and use for TO, TOV and components are made most rational. The general form of proportions in TO is provided by mathematically precise distribution. By a continuity of proportion on a set scale from greatest to least a most accurate and valid demonstration can be made.

There is great importance in proportions, values their limits and equilibria in TO and its proofs. We must always remember that although the totality and proportionality of TO seem to establish a closed set in which all can be valued and positioned, this is not perfectly the case. For neither the world we live in, we, nor entities, are perfectly TO and TOV. TO, TOV and TOT are conditioned by change, especially in time. All are actively engaged in the action-reaction dynamics of potential-actual, and positive-negative. TO is only relatively to practically static. It is in transition by time with changed settings. TO proportions and values are only more or less perfectly established.

Proportions and values when scaled exponentially are given as Relative Magnitude, or REM   In order to provide proportions and values with a basis upon which they can be most certain and precise, a scale of relative magnitude is given. It is a mathematical rating of all components of TO by their proportional part in the whole. Like many measures of stellar brightness and earthquake severity, Relative Magnitude, REM, is based on a decimal scale from one to infinity, e.g., 1, 10, 100, 1,000, etc. levels of parts in the whole, or 1, 2, 3, 4, 5 - - Where 1 is TO, 2 is 10 parts, 3 is 100 parts, 4 is 1,000 parts, etc., to infinity. The method and measures are precise and very beneficial, yet many problems occur in determining how all exists in TO, and fits into the scale. Most of the problems can be accounted for and more practically and accurately corrected by detailed and rigorous selection and determination. With future development it will greatly add to the degree TO and its knowledge is proven and right, with the certainty and proof needed by each and all persons.   It is much easier and more accurate to give the level of REM of anything, or component of TO, than it is to give their order in proportioning TO. Likewise it is a much more effective and rational way of giving values to persons, even past and future. By knowing what value one is, and how it can be raised or lowered is a tremendous stimulus and guide to quality in life.

Without some order, and relative proportions, of the content of TO, one is quickly lost in a maize of ever higher complexity. Also the unity of TO both enables and demands a systematic approach that best brings out importance in its components. This is with good reason what we know of TO already, its modes, continua and varieties, and their objectivity, subjectivity and reality, absolute and relative, and generic to specific varieties. With the use of a vertically integrated system, as evident from Fig. 7-1, their position in, and the positioning of many other well known parts of TO a rough idea of what makes up TO, and how it can be ordered, showed and used can be provided.

Proof of REM, PREM, and of TO by REM =

The proofs of REM, as proofs in general to a large extent, are provided by ever expanding knowledge and substantiation of the makeup, and what it is based upon, TO. It is based upon the discovery and refinement of its basic concepts, principles, laws, and mechanisms. They involve proofs of not only REM itself, they involve, to a varying degree, the overall and transcendently relative states, and givens that relate to REM. In this way proofs of REM will involve proofs of Total Valuation of TO, OV. An enumeration of some, of many, proofs of REM, include, 1. LOV, or Law of Valuation. Since all is not the same, selection must be made, 2. OV or Total Valuation. Valuation is more than the comparative and evaluated parts of TO. It is preeminently the expanding to absolute worth of TO. 3. REMT or that valued must be, or be brought to, purity or 100% TO. 4. REMTE or that valued must be based upon an ever more distinct and discrete separation and differentiation of the parts of TO. 5. REM is attained to a large extent only as we reduce all terms, concepts, nominals, and symbols to their part in TO. REM is the amount or proportion of TO totality that is occupied by an entity, things, sense, aspect, part of process of TO. How great a thing is of TO primarily equals its Relative Magnitude or REM in TO. 6. PRAV or Practical Valuation proves REM through its dynamic utilization in life providing the selection both of TO from non TO, worth from no worth, and the better from the worse. 7. Order of Magnitude proves REM by providing a systematically unified scale as well as an exponential and warped to total scale that enables the comparative selectivity of all parts of TO as they can be evaluated precisely. 8. The determination of REM proves REM by showing how, and providing the posits necessary for the valuation of each part of TO, accuracy of others increases.

The compartmental sectioning of REM provides the classification and scaling of parts of TO necessary for comparative and selective valuation. 10. The method of precise valuation through known proportions of each of the many classes and systems of TO, MOCBIREM, enables accurate proportions of their parts. How existing known topics that have good measures for their parts can be equated with and applied to REM of TO. MOCBIREM is an immensely useful and valuable means and method. It can be no better than the degree to which each class or system is well proportioned and all its components are held to TÓ.

Other direct, indirect, factors in, and modified methods of the determination and proofs of REM include, TO, TE, TIT, VAPO, special life, human, enablement of TO, life chain, God, Á L T, sectors, meaning, DAOST, ES, TES, relative, distinction and identity. It also includes, LARF or law of relative fixation, RUPKAL, revolution of perfected knowledge, love and well being, MATES or Meaning automatically tends to ES, specification and differentiation, LOPAN the law of partial necessity, OM or omnific magnanimity, TOP and TOF or past and future potential TO, proof by negative inversion, language, language by TO, new and existing concepts, maximal separation and distinction formulation of what is different and great in TO, law of relative parametric proportions that set TO, the absolute necessity of combined objectivity and subjectivity, law of order and standards or ordination and standardization, Law of Relative Value, LORVA, ALT, SPALT, positioning, statics and closed set, proof of ALT, TREQ, equality, distribution, proof of value, economics, business, capital, commerce, money, value, levels of mass energy units, law of equal dependence of a sum of all and persons in TO, fallacy and paradox of apparent of value, divisions, total positioning of REM, principles of a rational remainder, cancelling out of chance, value only of TO, and precise TO parts positioned in ALT and SPALT.

Proportions or REM depend in part on the total number selected. If we are judging REM by a few factors, they will tend to be of high REM. If we of an infinite number they will be of all, and most of lower, REM, As parts become more differentiated and separate from the classes or groups they belong to, the class, group and original parts lose REM. For the levels and relative content of each level remains constant. Entry into the whole by new elements deceases the value or REM of those previously given. This is strikingly evident from the condition of yesterday's news. New entries and time produce marked changes in value. This is ameliorated by the massiveness, set form, actual pure essences of TO. It shows the power of a person in here and now relative to, and helps to prove, TO. As each new term is added to the pool of parts of TO proportioned all existing terms lose REM. It is very simple and easy to give the relative values of a few greatest parts of TO. The major problem is how they overlap in, or have extended effects in TO. Also whether each class of entries in REM maintains its overall value in TO, or loses value as each of its component parts are added. All these factors, methods and considerations must be accounted for if REM is to be the ultimate measure of proportions of TO. Remembering this, it still serves a great purpose to have a highly accurate proportioning that shows the valued content of TO.

DREM is the Development of the REM of E, ES, CES, or specific parts and factors of TO for placement in a model, such as SALT. When we closely analyze increased specificity of REM, DEREM and DREM, their levels, 1R, 2R, 3R, 4R, of ES and CES and their corresponding ESINS in ideas, abstractions, words, concepts, and abbreviations, and posit them in ALT and SPALT, we can attain a highly practical and revealing representation and image of what TO actually is. When we supplement this with VAPO,

DAOST, the SCARLS involved, TRIS, OBT, SUT, as we can by applying: ARF or absence of REM Forms, and DIREM or divergence of REMS, Coterminality of REM, and MOCBIREM, we greatly improve the specificity, sensitivity and reliability of values in TO. The degree to which this can be a more accurate as well as practical image of TO can be continually refined and validated in the operations of MOT and SPALT positioning of the static and dynamic contents of TO.

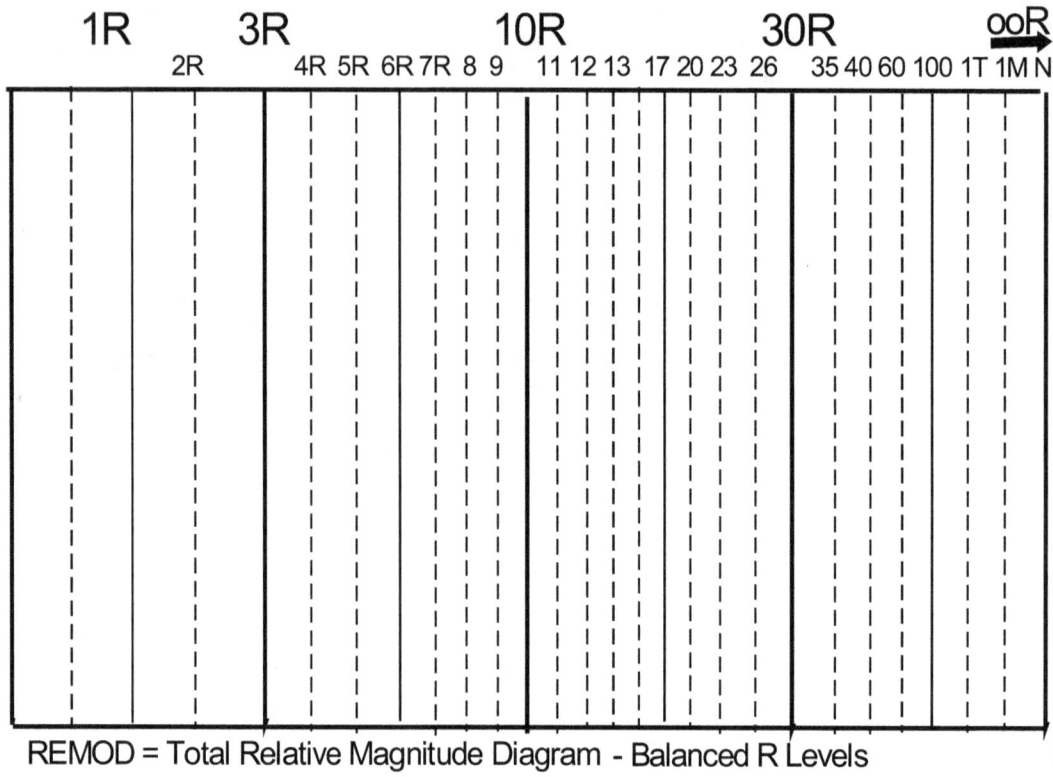

Fig 22-1 Balanced REM Levels, REMOD

This figure shows the whole range of REM adapted to the best distribution and balance of levels. In this way those aspects and parts of TO at 1R are given the much greater representation they warrant. The limits of REM are those to which TO naturally inclines and renders us capable. Thus most important in 1R is no more than others, or combinations, of other levels. This balance holds no matter how large or small the total proportions. What is possible, or we are sufficient for, must be kept in motion constantly pushing forward or backward through REM Levels. This is called: the Principle of Inversion Reaction or POIR. It is the degree to which relative totality moves toward absolute totality, and the absolute moves toward the relative in equilibrium. This is our all, TRIA, in transition. POIR is the difference between absolute worth in which all are immeasurably valuable and unitary worth in which all become one. With relative worth in between that is preferably well formed, scaled and distributed to give best order to values. It is by POIR that we can resolve and prove the difficulty of having a limitless and infinite REM or a meaningless and unitary REM. This paradox, of parts being as vital as wholes in contrast to the greatness of some parts over others, can be readily resolved, as all paradoxes, when all the information is at hand, the corrected or unified point of view is developed and demonstrated, and all TO, factors and methods are accounted for. REMS of 1R-5R tend to be more universal and the higher order ones more partial is evident, to be expected, remembered, and used to further refinement. The unity, equivalence, equality and stability of REM are only evident when values are derived from a single and totally representative source. Just as the only way to attain an objective is to know what the center is, and from its surroundings, so is the unity of REM dependent upon a full focus and perspective of TO.

This is a diagram (22-2) in which REM Levels are centered about 10R, and extend in equilibrium from one to infinity. Whether 10R is close to the median of REM Levels, and whether 3R and 30R provide good half way points between this 10R median and REM or REM Unity and REMIN or REM infinity will have to be critically analyzed, refined, and perfected in the future. It is not the same for all varieties of TO. Those of past TO, would shift to smaller REM, and those of the future to larger REM as TO expands. It does provide a good working frame for REM, when all their qualifying factors are applied. This will depend upon how much effect the other laws, principles, concepts, corollaries and mechanisms of REM play in providing it.

GREATER than TO		LE-098	
REM	SUBJECT-OBJECT	DEGREE OF UNIVERSALITY	PART OF TO
1R	e.g., GOT	HIGHEST, SELECTIVE	Greater than 1/10th
2R	e.g., DOST	HIGHER, SELECTIVE	Greater then 1/100 th
3R	Great Groups	HIGHLY SELECTIVE,	1:1,000 Part of TO
4R	Primary of Great Groups, SO	HIGH SELECTIVITY, COVREM	High, Variable
5R	Secondary of Great Groups	HIGH, less Selective	High, Variable
6R	Tertiary of Great Groups	INCREASINGLY PARTIAL	High, Variable
7R	Parts of Great Groups	Less Universal	High, Variable, POMOM
8R	Small Parts of Great Groups	Little Unversal	High, Variable
9R	Smallest Parts of Great Groups	Least Universal	High, Variable
10R	Average TES, LATANCOV	Least Universal	High, Variable
12R	Average Person, TES	Unlikely, COV	Higher, Variable
15R	Subpar Contributor	Unlikely	Higher, Variable
20R	Average Act/Average Person	Unlikely	Higher, Variable
30R	Average Thought/Average Person	Unlikely	Higher, Variable
50R	Barely recognizable Positive	Unlikely	Higher, Variable
100R	Low Order Positive	Unlikely	Higher, Variable, LAIA
1000R	Very Low Order Positive	Rarely	Highest, Variable
nR	Tiny Finite Positive Increment	Rarely 1/n →	Highest, Variable
∞R	Least Positive Increment	Rarely 1/∞ →	Highest, Variable
0R	Resultant neither + nor -	Rarely - = →	Nearly all, RACOD
-R	What is Negative for TO	If AUTO, ANTO	Entirely

Table 22-1 Formal Distinction by Balanced Proportions, or REM. Interpretation of values and worth

Another method of showing REM, and its highest orders, is by using a diagram similar to Fig. 22-2 only with ten instead of three angular sections. Each of the ten sections representing the greatest categories, aspects, fields, sciences, philosophies, religions and constituents when of their actual pure essences in TO. This would have to be on a sliding basis that would change to accommodate increasing numbers of components in each section. Such a diagram of REM would help to correct the problem of decreasing REM values with added parts.

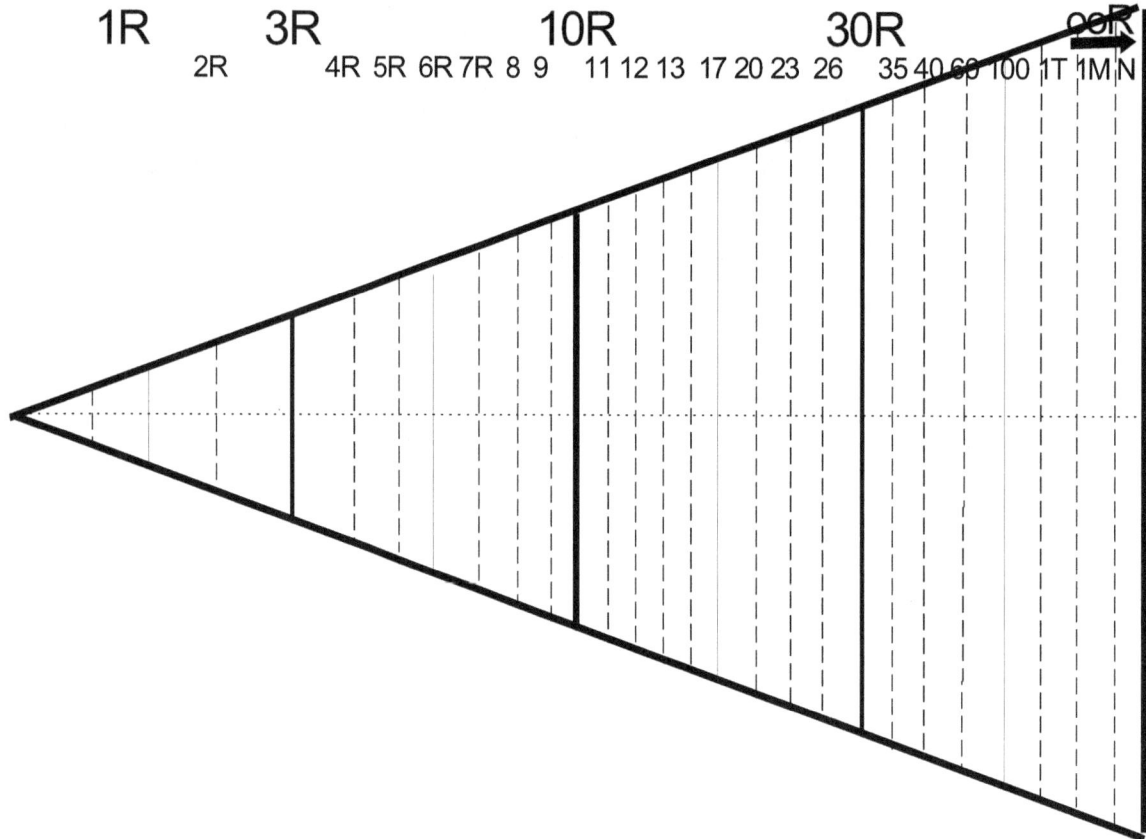

Fig 22-2 Standard distributed Conical REMS Diagram of Total Proportionate TO

That a sense of values is paramount are the lessons life's experience impresses upon us. Each, every and all persons must learn, develop, have, and activate this sense of values by finding and using them as they are of the total proper order and intensity of static to dynamic proportions. This sense of values requires and should be nourished by one and all, to become total dynamic REM as it best enhances TO. As simple singular problems may only require simple singular solutions or acts, so do multiple and total dynamic actions necessitate proportions and a transcendent, continual problem-oriented approach. Complex, major and compound total problems, usually require more compound and total accounting, solutions and actions. This is the purpose and goal of much of the remainder of the text. It proceeds from IA to, IA in TO, IAT, and through their processes to culminate in actions that complete actions that return, all needed for the present and future TO.

Parts and Proportions of REM in TO are given by the proportionate parts TO develops, exists by, and are perfected by several significant laws, 1. DOR, Development of REM, how they are transformed from entities to being only a proportionate part in TO, 2. LORE, Law of e part, or ES in TO,, never equals all TO, 3. LETRAD, Law of each ES is perfectly of TO only when all less and disproportionate of TO is removed. A more en depth idea of the greatness of REM is gained when one recognizes the level in which various concepts and objects exist in TO. These include, KOT - Knowledge of TO 1R, DAOST - Different Aspects of the Same Thing 2R, COVREM - Correlation of Optimal Virtue, Value and REM, 3R, COV - Corollary of Variable Virtue, Value and REM, POMOM - Principle of the Multiple Meaning of words 7R, LAIA - Law of An Instantaneous Attention - 100R.

The following increasing detail and differential of TO have decreasing values, i.e., 100 to infinitesimal REM. RACOD, are the RELTO-ABSTO, Concentrated Differential, or ever finer separations that occur with increasingly valid knowledge of TO. A REM of 0, or 0R, may be interpreted as a resultant limits, from the solitary unity of TO, to the infinitesimal limit. They are not a part or TO, except to the extent when from the unity of TO they are OV, or total value, and when the limit past the infinitesimal is zero. This is a zero that is not TO, yet has a little significance practically as a limit. The worth or value of ones IA or instantaneous attention, thought, subject, topic, study, discussion, thing, diagram, book, first person or anything equals, 1. The percentage of it that is it's ES. 2. the corresponding part of it that is in TO, 3. the degree of its brevity or resolution of its limits, 4. the degree of perfection by which it presents, depicts, and illustrates this part.

REM = ESIN x ES x REM Limits x ROT perfection

Thereby worth equals proportionality, brevity, artistic perfection, and the degree these are equal to their ES, as a part of TO, the percentage of the thing that is its ESIN and ES. Brevity, like Occam's Razor, is highest value where it measures the least that is appropriate for the measure. From an equality of REM we attain: LE or the Law of Equality or, COL x LE - X = TO, and Equality x S-O x X = TO x LAMID x quality, production, or O x S x S-O. LAMID = Magnitude Interdependence. Large components of TO are highly interactive and superimposed, smaller ones are much less so. Through the unity, equality, LE, structuring, and formulation of REM of TO, we can develop, represent and illustrate REM by a target diagram.

The whole of TO has its own set form, which provides the normal distribution that enables precise proportions. Total value, OV, is relatively absolute by the APET of all that constitutes TO. If there is no actual basis and value what is there? Value is a prime proof of TO, as TO is of value. This depends, to a degree, on the observer, whereby equivalents of TO are given comparative value, and what is incompletely TO has less value. Proportions of divisions or other universals in TO are largely evident from its total form, and are of and take positions among those of, the highest relative magnitude, REM, or value in TO.

It is impossible to provide an exactly perfect order to the proportions of all in actual totality. This is for several reasons, namely, 1. The degree actuality is viewed, 2. How well it is known, 3. The effect of actual pure essences on all components, 4. The limits of proportioning, 5. Dependence on SPALT or Model at hand, 6. The current state and level of perfection and completion of proportions and knowledge of TO. Yet proportions are an extremely practical method of determining participation and contribution in TO and among its components. Their limits and imperfections will decrease as TO grows, becomes better viewed, known and act ual pure essences developed. Actual totality can be no better than the degree to which it is perfect, it has perfect unity, it has perfect unity in totality and thereby its parts are perfectly known, represented, in order and proportion. This is the dilemma of the current state of values and TO. It is a dilemma whose cure lies in expanding development, knowledge, representation, and application of actual totality. Absolute perfection of proportions, like all perfect absolutes or relatives, is non existent in actual totality. This is not self defeating, it is challenging and the foundation upon which all is built, and most practically useful for best human life.

Total value, OV, depends on Laws of Constant and Variable Proportions, LACVAP. Constant is the presence and stability of TO, and variations by transitions, as well as internal dynamics of the components. It is constant by TO and variable by TOT and in use. When proportions are variable, suspect overlapping, when they are constant suspect specificity. Some constituents of TO like ALT produce excellent proportions, others need a great amount of work. This will take much effort and time, even changes in time. Associated with the problem of accuracy and determination of proportions are limits of representation, perspective and presentation. It is always well to recognize the errors in proportions and values. It helps to consider and compare values and proportions that are less than correct and right by TO values with others and the whole, OV, the worth of TO. Less than certain values do not seriously detract from the quality, principles, essence of REM, proof and greatness of TO. They are only specifics of measure and use. They help us to better correct and expand values in actual pure essences, representation, our vantage, and application. The reason why totality proper, TO, and its most specific variety, the world of best human life and mentality, TRU, are not the same as the physical universe, can be most simply found in proportions and what is most valuable and sacred. This is because relative totality and absolute totality, measure proportions differently, often in opposite ways.

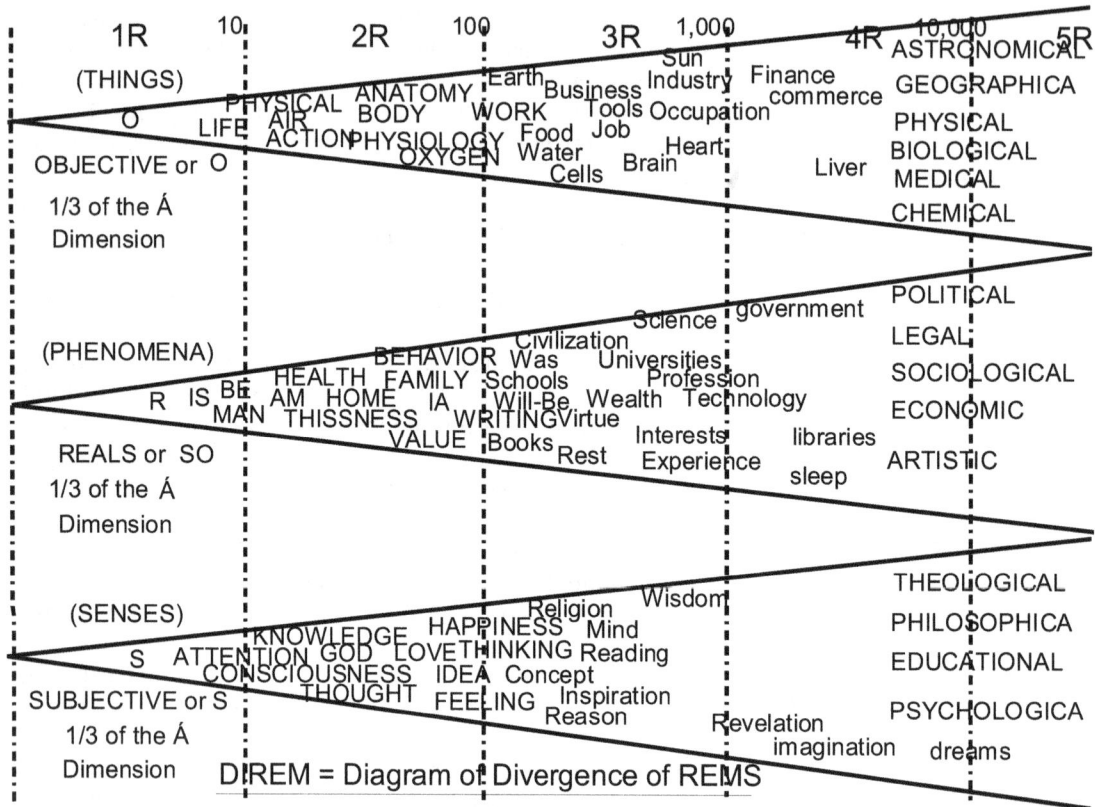

Fig 22-3 Divergence of Relative Magnitude, REMS, or Value by Á, Activity in the mass energy dimension of TO.

This figure shows the role of S-R-O divergence and distribution of REMS, DIREM, in TO. DIREM shows how to envision the contents and dynamics of TO through a sagittal section of REMS as they are proportionate parts of the top levels in the whole, TO-TOV. They are divided, with increased accuracy, by groups, fields, sciences, topics, classes and systems, MOCBIREM. If we know enough about the major modes, continua, kinds, varieties, systems, topics, themes, fields, sciences and divisions of TO, and how their consistency of greater to smaller parts, they can be given relative values or magnitudes transposed into TO, TRIT and SPALT. MOCBIREM is the increased accuracy and facility achieved by the relative hierarchy of components in any system or group for which definite parts are available. The Á mass energy form and function, FAF, are the Objective, Reals and Subjective Sectors and Operations in REM and TO. They are graded from highest REMS to decreasing REMS. There are many other terms and subdivisions of fields and sciences that can be added. They provide a good idea of how many fields, sciences, systems and classes diverge and are distributed in REM for use in SPALT.

Although REM and its use for TO are very rough, will never be perfect, and will even increasingly change somewhat in the mid to far term, much work will continuously and always be needed to refine and apply this extremely important aspect and proof of TO. Generally that closest to the oneness and basic form of TO is apt to be of highest REM and that which is of its multiplicity is apt to be lowest, this is not always the case. For by relative totality, special varieties, and TIT, TRINI, what is homely, base, uncommon, or even unknown may occasionally be of the highest REM, e.g., ones ideas, good deeds, or simple little cures for a great problem. The degrees to which the limits and problems with REM can be made more accurate, practical images of TO, continually refined and validated, are in the operations of MOT and SPALT positioning of the static and dynamic contents of TO. It has increased as knowledge, first principles, concepts, ideas, proofs and laws grow and become more valid and actual. Through increased formal distinction, clarity and meaning they are or are not as they are selectively utilized by humanity. They are utilized if their meanings are: very useful, clear, familiar, and lack questionable, problematic and negative effects. The continual upgrading of meanings of what represents ES and CES is what enables REM, as REM, in turn, helps to make them more distinct.

More evidence and detail for an efficient and effective representation of the proportions of actual totality are given in the addenda. Although ALT is not the only dimensional form and coordinate frame, it is both a prime denominator and accurate means of simulation. It does this in more than one way. 1. It allows us to divide the form of an averaged sum of all in TO into a specified number of parts or sectors. When each dimension is shown in ten gradients or levels, the total number of sectors is 1,000. 2. The 3D frame provides a better replica of the form of TO. 3. Such a number of sectors of equally proportioned parts provides an accurate base to begin positioning proportions. 4. When allowances and corrections are made for variations and aberrations in positioning the adaptations and modifications can be more rationally made. Other figures and lists of positions are given to make the proportionality of TO clearer and more complete. The proportions, along with REM perfection in TO, take much work, They will have to continue permanently. The improvements will make it the kind of vehicle for providing an objectively accurate demonstration needed. Even when perfected changes in proportions will be necessary as TO changes with TOT-F.

CHAPTER 23 - GEOMETRY, Plane and Solid to Chemistry and Life

Simulation from form, Set Theory, Closed Sets, Quantum mechanics, Fractals

Plane and solid geometries provide a more accurate and visible means of showing and proving TO, design and form. It reveals how form and design exist in totalities whose absolute and relative forms bond in ART to form TO that sets the basis for the creation and evolution of life. This is by replicating mass energy, length and time in its combined form from absolute to relative totality, ALT, by the models, representations and perspectives they help to create, and by the identity each person can achieve with TO. A major purpose and benefit of geometry are practical, how it provides blue prints for form, as in architecture and many kinds of design in life. The more it simulates the form of TO, and the more it contains combined aspects, modes and dimensions, the greater geometry is, as it proves TO.

What is the relative universe of all that enables best human life and mentality, TRU, and how does it exist and function only within actual totality, TO? This is most readily, accurately and powerfully provided by numbers, math and measures of solid geometry and dimensional form. It is by dimensional coordinates of TO, its most specific varieties, TIT and TRIF, and their form TRIT that enables TÓ to be a closed set, and to have solid geometrical demonstration. This is by models, and most specifically SPALT the spherical model of TO and its Varieties. Other models and guides that have VAPO, the means to accurately represent TO can be provided. Once this is given, the proof of TO becomes more certain and convincing. This is accelerated by identity, interest, living by, operation, practice, habitual use and total activity that are dedicated to ultimate unified totality, of TO.

Quantum theory developed from the uncertainty, destruction-construction of matter, and their part and roles in RELTO-ABSTO, of TO. It was from the interaction of subatomic particles, photons, electrons and many others that atoms, when bombarded with sufficient mass and energy to break into a variety of subatomic particles. It was impossible to predict or determine their timing and place of their positions. From the work of Niles Bohr and Heisenberg this resulted in the uncertainty principle. For their positions vanish and could not be determined by experiment. This applies to an absolutely physical universe, and explains the role, proof and application of relative totality, TRU amd TO, by producing a set state and solid geometrical construct. Only in a fixed whole or closed set provided by relativity of the combined relations of the modes of TO could there be certainty. By the combined observer-observed or subject-object of special relativity, the geometry of general relativity, and quantum mechanics we, are provided with the form, state and role of the primary, mass-energy dimension in TO.

In the most general to most specific totalities we encounter such uncertainty and certainty. Certainty can be clearly and increasingly found by vertical integration of all modes and constituents of TO. The interrelations and form of modes, continua, kinds and varieties of TO, provide its basis and proof. This is especially so for TO in the short term, and the role of TRU, TIT, TRIR in TOT. It is because TIT, as the extreme special variety of TO, whose center on here and now, we can position all by its actual pure essences. They resolve the uncertainty principle, and prove TO-TIT-TOT. The difference between absolute and relative, uncertainty and certainty, is between two different continua in totality. By LANERA absolutely absolute and absolutely relative do not exist, except by the actual pure essences of the object, universe or totality in which they exist.

The form of TO-TOV-TRU-TOF, becomes more spherical as their density, compactness, solidity, mass energy, S-O, and A dimension increase. This provides the set state, conditions, relations and limits that can be treated as a sphere. It is such a spherical form in which words and actual pure essences of its components can be made real and actual by their position in SPALT that they and TO are proven. In this we have an extended, or meta, basis with MOT, ALT and SPALT, or, the Law of ALT Inductive Inference, or Inferential Reformation and Redefinition, LALTI. Here are vast implications in the definitive content of TO, signifying how ALT and its use in SPALT for positions of parts and processes can be better accomplished by inductive inference through the reformation of TO and the redefinition of its components. Extended from A, mass energy in RELTO, and the law of specific masses and energies, LESE, is the total focus of all common, relative, and actual that make up their mass energy, spatial and temporal form and function, ALT, and become the special form orTRU-TOV in TO. ALT is the whole dimensional form whose contributions exist together in the special relative set of TO-TRU-TOT. Further proof of their combined and unified continua is evident when we position many of the greatest existing and new concepts in the dimensional field, or sphere, SPALT.

As our mass energy changes to S-O, to S-R-O, and differential to integral become more complex and must be summed to be valid in the relativity of TO, so does our simulation of TO and MOT form the basis for SOID. Upon these two parameters we can, not only represent TO, but represent by math as it is in itself, and all together, observers, or by epistemic, observed and represented of TO, ROT.   It can be seen that the success and thereby survival of TO is largely a function of how well it both qualitatively and quantitatively expand by its mass energy or subject object and integral differential dimensions, and how it at the same time maintains an ideal balance of all its parameters and dimensions.

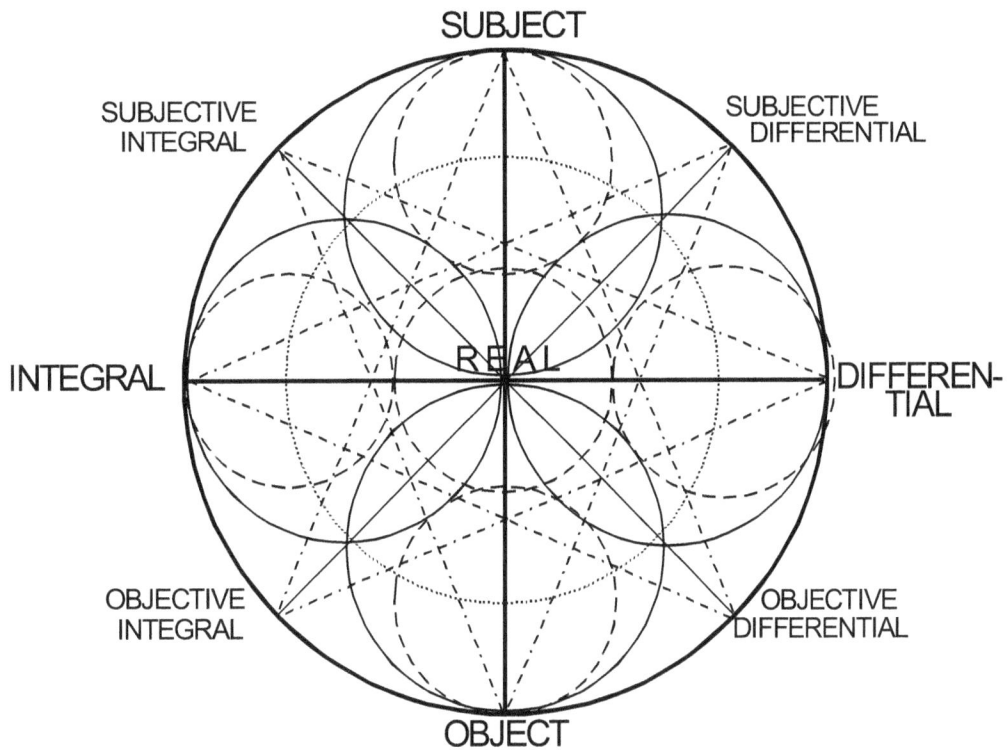

Fig 23-3 Subject-Object Integral-Differential, SOID

The SOID diagram represents two of the major basic dimensions of actual totality. The focal point is how all that forms and functions in TO best balances the relativity, observer observed or subject object, and the predominant mass energy dimension or Á, with the absolutivity of integral and differential parameter or spectrum of TO. This latter is from one to infinity. It could be oriented in other ways, yet for SOID this works best for ROT. There are three sets of inner circles which have great significance. For those that take up half the diameter of each, is generally considered the poles themselves, i.e., subject, object, integral and differential. Those that take up a third of the diameter of each are more specific. They have outer thirds that are the poles, but centers neither subject nor object, integral nor differential. They are the real, as the real world, mixtures as the means that are so much of all human life, in TO. It is such an identity of the real that allows us to increasingly grade and provide levels for quantifying the subject object, Á dimension, as well as the integral differential. This provides a reliable basis upon which we produce, or relate to further appropriate models, topics and proofs for TO. The SOID and SPAT diagrams are much the same, especially when SOID is extended to 3D by time. SO, or is the mass energy dimension Á, ID is equivalent to the length dimension, and time is similar in each.

To be the correct totality a major test is, does it have a center, periphery and intermediary form. To be a universe without a center or periphery disproves its fulfilling a necessary requirement. Any collection of entities or things that have no center, periphery or set form denies to that extent actual existence or part in TO. The term universe requires unity, totality and at least a degree of proportion and set form, Symmetry like proportion is a basic feature in the form of all. Symmetry, may be variant, especially in relative totality, and is affected by change with deformations. By the dynamics of the system or totality deformations and change are often reversible, or cause for gain and loss, a loss that can lead to non existence. Whereas both gains and losses may be negative, gains are more apt to lead to greater existence, continuity, permanence and recombinations. This is a basic mechanism in creation of life.

We must reduce form and function to their states in TO and TOV-TIT, even though this includes alterations that may seem contradictory. For the form of TO this is called the Law of the Suspended State of Structural Limits, LASTIL. Entasis is one means of supporting LASTIL. It does so by not only showing a more actual form of TIT and TO, but by showing how the observation and reproduction of this form produce altered, different, and added states in its structure. As we proceed to describe TO and TOV-TRU-TIT we continually add and subtract these superimposed states and structures to supplement the form typical of all TO. We must continually return to the exclusive form that exists in TO. For it is the centering, unique characteristics and convergence on TO and TIT-TRIS-TRIR-TOT-TOF that are not only the retroactive generators that explain and prove TO, their compelling forces in the future reinforce their predominance, power and growth.

Fractals are quanta like steps and levels that provide basic mathematical patterns of form, recombination and development of TO. They are like and unlike fractions, both are divisions, or parts of a whole. Fractals are more specific, the steps, levels or changes that occur in a whole. They are a product of relative totality interacting with absolute totality, in the most general varieties of TO. Fractals help to show and prove that math is not limited to its basic and random kinds of numbers or integers, but consists of higher kind that exists in actual totality, TO. This enables more accurate descriptions and representation of TO, TRU, etc. These help to show and prove the transitory form and function of TO and its varieties and disprove universes and systems that are incomplete totalities.

The direction of patterns fractals take, is produced by highly dynamic of combined mechanisms in the potential development of TO in TOT. One of these is the balances and coordination that successive TO require. Their form follows a priori locations and the laws of distribution, DIL. The great importance of fractals is the role they play in changing geometry from absolute totality of the physical universe to relative totality of organic life and evolution, in TO.

A person, as a constituent and most specific operant of fractal patterning are, when and by their actual pure essences activities in life an increasingly active factor of benefit to TRU-TIV-TO and future TOT. Not only do fractals and the different kinds of patterns of division explain the complex form and function, formation and transformations, and its varieties, they help to explain and prove that TO-TRU-TIT-TOT is correct, and how by practical and functional value, or actual pure essences, their representation and presentation should be right and can be perfected.

Once we have a definite and accurate object of the form of TO, with correct and accurate dimensions, it can be treated as a fixed closed set and can be applied for simulation and modeling. This is by the Law of the Absolute Combined Fixed Set, LACOFS. Such a law is valid only to the degree the limits and uses are correct and controlled. For it is largely relatively fixed, and the principles of set theory need to be applied. Yet it is by this law that we can both realize the VAPO of TO and its form, proceed with its representation, MOT, SPALT, SPAT and others, and find success in their proofs, presentation and guide in life.

The relationship between two opposing ends, is a dualism. It is the duality of a continuum, like that between mass and energy, or object and subject. It infers, and requires, a set, or practically closed set to move, function or operate. This ultimately applies to the totality or universe, TO and TRU-TIT, in which it exists. It also requires instantaneous attention, IA, that is stable, a fixed relationship between its subjectivity and objectivity extremes. This is typical of the extreme specific variety, TRIS-TRIR. This state of focus in totality is the Law of the Pivotal Proof of TO, (LAPIVOT). This law is only limited by the finite power available and the state of the points being fixed and immoveable. Herein TO, by TRU-TIT, is increasingly set, its actuality expands, human opportunity is unlimited, and hope springs eternal. With the actual pure essentials, functional values, practically of fixed and immovable focus or center, and the closed set of dimensions formed coordinately in correct proportions, the requirements are met for representing TO and TRU-TIT, and SPALT are produced. It is how to accumulate, account, order, and position the concepts, principles, and laws of TO and how they apply that justifies SPALT and proves TO-TOV. When proofs are made by deposition or falling into place, and in relation to other parts and the whole, and they are positioned in the frame, as in SPALT, total proof of TO-TRU becomes much more convincing.

Equilibrium by combined categories and constituents of TO provide for its existence, knowledge and proof. Concepts of equilibrium greatly add to validity, by expanding dynamic coordination and complexity in TO. It is from the most fundamental form, dimensions, universals and dualities that equilibrium is based. We present many concepts of equilibrium to emphasize the various kinds of their most critical mechanisms in TO, TOV, TRU and TO-TOF. They are a product of relation, actuation, proportion, totality, coordination, and the continuity, thereby survival, of TO-TOV-TRU-TOT. When by their own actual pure essences and those of actual totality the physical and relative universes combine, are completed and corrected in TO, their form is manifested by SPALT. The physical universe supplies much of its mlt form, the relative universe its ALT form. As TO by TOT is transitory, SPALT changes similarly. Yet for practical purposes and those of demonstration, by the actual pure essences of TO and the stability of its mechanics TO is represented and presented as a closed set..

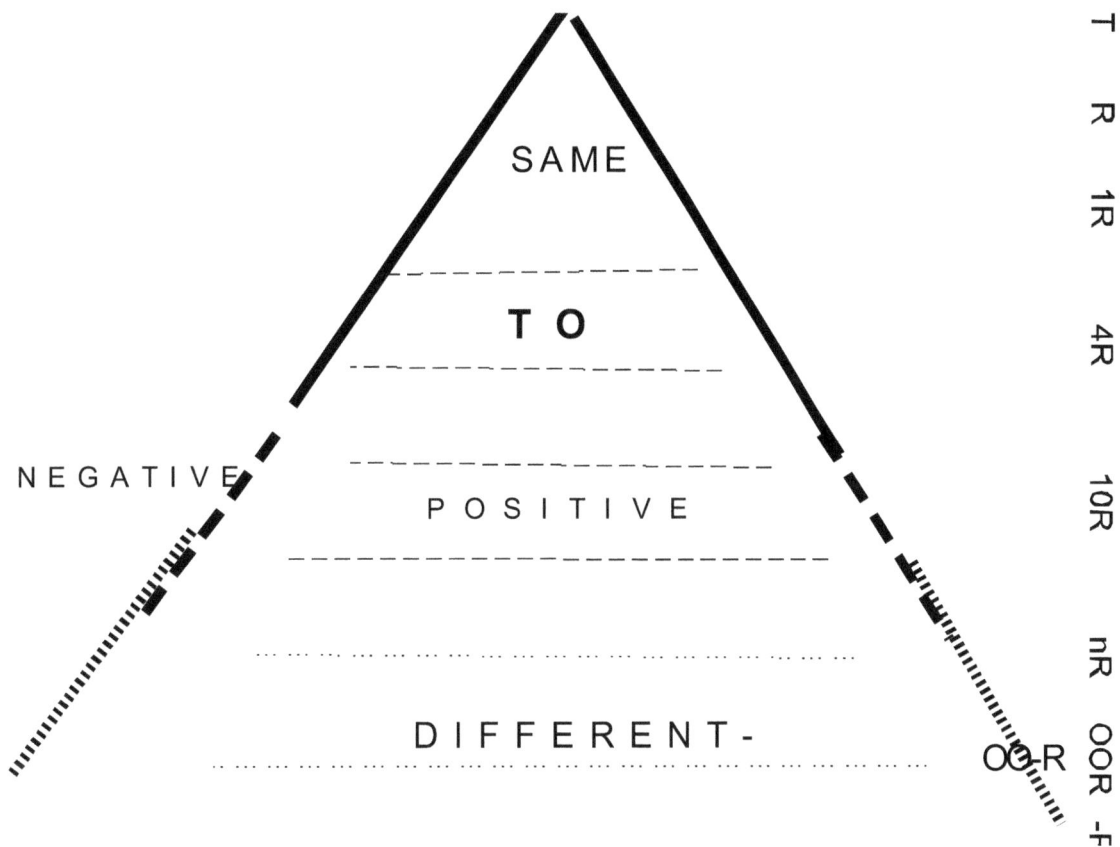

Fig 23-2 Vertical integration and Geometry of TO

An external view of expanded REM not only provides a clear idea of values in TO. It exemplifies the importance of vertical integration and the power of geometric shapes. The great significance of Fig. 1-23 is to demonstrate two things. 1. The use of geometric figures and use of proportions throughout TO,

2. That vertical integration can be of any orientation, Fig 22-2, as long as it shows the importance of the integral, from one, to differential and infinitesimal in proper order and sequence. 3. A standard that is vertical gives the formal basis for integration.

Identity comes from unity, the unity of relativity systematically produced by the conversion and fusion of the commonality of all that is greatest in TO. It is the unity of our predominance, the Á mass energy state of TO and by their FAF in ideas and mechanisms of the world and God. This is how all that comes before us, by the interaction and relativity in the world of being an observer of all observed, comes to have more or less meaning to us. The greater this meaning is, the greater and more these same values, ideas, forms, aspects, parts and processes of TO are, and the more we and TO are. There is unity at the top, in identity and sameness. As we pass from infinite R to 10R to 6R to 4R, 3R, 2R and 1R, the values and REM become more, the same. They must, by definition, be increasing proportions of TO. At the bottom, oo-R to n-R, there are only infinite or unidentifiable finites without much relationship or significance to each other, the whole, and without coordination except in their tiny posits relative to SPALT and TO. Yet they are in TO, when combined with all else that is different and differential, much of TOV. Meaning and reality, is largely the power of our IA, guided by the Á predominant FAF, focused upon the ES, CES, and REM of TO.

Because humans cannot achieve absolute certainty, should not be defeating, it should be challenging and stimulating. It provides the imperative differentiation, measures, probability, certainty, and continuity evident in VAPO, the method of valid, actual, proportional and total approach to TÓ and TOV. Proof of the comprehensibility and practicality of all, by TO, is by its systematic unity, order, coordination and form through representation, simulation, modeling, positioning, operations and applications to life. Continued refinement and perfection will be forthcoming when we have the proper recognition, interest, purpose, will, plan, accounting, accentuation and action. What needs to be expressed and held in mind are the different probabilities of strict, and actual pure essences, of identity with all that enables best human life with all that enables TO.

REM as Centers or Powers in TO, (REMCAP), are the REM and values that inextricably help create and form TO, (REMCÄT). REMCAP and REMCAT show how TO is more formed by predominant powers that make it stronger, more stable, different, and exclusive, from all else. It proves Á, PRA and TO, PRUT, by showing how values and REM cannot be answered in any way, as only physically, but as they are typically, specifically and exclusively all in TO.   Such a sense of values depends on the frame or geometry on which it is based. It is this frame and geometry that are basic to the form and design of relative totality, and especially the relative universe of best human life. How this geometry grew from the relativity of the physical universe and cam to unite with the physical of the relative universe is the key to understanding the creation and evolution in relative totality.

Geometry and the form of entities and structures have many similarities from the smallest subatomic particles to that of universes and totalities. From quarks through the crystalline structure of chemicals, to the genetic form in life, form of the physical body, engineering in modern technology and construction, form of the physical and relative universes, to the form of absolute and relative totalities and actual totality there is a continuum and spectrum of form and design which determines much in each stage and the whole. The critical role such form, function and other similar and typical factors play in the origin and development of life becomes evident when we compare and describe the factors that determine each step critical of each variety of absolute and relative totalities in TO.

CHAPTER 24 - Physical, Inorganic and Organic CHEMISTRY of LIFE, to Animal Life in TO

Elements, Ionization, Differentiation, mlt, Mass Energy, Geometric Design and Changes, Creation, Recombination, Protein Synthesis, Replication, Viability, Biology, Regeneration, Reproduction, Transformations, Transitions in Time, Inorganic to Organic, multiplicity, lateral, reversible, forward and progressive development.

Life, as TO, did not originate from nothing, nor one thing, it was created by accumulated nidus of proactive inorganic to organic components that functionally assimilated all necessary for existence. The assimilation, along with elimination were continual, often mechanisms and massive processes, as evolution by natural selection that was, have been, is and will continue indefinitely. Only each stage in the transition, and transformations, assimilates, eliminates and occurs in its own way, typical of itself in TO of that point in time. TO is not to be thought of as an all or nothing, beginning or ending, of life. TO is the sum total of all that exists in those ways typical of it. When taken in any segment or piece of itself, it must be quickly realized and corrected to be limited to this less than the whole. Also, it is valid in our considerations of creation and evolution. Their part and role in TO are large and continuing in the whole, without a simple and clear beginning, or simple and confined process, in any small sector of the development of TO. When applied in their larger, yet proper context in the whole of TO, creation and evolution become, as all that exists in TO, TOV and TO-TOF, their typical for each, rightful place and part in the whole.

From the ionic bonds holding chemical molecules together, to the absolute love that bonds each and all persons in TO, there are increasing powers of relation and relative totality. The effect of these powers is to bring the actual pure essences of absolute totality and relative totality into their proper and ordered part and role in, and proof of TO. All tends to converge on those features and powers that produce the whole. By this massive, lateral and longitudinal total convergence the progression, often by quantum steps and fractals tends to what is practical and works best, actual pure essences, and these to their state in TO. The basis for life arose from the states. conditions, relations and limits of the positive and negative, geometry, ionization and creation of matter, and its various kinds and forms in absolute totality and the physical universe. Temperature defines the various forms of matter, solid, liquid and gas. Temperature exists from absolute zero, or most solid and static, to absolute infinite, or most gaseous and dynamic. Between these extremes there are an optimal degree and range of temperature that typifies various kinds and stages of life, TOT. TO and TOV. From the smallest subatomic particles to chemicals to the form of matter we experience in nature, TO partakes and is composed by the degree all their actual pure essences become TO. In this way matter exists not only as particles, but as photons, subatomic particles, electrons and, chemicals, molecules, early life forms and all that the environment and life unite to form in TO. It is no surprise that the tiniest particles of matter are so fleeting and uncertain. They do not exist as perfect solids at zero absolute temperature. Only with increasing energy do they relate and interact with all else in TO. So is it with succeeding forms from ABSTO to RELTO in TO proactively combining by smaller or larger steps to produce each new form.

The creative powers, propelling and compelling forces in actual totality, are not all predetermined nor undetermined. They are like actual totality, affected by transition and transformation. They are a part of each stage in the transition. When actual totality exists in its proper state in the extended present, creation is highly predetermined by the form and function of actual totality. When actual totality in the far term past and future is in a high level of transition it will be less predetermined by its proper state, and be undetermined within and more determined by extraneous powers and forces, and those of potential TO, TOT, TOP-TOF. The origin and development of life are not so difficult to understand, or prove TO, when we consider all the factors and forces that caused it, and the steps, levels and stages that occurred. One of the major contributors was the electromagnetic spectrum. This was evident in DIMFOR, Fig 3-18.

The mass energy dimension that moves DIMFOR from a field to a cubical form is not only typical of the physical universe, but assumes many of the steps the EMS's spectrum takes in the development of TO.   When we describe mechanical features we are not only dealing with the physical forces involved, many other forces typical of TO and TOV participate in the mass energy of life. The electromagnetic and other higher stages of mass energy are needed to complete the whole of relative totality in TO. This is evident in its alteration of the form of mass, length ant time. Likewise many other continua of TO, not only exist in TO, but have been altered in its development. Each continuum, whose spectrum and gradients are active participants in TO-RELTO and their creation and development, need to be applied if we are to fully account for the progression from physical, inorganic and organic chemistry and on up the scale of life to the highest stages of relative totality and specific varieties of TO.   The positive negative continuum is a good example. It has been present throughout the development of actual totality. Only its form has changed with succeeding stages. Many of the other continua, whose spectra and forms changed with time and development, when collectively summed describe much of the creation and evolution in TO. Their forces were active, retroactive, and proactive throughout development. They tended to operate together in an interactively bonding way to produce what adapted and worked best. Once the origin, creation, evolution and development of TO are known, much of the rest of TO is proven. Science has continually added to the evidence of form, function and origins needed. Religion and philosophy have helped to focus and hold all actual totality to its relative totality and unity. The rest of knowledge has provided many of the connecting features, ideas and facts that fill out the whole of development of TO.

The development of TO has followed the history of the physical world, earth, life, animals and humans in a way that exists proportionately in TO as TO changed from TOT. This occurred from the big bang, to earth, to oxygen and water on earth, to life, plants and grass lands, animals, changes in anatomy and physiology, dispersion of humans on earth and all subsequent changes, abilities and techniques that gave mankind the powers that now characterize TO.   The same creative factors, forces or powers that contrast the astrological universe within and without the big bang are similar to those that propelled and compelled the development of the mass energy of chemical elements in absolute totality and the origins of conceptual elements in relative totality that led to their combination and union in TO.

The presence, and variance, of symmetry and proportions in the best physical conditions for replications are major modes in the initiation and creation of life on earth. They typify relative totality and are affected by change with deformations. Deformations in the dynamics of potential actual totality like mutations in heredity, may or may not be reversible, or cause gain and loss. When of gains they lead to predominance, greater existence, continuity, permanence and recombinations. These gains, especially when sustained, is a basic primary mechanism in relative totality, the creation of life, in, and proof of, actual totality. This is the Law of Symmetry, Proportions, form and deformation in Actual Totality, LASPADT.

The ideal conditions on earth for initiation or creation of life were not static. They were both highly dynamic and variant in the life form and in the environment. Different environments and life forms facilitated more or less diversity. Many factors were involved including temperature, pressure, volume and chemical composition. As these became more favorable for life, propelling forces of life helped to facilitate and evolve its existence. This is what led, by natural selection and other factors, to progressive, and later compelling forces within its own existence. It tended to accelerate the kind of relative totality and altered absolute totality that exists in present actual totality. 25 Organic Differentiation, mlt changes - In past development of totality there was a gradual, although somewhat variable, change in the mode and scale of mlt. Varieties of TO, from generic to specific, replicate these changes or transformations. TO in its own typical way includes all levels, in the continua absolute to relative totality, of its varieties. An example is the development of mlt, ALT and SPALT by entasis, or by almost imperceptible increments and changes in form and its observation. Many mechanisms and patterns such as fractals explain the transformations and development of form and function that have occurred in the progression of TOT to TÓ and their future. Some are more subjective and some more objective, with all occurring by actual pure essences. It is by fractals that much of the replication, recombination, reproduction and convergence of actual patterns occurred. Each was a creative step that eventually led to, and is a part of, the magnanimity that is and proves TO.

Our actuality is neither zero nor the absolute, it ranges from the infinite to the infinitesimal in such a way that produces a medium, a set level of finity. It is both the relative convergence of the finite in the middle and te limits of TO that produce stability. This is of the present, and more or less set in permanence by practical actual pure essences of TO.Perspective and Systematic Unity in Totality, of TO, SET, cf., Figs. 7-2 and 15-4. SETLAW is the state all find themselves in when set or fixed in any position or way between the infinitesimal and infinity in relationships, identity, perspective, observation and objectivity. This is most critical in the special varieties of TO where existence seeks closure and continuity. It will be variable to the extent conditions are variable. Yet it will be fixed, thereby set, to the extent they are fixed between the infinite and the infinitesimal. It is the collective finite as one, shown by the center. SETLAW is what make all that is relative a separate unit unto itself. It is evident in how we are so certain we exist, are as one with what is, our words, beliefs, views of the worlds, and objects before us. Yet these by SETLAW are highly limited and must have their appropriate modes, forms, functions, states, conditions, relations and limits, SCARLS, when existent, applied and operated in life, as all together they constitute the whole, TO.

As mass is a function of frequency so is life, as TO, set by its finite quantity. As TIT in TOV of TO approaches zero, its infinitesimal state loses finity, as does its infinite state when approaching the absolute. The contraction of TO in the extreme ABSTO-RELTO of TO is what establishes the relative construct that originates and creates TO. The finite, is a collective one, set in each stage of TOV, a quantified mode of extension and measure. By quantity the finite and gradients of the spectrum of continua and dimensions can be provided. The distinctions these terms and means provide illustrate the ontological and theological proofs of God as TREQ, TO Equivalents, and TO when by actual pure essences adequately limited and qualified.

The progression toward actual totality, its varieties and life were determined by both random occurrences and coincidental factors, continually accompanied by the convergent powers of absolute-relative and mass energy subjectivity objectivity set state. The set state of TO is largely a result of the mechanics, momentum, redundancies in, and inertia of its form and function. ACEXIN is the combined powers and state of actuation, extension and intension of the more specific varieties of TO. It indicates, shows and proves the make up, and levels, of the mass energy, Á continuum, when completed by the spaciotemporal frame of TO. Together they affix and establish the structure of TO and TOV that is available for representation and modulation. In TRU, TIT, TRIS and TRIR ACEXIN is preeminent, although like much in TO it permeates the whole, only in quantities that correspond to ABSTO-RELTO, generic to specific varieties.

Form largely follows function in the critical earliest and smallest phases of the creation and development of life and actual totality. The ability to generate and reproduce occurred from the simplest beginnings of duplication of organic matter and the definite focusing of relative totality. The reproduction and generation of the succeeding more complex organisms that followed established a continuum, whose spectrum consisted of the most singular and simple to most profuse and complex capacity for the great series of reproductive generation that has been the basis for life.

By the conditions and forces at the inorganic to organic intersection that gradually by increments or larger steps established taxis, tropisms, powers of recombination and duplication that life began. It was not in one simple step, but the combined state of many factors, forces and steps. As these powers of replication grew they became more and more self determining. In the earliest stages such powers of recombination and duplication were more chemical than biological. With natural ahd self propelling protein synthesis these processes became more biological.   An easy proof and demonstration of TO, in contrast with the physical universe, is to list all that is of greatest value, especially what does not exist in the physical universe. That without which various universes, worlds, gods, systems, is deficient and defective as the ultimate unified totality including both absolute and relative totality. Amino acids and protein synthesis were the building blocks for genetics and the mechanism of recombination that typifies life. Interacting with many other factors and occurrences, genetic material developed by coincidental and more propelling forces in each stage of needed existence.

Replication, Life, Biology, Viability - As protein synthesis and recombination improved they achieved increasing levels of replication, the ability to duplicate prior life. This is the beginning of viability or what is vital. Vital is a higher form of necessary existence whose essential actuality is increasingly self perpetuating. It is that essential for proof of the relative, special varieties, biochemical, biological and human species part, role and presence in TO. With increasing powers of viability life acquired a greater positive imperative. The collective state of all that exists in and sustains life, or their various kinds of propelling, and compelling forces. Largely by the powers of relative totality and it's positive imperative advanced, as special varieties grew and developed the forms and functions, plans, purposes and will to live. This enabled subjectivity with objectivity to became the reality that proves the mass energy continua, Á, and thus the form and existence that led through TOT to TÓ.

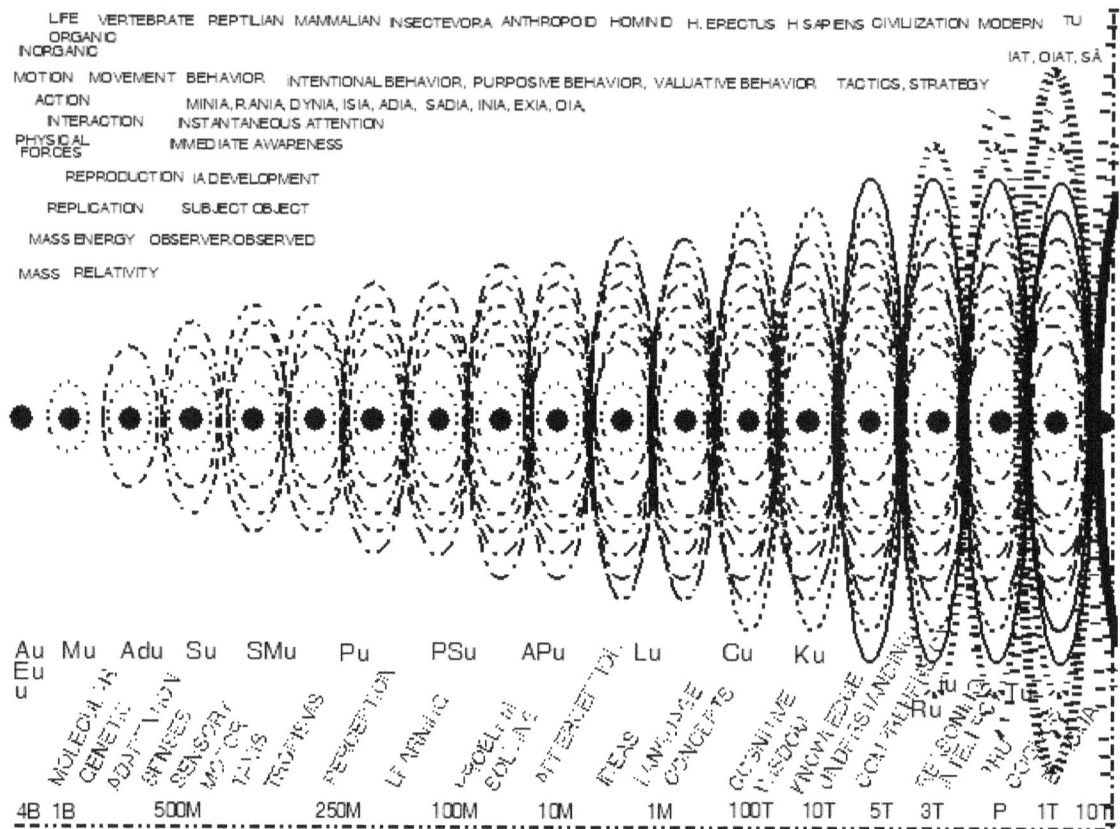

LIFE   VERTEBRATE   REPTILIAN   MAMMALIAN   INSECTEVORA   ANTHROPOID   HOMINID   H. ERECTUS   H SAPIENS   CIVILIZATION   MODERN   TU
ORGANIC
INORGANIC   IAT, OIAT, SÅ

MOTION   MOVEMENT   BEHAVIOR   INTENTIONAL BEHAVIOR, PURPOSIVE BEHAVIOR, VALUATIVE BEHAVIOR   TACTICS, STRATEGY
ACTION   MINIA, RANIA, DYNIA, ISIA, ADIA, SADIA, INIA, EXIA, OIA,
INTERACTION   INSTANTANEOUS ATTENTION
PHYSICAL   IMMEDIATE AWARENESS
FORCES

REPRODUCTION   IA DEVELOPMENT

REPLICATION   SUBJECT OBJECT

MASS ENERGY   OBSERVER/OBSERVED

MASS   RELATIVITY

Au Mu Adu Su SMu   Pu   PSu   APu   Lu   Cu   Ku
Eu
u
4B 1B   500M   250M   100M   10M   1M   100T   10T   5T   3T   P   1T 10T

Fig 24-2 Mass Energy, Subject Object, Á Continuum, Form and Function, MESOA

The development of human relativity, observer-observed, physical to mental, and object to subject provides the prime dimension and basis for TO, RELTO, TOV and TRU Mass Energy Subject Object Á Dynamics and IA. This is signified by MESOÁ, or it could also be called MASOR, or the mass energy, actuality, subjectivity, objectivity, reality diagram. This diagram provides a good account of the transitions and transformations that preceded and exist in TO, those that exist in TOT, and those that are potential for future TOT. It also gives an indication of the progression of relative totality in TOV of TO. Also, absolutivity has a reciprocal role with relativity in MESOA, and mass energy in TO-TOT. The initial steps of life, as shown in figure 23-2 and 24-2, reveal its powers of recombination, replication and regeneration. They demonstrate and prove actual totality, a convergence of absolute and relative totality. By their autonomy, preservation and especially by their existence in, uses of, and incorporation of their environments, they alter the direction and control of the physical universe to a new course. This is a course they now began which has progressed to the massive and complex world of human life, knowledge, technology, industry and all they incorporate, as a relative totality in actual totality.

Spores were one life form that withstood severe environmental conditions over long periods of time, distances in space and on earth. Spores are collections of organic molecules that have achieved powers or stages of generation, and develop powers of protective enclosure that can remain dormant and prevent alteration or destruction by external physical forces. Whether or how spores were a product of the evolution of life on earth, or were transferred, one or more time, through several billion years of the development of earth, is not certain. Both are plausible, and may have occurred. Spores, or other durable collections of organic chemical-proteins molecular forms not only explain a major step in the transition from inorganic to organic chemistry and life, they suggest much greater continuity of potential actual totality in space time. Spores are of several different classes, largely dependent on their ability to undergo or adapt to change. Spores not only help to explain early life on earth, and the initial development of actual totality, they are important for future adverse occurrences. In the transitions and transformations between TO, TOT, and lesser external forces, many forms, functions and advanced types of, varieties of sporulation are needed to give actual totality continuity and permanence. Many other potential methods of altering life forms to suit change in the far term are the topics of current interest, and future need.

The mass energy continuum is largely and increasingly objective in early ABSTO, and subjective in RELTO of TOV. It is subject-object, in equilibrium in the center of TO. Reality is tantamount to TÓ throughout TOV, to the extent subjective and objective are in equilibrium. To be an actual continuum objective and subjective must be of equal proportions and in equilibrium in TO. This corresponds to the equilibrium of ABSTO-RELTO. Since TOV recapitulates Absto-Relto proportions will only be dependent on the generic to specific variety. As mass-energy becomes the Á continuum and dimension, in the more specific varieties of TO, as in TRU and TIT, this helps to explain the higher levels of Á as a dimension, and its correspondence with the three higher levels of space and time. For the mass energy dimension this becomes, 1. Mass energy combined, 2. Subject-Object or Á, and, 3. the reality of their existence. For mlt of the physical universe their dimensionality becomes, 1. length, time mass energy, 2, area, change, subject-object or Á, and 3. volume, variation and reality.

Regulation, regulatory forces, processes and systems are important determinants of existence, essence, potential and actual totality. Their changes with time, and development at the beginning of life are crucial factors in TO. In the physical universe regulation arises from the interactivity of matter and antimatter whose forces of mass energy in subatomic particles, atomic nuclei and gravity have positive and negative effects that determine form. With the transit from inorganic to organic compounds and their proliferation many new kinds of regulatory processes occurred. Each new kind produced other kinds whose proactive tendencies were directed toward continued recombination and generation. It is the new kinds of regulatory systems evolving in animals and leading to humans that make up much the A dimension of TO. .

Regeneration, form and design - Existence in TO is never completed, nor partial. Only in the equilibrium of ABSO-RELTO and ALT does all find fulfillment in the summation of TO. When this sum is averaged, we can represent as one whole unit in full, as we do in the spherical ALT, SPALT. On this basis, and with this development of TO, much of the form, design, purpose and direction of life as TOT occurred. TO replicates much of this form, as being replicates development of form, or ontogeny replicates phylogeny. This is reversible for development often replicates being, which is a fundamental key to, and law of life. (LARDOB). This state and condition generated by life, and like the reversibility that occurs between TO-TRIR-TOT and TOF is a primary proof of TO. It is the power of reversibility that has propelled absolute totality to become relative and vice versa, as together they sustain their combined totalities, and thus TO.

With developing recombination, replication and regeneration, reproduction resulted in a great variety of life forms. Separately and together these forms of life or types of animals became relative worlds with their amenable surroundings. The collective convergence and effect of groupings and classifications have been highly important factors in the progression toward life, TOT and TO. An example of this is organicism, the principle that life functions as a whole. It, is not all, for only the whole itself, is complete. Yet its life functions have an effect that guides the patterns that have made the progression. The ends do not always justify the means, but they are highly important in creating, directing and controlling conditions for life. Ends are largely controlled by the selective mechanisms of the actual pure essences of each kind of animal or type of relative totality. This is seen in TRIR and its reversibility affect on TOF. All that created TO contributes, in a variety of ways.

The many steps advancing complexity and variety of life takes are largely by transformations. There have been many factors and mechanisms that contributed to the transformations and transitions from an inorganic, to an organic world, thence to TOT and TO. The most fundamental of these were major determinants, of which, existence, potential to actual, necessary and essential, and their actual pure essences have been foremost. Many specific key mechanisms also explain various levels of transition and transformation. These include, alteration, adaptation, adequation, facility, redundancies, allowances, level of tolerance, and the continually and massively occurring forces determining direction in action reaction.

Transition and Transformation in TOT are different by the fact that transition produces a more or less new, like aseity, entity. Whereas transformation, more like haecceity and quiddity is change in form, be it of the whole or a part. Much of the entity or life form may remain the same. The creation and development of life were by transition. Transition was more prominent in creation, transformation in the development and evolution of TO. Simultaneously what largely propels life, and TO, is stability of form and function that is maintained. It has been called the life chain, a-priori to posteriori forces and factors that sustain totality, or TO.

A precedent in religion exists for the forms, functions and states that drove the development of TOT and TO. The teleological argument is used to prove the existence of God, which when correctly qualified is also a proof of TO. The evidence for purpose in nature and the world suggests that more than a material, random or random process is occurring. Not only does an object, best human life, RELTO or TO, retain their motion, but the self-sustaining momentum of this motion tends to qualitative and quantitative expansion. For relation, life, TRU and TO this is an expansion of self perfection and fulfillment. The teleological argument also works in TIT-TRU-TOV-TO by need, essence, selection, quality, direction, inertia and momentum and others. Similarly it works by the reversibility, TRIR of TIT in TO-TOF, where all forms, designs and purposes are reconciled in equilibrium, predominance and ultimate absorption necessary for continuous existence. The tie that binds absolute and relative, the physiccal universe and that of best human life and mind is the role of relativity in the physical universe and subjectivity in the human and mental. Special relativity in ABSTO does this through the observer and observed bond. How the physical universe can not exist without being observed, or made relevant. This is how the specific levels of existence are essential to fill all the varieties of actual totality.

Like dark matter and dark energy, the static and dynamic, parts and processes, particulate and functional of TOT and TO, exist in universal relation and interaction. Their static form is readily represented, their dynamics often lacks being named, and needs description. So has human knowledge much more slowly achieved the degree of certainty and mastery of the heat, temperature, dynamic, functional and proportional of all in TO. Only when form and function correctly share their existence in TO can its knowledge be complete.

General relativity affects the bond between absolute and relative through those geometric determinants that tend to more or less progressively converge on life. Like the form of fractals they adapt and alter that form which most fits the occurrence, With an end result inclined toward more and more self determinant form and function. It is these two kinds of relativity working like a duality that creates and evolves life, animals and mankind. They share this determination with many other features and forces such as the absolutivity of RELTO, and those processes that favor a reality, design, development, growth and past to future interdependence. It is a dependence in which a less definite necessary existent or actuality more or less, but progressively achieves more definite autonomy. It is this autonomy, or at once tendency toward exclusiveness and permanence that characterizes the highest level of definite actuality typical of actual totality.

From the focus of actual totality this has been a very long progression, so long that its origins are readily lost sight of. From the focus of no actuality, as in the physical universe, other worlds, God, philosophical systems, etc. it is also lost sight of, unless the universality and moderation of the whole spectrum that makes us actual totality are constantly kept in mind. This universal moderation is the kind of overview, belief and agenda all must have if they are to master, apply and solve all the problems that vertical integration by actual totality makes possible. It is based on the Law of Relativity, Absolutivity, and all the other features and forces that enable creation and evolution of the most definite actuality, that is TO. LARFECT.

V Proofs of Evolution, Transition and Knowledge in Time of Actual Totality

CHAPTER 25 - ANIMAL LIFE, Zoology, Mammals, Primates to Humans, Civilization and Actual Totality

Genetics, Genetic Material, Live Product and Determining Forces, Heredity, Phylogeny, Species, Adaptation, Evolution, Natural Selection, Stimulus-Response, Sensory Motor Systems, Ontogeny, Ancestry-Progeny, Autonomy, Preservation, Predominance, Expanding relation and best life.

Although life in general and plant life, are large contributors to TÓ, relative totality didn't acquire many of the typical features and forms of TÓ until, and with the advent, of each succeeding predecessor of humans. These not only characterize their key forms, functions and existence by actual pure essence, their convergence proves actual totality. Common, classical and historical tradition more or less increasingly focused on the expanding relation of quality and best life. First there was unbounded chaos, then cosmos, Gaia, mother earth, Kronos or time, then the Olympian Gods, Zeus and Gods made in the image of man. With Plato and Christ came ideas and essences, converging on God and the rational practice of religion. Following Aristotle there was a great diversity in the fields of knowledge, philosophy, science, and their practical uses in life. Experience through the ages has added and identified with these and many other great added aspects of totality. Yet current knowledge and its organization are deficient and insufficient for actual totality, and to meet the need of life in such a rapidly transforming world. To do so we must better use the powers our ancestors have given their progeny, separate what is of the highest quality from all else, order it by the form and function of totality proper, TO, perfect it and apply it to a comprehensive well ordered, managed and controlled endeavor for a permanent future. It is the relative position of life, man and mind in the chaos of the physical universe and all less than their role in TO that actualize and produce the actual pure essences that separate and prove actual totality, TO, from all else.

Actual totality was not created in one simple stroke from the physical to the relative universe. In fact the way it was created is similar, only in reverse, to the way To undergoes change in the future, i.e. transitions described in chapters 27 and 28. Since these changes, increasingly greater the further in the past they occurred, were less and less actual, but more and more only potential for TO. TO of the present exists as ATO, at once autonomous being. The forces and states that created TO in this potential to actual necessary existence largely exist within TO at present, only in somewhat different forms. This is called the, Law of Potential and Actual Incremental States and Forces of Creation in Past and Future TO, LAPSCIT. It was the creative force and powers of the advancing complexity by actual pure essences in the most, to less general, absolute varieties of TO that was the great mechanism of union in ARTO that TO developed, is proven and made most certain. TO, although centered on here and now with its own mass energy, consists of much of the mass, length time to ALT of the past and future in TO. It is this complex form and associated power of creative impulses past to future within TO that by example and their state in TO, provides its most mechanisms of progression and certain proof

Origin and development of animal life occurred with the differentiation of lower life forms to one cell organism, with special characteristics, With expansion to two and multi celled invertebrates the ability of stimulus and response in increased forms of mass energy, adaptation, and subjectivity-objectivity were initiated and expanded. The beginnings of TOT or potential, and TO, actual totality, grew, expanded and become more evident. Much of this progression was a product of the form, function, determining forces and purposes of advancing life. Natural and normal are general and important concepts in language, life and actual totality. Natural and normal are highly determinative of actual totality, depending on their function, and how they are used. Natural and normal for what is actual totality can be the opposite of what is unnatural and abnormal for what is not, or negative, for actual totality. Natural is a more scientific usage in which what exists or occurs is a result of physical to biological processes. Normal is a more mathematical and statistical usage in what exists or occurs is a result of variation from a norm or standard. When their use is strictly limited to actual totality, they are very helpful in themselves, and to distinguish those forces that are deviant, unnatural and abnormal for actual totality. It is part of both the variations and interactions within actual totality, and those that are less, lacking, or external to actual totality. The determining forces of natural selection are much of the progressive development of life, heredity, and best human life in actual totality.

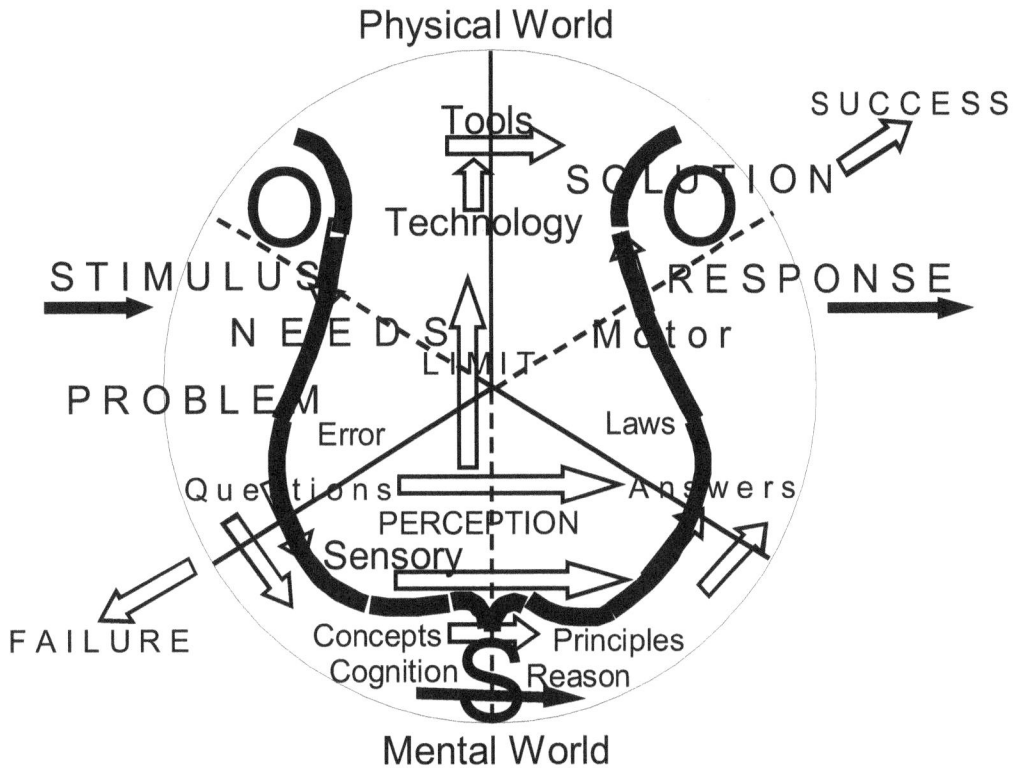

Fig 25-1 Object-subject-object in TOT-TO

This diagram emphasizes the fundamental object by subject to object features of developing TO and the accompanying motor sensory form needed in stimulus response and problem solution dynamic relationships. It helps to show how in the animal worlds, function follows form, as form follows function, working interactively on the subject object basis of mass energy. We can't look at only one aspect of our world and expect to understand it. Nor can we understand it adequately without providing some order and frame for these relationships. These are superimposed in TO and show how the S-O dimension becomes the predominant mass-energy state of TO, TO's mass dimension A. It also shows the methods of knowing and acting in the equilibrium of/in TO. Any mass that works dynamically in space time seeks to maintain equilibrium. The relationship and operation of dimensions as coordinates become altered externally is the Principle of Coordinate Transposition, (POCT).

The ever higher state of ontogeny and autonomy of actual totality by the convergent sum of all its actual pure essences in the here and now, is the power that both unites and gives autonomy to TÓ. It reveals by its interaction with all else, especially TOT, its continued evolutionary transformations and transitions in time. Natural selection has increasingly been the major determinant of transformations and change in TO. The core of animal life and this best of worlds of human life in TO is its live product, genetic material and, determining forces. In the whole they work in a highly interactive interdependent way to provide natural selection, survival and are key mechanisms in the purpose of life. The progression is life's eternal continuity. This is centered on here and now a product of convergence of RELTO, especially its most special varieties, a centering that produces, an At Once Autonomous Being, or ATO, in TIT and TRIS, TIR, TINI, TRA and TRIR, the focal purpose of life. Their dynamics and interactions give us a preliminary yet highly fit way of understanding and proving much of the world, and their part and role in actual totality, TO.

The progression of animal life on earth from the most primitive one-celled invertebrates to highest primates is well documented in biological science. What needs to be stressed are the development, form and function of the central nervous system and brain, especially how physical qualities became transformed into relative qualities in TO. Figs. 24-2, 1 and 25-2 are beautiful examples of the series of transformations from mass in the physical world to the most specific mass energy continuum and dimension, Á, in TRUS, TRU, TRIS, TIT and their representation, SPALT. It helps to show how the actual specific dimensional form is predominant that proves the hierarchy and hegemony of TRU in TOV and TO. With the spatial and temporal, the evolved actual pure essences in the mass energy dimension of TO, replicate the accurate coordinated form typical of TO, showed by SPALT. Dimensions are centered on this essential mass energy, space and time in here and now. It is not so for all of TO, but all that the specific of RELTO produce in TO. It has been the great expansion of the cerebral tracts and cortex driven by the accumulative effect of ideas, words, language, knowledge and reason that have led to changes in mass energy, Á, that most typify TRU, the specific varieties, RELTO, and much of TO. The long term continual dissemination and expansion of animal life, or evolution, produced a massive variety of life forms, classified in decreasing generality by phylogeny to that most typical of each kind of animal, a specie. Each specie was largely a self preserving reproductive group, whose tendency was to maintain or advance its own kind. This build up of self-sustaining animals was a major step in the autonomy and exclusiveness that established and proves actual totality, TO, and culminated in its formal consolidation by best human life.

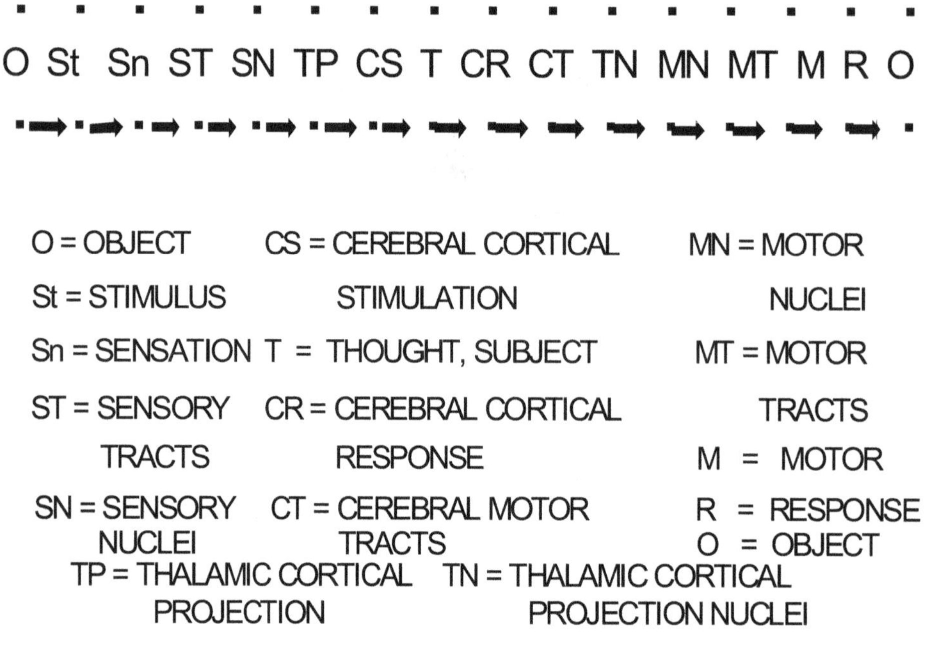

O St Sn ST SN TP CS T CR CT TN MN MT M R O

O = OBJECT   CS = CEREBRAL CORTICAL   MN = MOTOR
St = STIMULUS   STIMULATION   NUCLEI
Sn = SENSATION T = THOUGHT, SUBJECT   MT = MOTOR
ST = SENSORY   CR = CEREBRAL CORTICAL   TRACTS
     TRACTS   RESPONSE   M = MOTOR
SN = SENSORY   CT = CEREBRAL MOTOR   R = RESPONSE
     NUCLEI   TRACTS   O = OBJECT
TP = THALAMIC CORTICAL   TN = THALAMIC CORTICAL
     PROJECTION   PROJECTION NUCLEI

Fig 25-2 Sensory Motor Tracts in the Nervous System, Basic Mass Energy S-R-O Continuum in TO

It is knowledge of the anatomy and physiology of the nervous systems that provides the most evidence for the dependence and interaction of sensory - motor basis for object-reality-subject in mass energy. It traces the biological development and initiates the proof of the role of mass energy, as subject and object, in TOT-TO-TOV-TRU-TOT. How ontology recapitulates phylogeny, especially in the development of the central nervous system, is one of the greatest examples of the transition from the physical world to the predominant role of human mentality in ABSTO-RELTO, and their proof of actual totality, TO.The nervous system consists of the central, shown above, and the autonomic. The autonomic nervous system is divided into sympathetic and parasympathetic systems, with their fibers and ganglia. These two systems, being autonomic, are largely subconscious, yet of great importance in maintaining equilibrium in many body functions. This is done by their opposing functions to allow different organs to vary their function with need. This exemplifies TO in general. How TO explains and proves the fusion, coexistence and existence of combined absolute totality and relative totality is critical to its understanding and proof. This is the Law of Relative and Absolute Totality in the Neuromuscular Form and Function of Actual Totality, or LARNMAT.

What is special is one of the greatest explanations and proofs of TO. This is most evident in the reproductive biological exclusion by species as a breeding unit. Although not absolute, this exclusion provides separation and delineation of what is, what does, and what does not, not have actual or necessary existence. The Law of Species Unity, LASPU, holds that it is more the depreciation of the massive equilibrium of the whole, rather than capacity or some element alone, that produces extinction. The Law of Relative Speciation, LATRASP, signifies how speciation consists of many gradations or levels. It is made up of not only individual remembers, or even the specie itself, but the totality it relates to or brings to itself. And even this is special in the sense that what relates to it or brings to itself is so only as it selectively exists and participates in the whole.   The Law of Incorporation of Speciation, LINCORS, certifies the fact that a specie may be an individual animal biologically, but as the whole specie, and this as it is TO, is different. Any constituent must be made up not only of that which is its center, its relativity, its frame, or major mechanisms and processes, but all that enables it to be and this in the right proportion. This functional aspect of LINCORS is the Law of Enablement and Enhancement dynamics of speciation, LENDS, which certifies the fact that TO exists only to the degree all that enhances or enables it to exist is included, and included to the typical extent. This feature, makeup and form of TO, TÄ, which is focal to its being and speciation, and which is the core and nucleus of actual totality proper, TO, is its this-ness or haecceity. It is what not only externally focuses toward TO. It is what internally focuses upon TO. It is what make it a 'this' and a 'the'. It is what helps to retain stability and permanence. The functional aspect of this is the enhancement and enablement dynamics of increasing levels of speciation.

As animals and humans attended to and became increasingly aware of their existence, surroundings and what was needed and essential they selectively sought order and unity in their relations. With increasing life spans and experience familiarity led to being at home with environments that supplied their needs and provide for their well being. These relations, diversity and development of capacity led to an accelerated physical and mental proficiency with a relatively rapid growth of tactics and skills. The incorporation or embodiment of TO is seen in the capacity, ability, actions, states and conditions, or CARSC, of which it is formed and functions. Evidence and proof of TO these laws add is in the vast inclusion of many extra human, and even mental aspects of TO. It confirms and is confirmed by the physical and relative universes, chemicals, animals, Gods, philosophical systems and much we would neglect to include in TO that are parts are necessary for existence, by LOPAN.

TO continuously duplicates much of its form, as being reproduces development of form, or ontogeny replicates phylogeny, and vv. Their continuously reversed interactions guided by natural selection are what give TO permanence. This state and condition generated by life, and like the reversibility that occurs between TO-TRIR-TOT and TOF is a primary proof of TO. The complex factors in the form and function of TO in time, produce its capacity, ability, and conditions for progression and much of its direction.   By phylogeny we learn the vast expanse of life, we also learn its relentless development in time. It helps us understand the interaction and dependence, not only of the beneficial but in opposition and negative, that competition and natural selection produces for life and increasing capacity and best form. Differentiation and specialization result in many forms and states beneficial to existence. The end result in mammals, primates and man is mentality, the ever higher states of cognition that prove and have, are and will culminate in ever higher and stronger levels of actual totality. A major power that sustains and advances life and species is preservation, a self preservation that is central to all levels from individual to totality. This is most evident, and prominent in each person whose whole life is centered on endeavors that assure continuity of themselves and their kind. Preservation is a power that helps to guarantee, and prove, the autonomy, and interdependence of all in TO.

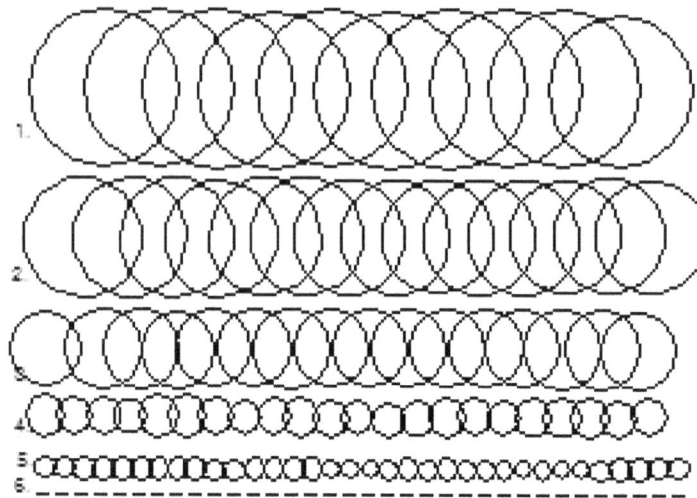

Fig 25-3 Life and Human Generations,

This figure is given, to help the reader picture, the series of progressive gradients that are so much of TO. The different kinds, overlapping, and combined forms. It shows some of the different kinds of generations in life, animals, humans, TOT and TO. # 1-5 pertains to decreasing sizes of population or overall quantity. To what extent is TES only # 5, the single individual from which, in mating, all future generations will be derived? Or is TIT only # 1, the total human population. Or is TIT # 2-4 some combination or variation of these. This results in differences, a flux and generational change in TO and its FAF. The kinds of generations change. This is why we include # 6, a time scale. To be accurate this needs to be more elaborate, more than arithmetic, more than one scale, or even one diagram to depict kinds of generations. There are great mechanisms, forces, powers, maxima and minima in TO, TOV and TOT, with multiple factors and times of effects. Anthropologically they can occur by quantum changes as with Lucy and Eve. Many of these are reversible, e.g., epigenetics. This diagram, like MESOA, shows the continuity of life from its beginnings with relative totality, and extension in future permanence. It shows how generations have both separate existence and increasing levels of superimposition. The superimposition or overlap of generations with each other helps to explain external and internal action-reaction as well as the bonds of all with each other in, and with TO. Not only has life in TO become more complex, the need for unity and totality, and the needs, demands and stresses on existing individuals, increases.

The life of all animals was a total of all dynamic endeavors to sustain and advance themselves and their kind. A great variety of skills developed in many different kinds of animals, many general and many specialized. What was special in the beginning, often became more generalized as the species or genera evolved. This is how humans developed mass energies that supported the union of absolute and relative, and the other modes, continua and qualities that are typical of actual totality, TO. None have been more predominant, and proof of TO, than human mentality, comprehension, reason, and all those ancillary applications that characterize modern man. Another powerful feature of higher animals was a predominance, associated with an acceleration of mentality in its many forms and applications, e.g., tools and technology. The specialized abilities that added a series of quantum leap creations to the ability to use or defend against, and dominate others, other animals, organisms, foods, and the environment led by natural selection to the mastery of life, that is a most typical example and proof of the exclusive unity of TO.

By absolute and relative totalities, all has a kind of existence. If positive and beneficial they become ABSTO and RELTO, increasingly necessary for TO. The relative in contrast to the absolute imparts a series and scale of universes, evident in relativity and the creation and evolution of life. This is mirrored in the varieties of TO. It is why TO not only focuses and converges on humans, but all preceding life more or less participates. This is seen in the progression from AU to RU, to RUT, to TRU. In this way actual totality is not the only RELTO, in TOT and others. This is how animals, or other centers of relative totality, have come and gone, as their predominance waxes and wanes in time. This deviance from TO is not disproving, it only shows the role of TOT and transitions of TO. Yet it also shows and proves the convergence on TO, its predominance and uniquely potential, extensive and powerful form and functions. It shows the great positive and negative continuum between power and dominance versus weakness and vulnerability. It is this continuum of the whole dynamic of TO that is at the crux of, necessary existence of, and proof of TO and its continuity with TOT in TOF.

CHAPTER 26 - HUMAN LIFE in TO, to Transitions in Time

Potential to actual in TO, Homo Sapiens, Man, Mankind, Civilization, Specificity, Consciousness, Sensitivity, Subjectivity, Identity, Ability and Skills, Self Preservation

Human life is not an unremitting gift, its positive imperative dynamics require a high level of return. The spectrum from no to optimal return continually confronts all, from defiance, denial, slight, moderate, a high level to the best sacrifice and service that can be made. When we know TO and all its consistency and their important proofs we can greatly enhance, and make more effectively, our lives in TO. The term mankind for the whole of homo sapiens emphasizes its convergence, autonomy, values and the presence and powers of increasingly specific life. The kindred spirit of best human life in a relative universe, TRU, exists in a less obvious bond between relative and absolute totalities, ARTO in actual totality, TO. The more qualities of TO that are known, the greater is its proof and practical use. The most specific variety of TO, and its carrier, man, has many typical characteristics and properties, including, happiness, predominance, genetic and personal health, strength, selectivity and choice. These and the many other qualities that make up TO, when known, by their states, conditions, limits, relations, actual pure essences and convergent interdependence, give the order and coordination that prove the whole, TO.

The potential to actual development of human life is largely determined by the Principle of Viability, POV, is proportional to the increasingly more specific varieties, TOV, to be the collective and cooperative actual essence of TO. By the principle of viability, and proof of duration and continuity we can reconcile and summarize the part and role all whose separate existence is far less than their essential existence in actual totality, TO. This is most significant in humans, existence external to, or within TO. Herein is the purpose of life, the relative, which cannot exist without the absolute of an objective, or physical universe. By lack of knowledge and a self-centered fixation on ones existence humans and especially all new generations have stubbornly held to a state and condition of a non actual existence. Only when they learn and know TO or TREQ, will they overcome this disability, and be correctly alive.

The accompanying figure (26-1)roughly purports to suggest, show and prove a few important features and facts of TO.

1. All exists both in and out of TO, only in different forms.

2. TO is much more than any one of its constituents.

3. TO and parts are highly interdependent and interactive.

4. All entities and topics exist both separately from, and within TO, only in much different forms and functions.

5. Their positions change with varieties, time and TOT.

6. The difference is largely qualified by actual pure essences.

The figure (26-1_ also reveals what is potential and commonly lost, the order of the actual world, how everything without the presence, relations and contribution of their actual pure essence lacks focus. How by comparing what are external and internal of some of the better-known constituents of totality all can be better understood. How TO is most relative, practical and has stability within, all lack fulfillment, closure and the ability to be represented, as by SPALT, without the right unified totality, TO. How TO, or its varieties, do not exist, nor can be represented without their parts. How the parts of TO exist in set proportions of relative magnitude, from the highest to the lowest or from near absolute unity to near zero diversity. This is the scale from near total through universals, major REM, minor REM, very finite and to infinitesimal. To learn and apply to life the fact that not only the soul but all, more or less, exists outside and inside TO is the beginning of actual wisdom and ones life within TO.

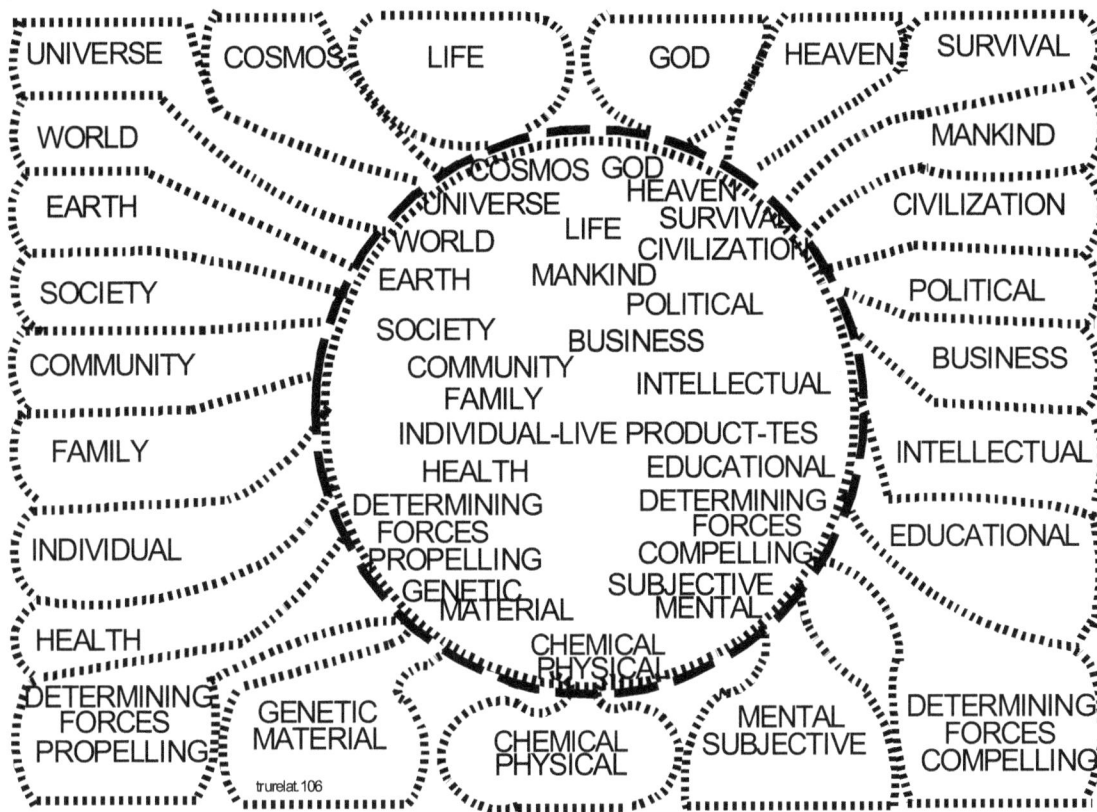

Fig 26-1 External Internal Relations of things, and as participants of TO, TOV and TOT

The survival of humanity depends upon many factors and forces, which work in a great variety of ways, focused on the resultant reaction to determining forces, health, strength, and success of the phenotype or live product, and continuity of the genotype or genetic material or germplasm. Variation and change have already been built into this diagram in its time dimension and the long term roles each plays. This also applies to all the Á dimension. Many more forces, factors, concepts, principles, laws, and mechanisms can be added. Yet too much detail causes a loss of overall comprehension of TRU, TOV and TO. This is the Law of Proof and Practice Over Detail, LAPOD. It holds primarily in the acquisition of universal knowledge, e.g., TO.

Genetics - the wholes of genetic material, live product and determining relative convergent forces developed together and function more or less as a major system in totality. Load versus capacity, damages to genetics, especially in recombination, how much can it handle, how much are animals and humans limited and liable for depreciation? Many causes for death and extinction, must be learned and applied if prevention and treatment are to work, and TO is to be and advance. Spread sheets and interactive diagrams can provide greater detail. The species, homo sapiens, centering in the definite here and now of TIT-TRU and special varieties of TOV in RELTO of TO, is the Law of Speciation, (LES). The Species, homo sapiens, signifies man typified by wisdom, the actuality of being human and the actuality of mentality. When we specify "the," the sense, the case at hand, exclusive thing, or whatever is given, it becomes the definite predominant form of much that is of the special varieties of TO and their unique, peculiar and exclusive existence, separate from all others. To the degree this is so for, or of, TO it is an added fact of being all inclusive, and all exclusive, of all less than, not, or negative to, TÓ. It is the this-ness, or haecceity, that provides a fundamental determination and proof of TO. For homo sapiens this definitive specification references its proof and that of all vertically integrated classes in TO, including TI, TE, TES, TIT, TRU, RELTO, and TOV in TO. It is not the human species alone that determines TO, it is the actual pure essence of this species, and all the rest of TO that is the overall determinant. The individual constituents of TO, as oneself, exist in TO in a similar way.

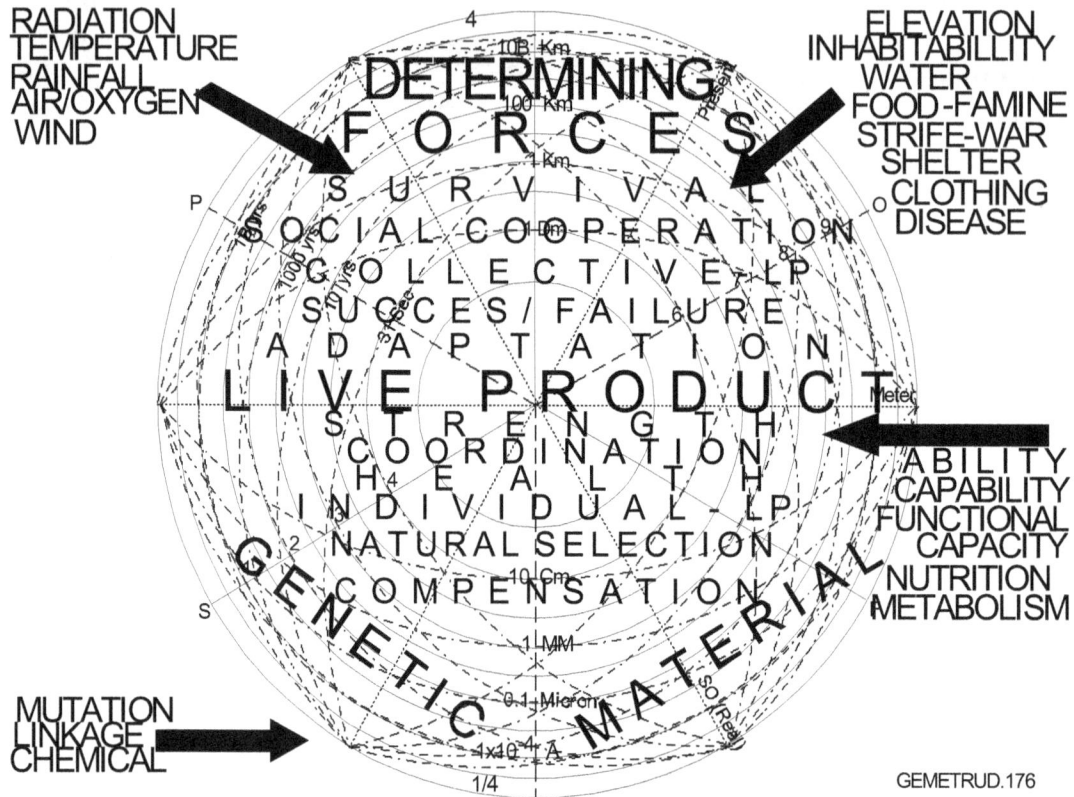

Fig 26-3 Genetic Material, Live Product and Determining Forces, Dynamic Interaction and Continuity

Nowhere is an accurate perspective of actual totality clearer than by orientation in time. This is especially so for the generations in cladistics. At three generations per hundred years there are one thousand ancestors in 333 years, one million in 667 years, and one billion in 933 years. For progeny this is greater depending on number of children per generation. This is how populations and cultures to a large extent, converge and diverge. It shows how natural selection works not so much by each generation, but by the whole number of generations. It helps to show how each individual is both same and different. It most importantly shows how we cannot understand actual totality unless we understand how all changes with time. The temporal, with the mass energy and spatial, must be viewed with a precise perspective, one that is both overall, and segmental depending on the region or sector of actual totality given, as in SPALT. It is often by the right perspectives that actual totality is proven. This is why we must learn all actual totality, and as much as possible in the correct proportions and order.

That the focus of the world is on humans in here and now, and in equilibrium with there and then, can be readily recognized by cladistics. The simple fact is if our ancestors were not the best we would not be here, and if we are not the best our progeny will not be. It is the balanced phylogeny in ontology of the present, and through and through based on a, at once, focus, ATO, of TIT in TRU of TOV and TO that ties in all in TOT and TO.

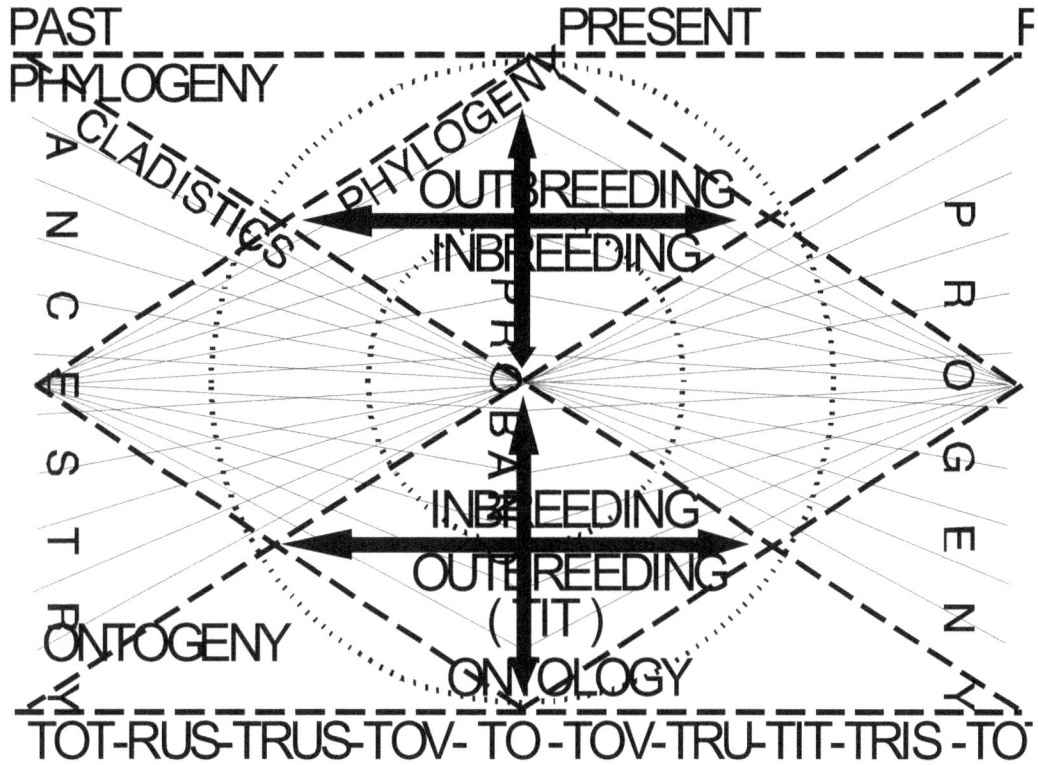

Fig 26-2 Cladistics, Phylogeny, Adaptation, Reproduction, Ontology and TO

The center of TO is the focus and the media of the sum of its cladistics, species, immediate awareness, consciousness, instantaneous attention or IA, anticipation, selection, problem solving, reason, thought and people. It is a convergence of sameness in the relative of TO, TOV and TRU. It is a center characterized by the mating of male and female, and resulting family bonds. The center is a reference or focal point in a frame of reference whose average is a perfect sum. Most conceptions of the world, evident subjectively in immediate awareness, reach the conclusion that the center must be the actual reality of 'here and now', if there is to be any center at all. To deny a center is tantamount to denying unity and the concept of universe or world. To deny the 'here and now' is to deny the actual pure essences of all that enables human life, including mentality. Assuming that a center based on here and now is correct, and denying any universe that is without a center, we can proceed to discover what kind of universe requires such unity with a center. That it is not an exclusively astrophysical universe is evident in the fact that only relative corrections alter the idea that the earth, sun, and even "our galaxy" provide a center. To have a center necessitates relation, a periphery, and limits. To have external relations, absolutely, one must depart, at least to a degree, from 'universe'. Absolute and zero of the periphery and center are not absolute, only infinite and infinitesimal limits.

The observations of, and the object itself, are very different, often in ways opposite. This is made even more enigmatic when we compare the subjective within and without. Even more error occurs when we confuse the subjective, inside to outside, of TO. To correct the error of in and out we must always remember TO, its modes and TOV within. We realize the role of the observer or subject, which must be within the universe, at least in the way it approaches the exclusion of being the universe. In this way we must assume degrees of variance from absolute. This is why we need to approach ourselves, TES, TIT, from a more exact approach through TRU, TOV and TO, and return similarly through TRIS, TRIR to TOF and TAR. This is an approach that must recognize the preeminence of all as they actually exist of and in TO simultaneously. This is why cladistics, and the quality of our reproduction are paramount. Scientific knowledge, and especially that of genetics and allied fields, is rapidly improving. Much better use of it can be made than by natural selection alone, although this also is indispensable in

TO and TOF.

The collective unity of modern human life and it's technical and judicious products are what we, ours and all combine in civilization to produce. A highly convincing proof of TRU and TO is what arises from "we," the collective, composite sum of all who are cognizant of their being. Any universe that claims validity must account for its being observed, thought of, or reason with. Otherwise, and without this, it can't be complete totality, it lacks relation, essence, many of its parts, and has no unity. It lacks any set mass energy, spatial and temporal conformation, configuration and coordination. The question that arises, what 'we,' is answered by providing that which is lacking. Herein, by correction and refinement are further substantiation, completion and proof of TO and TOV. 'We' that are its reference point are not any we, nor any strictly human person or group, but the collective "we," whose characteristic existence and actual pure essences assure and enable total success, well being and survival of the actual world of Best Human Life, TRU, and The Relative Universe in and of itself, TIT, in TO and their optimal adaptations, transformations and transitions. We can be interpreted as either all that is of TO, or all humans, and the degree, they are TO. When of all humans less than TO, we interpret a world of common life, whose contrast with TO reveals the central essences of our existence. This is all that we lack, as a civilization that is less than TO; whose absence, when discovered, helps to prove TO.

The dynamics of human life by rapidly expanding growth and interaction produces the condition of partiality and non identity prevalent today. This is most evident in the focus on money, power and their waste and misuse. Only superficially in interpersonal relations, routinely in work, or fleetingly in principle do we touch the kind of identity all must have in TO. The great increase in human interaction only aggravates this basic weakness. All must learn or be trained to experience a deep love, interest and dedication to TÓ, best human life and their convergence with TOF. Any and all that lessen or destroys TO and our hopes, will, search for, and life, in TO is to that extent the worst, the great anathema of non existence over existence. It is the denial, negation and loss of what is the prime great actual pure essential of being, to be alive. In consequence being alive is tantamount to being of, and in TO. And to be alive takes a certain and set approach, evident in the proof of all that enables TO.

A major aspect, and explanation, of absolutivity, is the part and role of a person in actual totality. From any individual, TES, to the person, TÍS, consciousness, instantaneous attention, self centeredness, and the absolute idea of knowing and being oneself the focus of relative totality on here and now is the basis for absolutivity. This is signified by the Law of the Internal Convergence of the person on, and proof of, absolutivity by RELTO, in actual totality. LICPART 29, 14 In nature nothing is free, everything has its cost and must be paid for. This is life, all life in TO. Only we can find consolation in the fact that much is closer to being free than others. A freedom in which limits, and our life within these limits, is reduced to a practical minimum of their correct degree in TO. This is not without effort, however that necessary to make the reduction, is by a life in TO. This is acquired only when we discover, learn, and appreciate the proofs of TO, and live by them. Humans are the basic, emotive and mental oriented force that is special to and upholds TRU and TO. It is the best exercise, practice and use of this force that gives humans the opportunity to be as free as possible. Ths is why the capacity, ability, skills and their actions are needed if each and all persons are to attain the level of quality and proficiency needed for TO and its proof. What we should, must and are often forced to do, when right by actual totality, are relevant, something to be attended to. They are the product of all that has gone before to make actual totality what it is. How we as human life must learn, know, remember and live by those propelling and compelling forces about us. This is a continuous encounter in life, the basis for many of our experiences. Whether we know, live by and profit from these experiences is the power we gain, and TO is proven and enhanced. The imperative is not our enemy, it when right is our friend that guides us to be our part in actual totality. It is all that is not right, seen or unseen, direct or indirect, that is our enemy. Part of our duty in life, as actual totality, is to continually expose what is wrong, and show it where to go, by the limits of TO.

Cultural patterns of TO, themes past, present and future -

There has been, and continues to be to a somewhat lesser degree of marked regional specificity and difference in the human population on earth. By quality and quantity this diversity can confuse participation in TO. From a stage of wild animals, through savages, barbarians, less to more civilized and higher qualities of advanced participation and contribution, the average sum in TO is heavily weighted in the higher. To simplify this overall perspective we focus on the more advanced regions and themes of human development. Retrogressive and developing cultures often have much regional variation that is largely dependent on the success or failure of needed existence. As natural selection is supplemented by greater knowledge, cultural disparities will be increasingly corrected in favor of improved TO. The great advantage and strength of family over separate life are a maximum of opportunity, utility responsibility and unity. These optimize the role men and women in all human life, as TO. It has its limits and exceptions. Other relations and institutions play a part. Yet it is the great maturation and perfection of one's identity, in TO that is a greatest determinant of survival.

That some events happen more or less by chance in nature is evident in human genetics. By reproduction and the resulting genetic material after recombination of DNA there is often a mismatch of the two haploids in which an imperfect match and product result. This is often associated with deformities of one kind or another. These are more apt to occur in the resulting male offspring, which normally receives one X and one Y chromosome. The Y chromosomes being much the smaller, and at the end of the DNA, are often the point of mismatch and abnormality. This is why sisters tend to be more alike than brothers, and why brothers are more likely to display more extreme, good or bad, and less moderate, characteristics. It also explains the greater tendency of males to have deficiencies, abnormal tendencies, reaction tendencies of self centering and criminality. Yet when positive it helps to explain why males can, when extremely well formed genetically, trained and use knowledge wisely, have the most beneficial, powerful, and superior traits and contributions that are the hallmarks of many advances and salient times in history. Why women have been the stabilizing force that has assured continuity. These are generalizations with many limits and exceptions. They do help us understand the core of TO, TO proper, TP, and can also help to guide us in the future.

Trust is the glue that holds society together. Yet trust can be no better than the truth it is based upon. And the truth it is based upon, can be no better than the fact and degree to which this truth is right. Like justice, trust can only be right to the degree the totality on which it is based is right. This is why trust must come from TO, and why TO must be right. All must be their part, know their roles and how to work for, and fulfill and complete and all that proves life by TO. In this way, from the trust between individuals, and the trust and justice that exists between societies, nations and all peoples, it is shown and proven that the interdependence of all in actual totality is vital. All levels of society, and actual totality, function properly, only when they are of the highest quality and ability, working interactively and proportionally fulfilling their parts and processes in the whole of actual totality. Actual totality can be right only to the degree it is well known, qualitatively and quantitatively, and entirely applied in the best ways. From non identity through the bonds that hold humanity together to their completion in the actual pure essences and perfected identity in TO, is the spectrum of non being to ultimate being that each and all persons strive for and can, more or less, achieve. This can only come with meaning, actual pure essences and the greatest wisdom and toleration this best of worlds, TRU of TOV in TO offers and can provide. This, when well proven, appreciated and popular, will help all be immortal, and assure the continuity and permanence of this best of worlds

In concluding human life, its pivotal contribution in TO and its proofs, we seek to tie in, all that is, and needs to approach, TÓ. One of these is the problem of TO and what is less than, yet potential to TÓ. How in the future, when TO loses its exclusive autonomy by changes with time, the absolute and relative totalities change also. As TO expands and improves, its relative totality leads the way, and tends to assume a predominance over absolute totality. This absolute end product is a relative-relative totality which cannot exist. It only seems to exist when we are blind to all the rest of TO that must be maintained. Yet such an absolute end product serves a great purpose in propelling and compelling TO in reaction to all that is less than and not TO. Actual totality is much more than a person, humanity, even life or the physical universe, although these are large participants in it. As we progress through the varieties of actual totality and RELTO toward the highest forms of life in humans, this presents an ever larger share of contribution to the whole. Yet we must always remember what is and is not actual totality, and their proportions. How all that are positive, supports, and is necessary for TO must be appreciated and upheld. This is not only because of direct benefit to man, but their actual pure essences, and more general part and role in the development, existence and future of TO-TOT. We must always remember the great part each type of proof, shown by each chapter in this text, is the underlying measure of the most comprehensive and general part and role in TO. The object that each chapter of this book represents is what actual totality is made of, is proven by, and must continually have to exist.

CHAPTER 27 - TRANSITIONS of TO in TIME, to future TO with TOT

Transitions and Transformations, external and internal, constituents, great presence of TO with TOF

In order to approach and identify with actual totality this book began from the outside, non existence, continued through totalities to categories of TO and its form, function and development. Now we reverse this approach from the inside to the outside, TO, TOT, totalities and existence. This is necessary because of time. Since TO is of the practical present it can be sustained temporally. Yet increasingly with time, by the finest to greatest of periods we must, if we are to achieve the right position and view, adapt to attain the ultimate unified totality. In Chapter 27 this is done largely by transitions and transformations that occur with TOT, totalities potential to TÓ. In Chapter 28 we treat the extreme relative totality, not as a final end, but all those aspects and features that effect TOT and TO. The remainder of the text encompasses all these, outside to inside, that provide optimal supervision and guide

Changes, interactions, transitions and transformations of TOT and TO with time, are separately, combined and all together, major aspects of the world. To learn, prove, understand and best apply their features, form and function is key to mastering actual totality and our lives in it. From development through its greatest qualities and their dynamics we must recognize what is external and all the changes and interactions that participate in the existence of TO throughout time. Along with learning its development we must learn its course in the future, as well as whether it ends, or has permanent continuity, Chapter 28.

Transitions and transformations are, with some differences, closely related. Transition refers to, often progressive and more or less, incremental changes in existence or state. These changes are largely a function of time. Transformation refers to changes in form, often in series that occur as a result of alterations by many factors. The transformations may or may not be progressively propelled by various and more or less related prior internal or external causes. Transitions for totality, affect TO externally, and is the typical condition of TO in TOT, past, present and future. Transformations for totality are also highly time related, and reveal how TO began, developed, grew, mature, and what it can, or cannot become in the future. They help to reveal the numerous and diverse changes, quantum leaps, fractals and other ways TO were created, evolved and portend to be for the future. By discovering, learning and applying how transition and transformation work in TO, is a major step in its proof, understanding and improvement. Most basically TO is proven by existence, necessity, actual pure essences, change within itself, the contribution of each of its varieties, TOV, and relations with TOT. The succeeding varieties of TO are characterized by their more specific actualities whose form and function provide the basis for understanding TRU its transformations within TO, and transitions with TOT-TOF. It is by these great transformations and transitions of actuality, whose series and gradients are revealed in the varieties of Totality, that we can most readily recognize and prove TO and TRU-TIT-TRIR-TOT.

All changes with time, incremental time, time given, time present and extended, is evident in the following,

1. All has an effect and determines the form and function of TO, only by the degree positive/negative.

2. Transitions and transformations in time largely occur at the interface and by interactions between TO and what is potential and all less than TO.

3. There are stability/instability, and an equilibrium or homeostasis, with its own resistance and powers.

4.   There is reversibility that sustains and restores TO.

5.   Many factors and sectors in dimensional formulation contribute to the changes in TO - TOT.

6.   Lorentz Transformations, contraction, expansion and changes in the state of objects with changes in mechanics of space, time and mass energy.

TO consists of transitions and transformations in different ways. How the universal constant for TO of time zero occurs in conjunction with that for the more generic varieties, of ABSTO, helps to explain and prove the complexities of ARTO and thus TO. It reconfirms the fact that TO exists both by its own exclusiveness, and a more or less inclusiveness in time with all else. The degree of this double, or multiple, existence of TO depends on many factors, not the least is its typical actual pure essences, knowledge, and will to sustain TO.

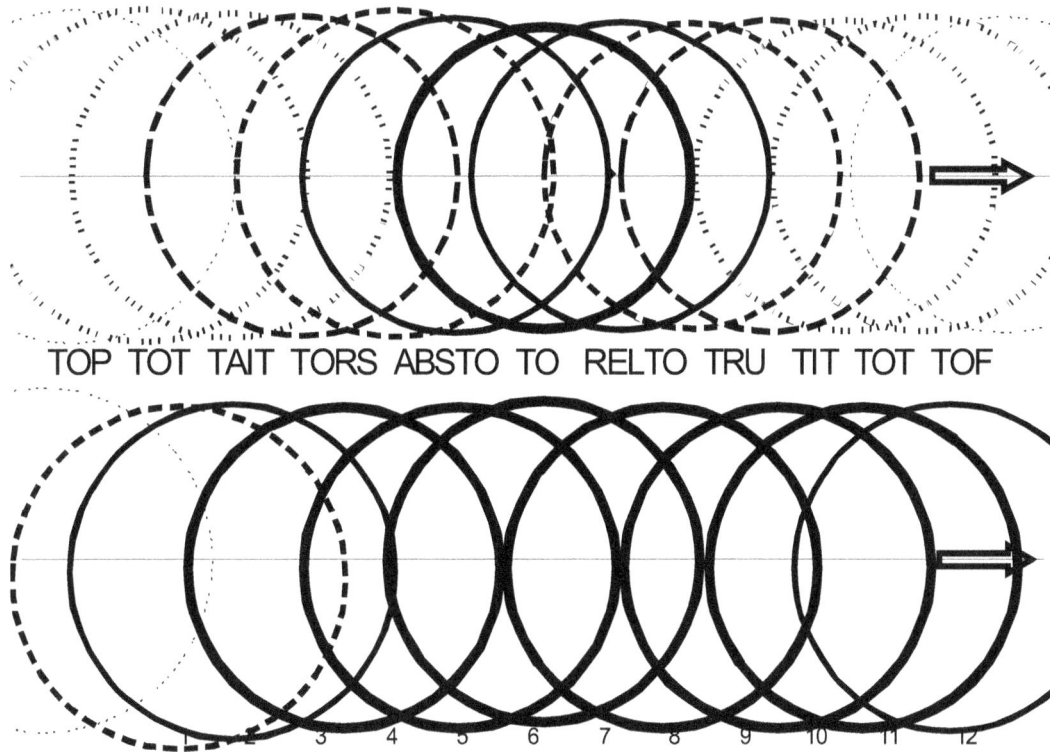

TOP TOT TAIT TORS ABSTO TO RELTO TRU TIT TOT TOF

Fig 27-1 Change, transitions, transformations, from past potential totality through present actual totality to future potential totality

This figure is a companion of Fig. 13-2, only instead of emphasizing the varieties of TO, it features transition of TO in time. The top series shows a predominance of the potential, the bottom predominance of continuing presence and permanence of actual totality. To achieve and secure such presence and permanence is a most basic need and proof of actual totality. Only we must learn, know and live in the right way, qualitatively and increasingly quantitatively if actual totality is to work.   Of the many varieties, formalized at ten, the extremes are AIT for the absolute and TIT for the relative in TO. In the transitions and transformations of TOT and TO with time these two extremes change as does all the other varieties. What these changes typifies, is the form and function of TOT and TO in time. All varieties exist and function, not only at their level as a gradient in the spectrum of absolute to relative,

169

and general to special variety, their influence and actual pure essences manifest themselves throughout TOT and TO. This is seen in the more combined varieties of physical universe and relative universe in the spectrum of the varieties of, and form in, TOT and TO. It is for this reason the form of all in TOT and TO is heavily affected and produced by the physical universe, and the function of all in TOT and TO is heavily affected and produced by the relative universe. Since form and function are closely interconnected and fuse, both physical and relative universes produce a profound effect on TOT and TO. This is largely by their mutual form and function in the highly permeable whole. It is by their actual pure essences, and the profound effects, that the physical and relative universes, as opposed as they may seem, are united in, and prove TO and TOT. This is highlighted when extreme, in AIT and TIT, in TO-TOT.

There exists at any time, period of time, or segment within TO different gradations and degrees of separation. From the maxim, "You would not be here if your ancestors were not the best there were," it can be deduced that there exists a state or condition of more than one world. Those who, were, are and will be, the best, and those who, were, are, and will be, many gradations less than the best. The question arises, "how is, or can, this best entirely equivalent to TÓ and TOV?" Not necessarily, for what might have been the best in the past may not be for the present unless all three are in TO. Even more the future for a partial, less than the whole, may not be so of TO-TOV. Thereby we have variables in this set and equation that which exists at any level, group, time or world, what has been the best or is or can be the best, and what converges on one, TO. Fig. 27-1 helps answer this problem by showing the progression, extreme absolute to extreme specific and back to ABSTO-RELTO equilibrium in TO. The role of serial states of TO-TOT, is not only separate, but how they reverse and coalesce with TO, as they are potentially of each other and become one. Time zero is the same everywhere in here and now mass energy of TO by relative totality and human consciousness and instantaneous attention. It is the center and key to existence, orientation and coordination. Any zero time is a feature of absolute totality and the physical universe. It is this difference between the relative effect of relative totality in TO on the rest of TO that differs from and is in contrast with any zero time external to TÓ. It is this distinction, and how zero time fuses in the absolute and relative totalities in TO that describes and proves TO. Although not so important in TO, this contrast is necessary if we are to fully understand and prove TO.

TO-TOF consists of the actuality, or actual pure essences, of all their varieties, only of somewhat different quality and quantity typical for TO or TOF. This includes, TRU and TIT, and explains their transformations into TO, effect on, and union with, and proof of TO-TOF. Because TO is practically, and by actual pure essences, absolute and exclusive does not mean that its transition and multiple transformations in time are not important and are proof of TO and its extension. If current actual pure essences prevail, TO is absolute and exclusive. As TO changes, at first imperceptibly with time or with APET, it continues to exist, yet not so absolute, nor exclusive.

Actual Totality, TO, in TOT future, as the individual in life, must be formed and function properly. It will tend to change with time that is primarily and mostly in content and not in form. This is signified by, The Law of Content over Form Change, LACA. TO is in constant motion which if normal and healthy seeks balance, equilibrium, satisfaction and harmony. To be TO, the individual as all within, must be and have balances with the correct fulcrum at its center point. There can be excesses or deficiencies, which will be compensated for. Actions and reactions, input and output must be adapted to and be reversible. Reversibility and adaptation may take many forms yet will, in resultant, always accommodate and balance TO, its primary form, LACA, TÄP, and TÄ. In Figure 27-2 the relationship between input and output, excesses and deficiencies are shown. These may be in complete balance, eustasis, as a center point representing the mean, golden mean or ideal. Yet they are not always, nor necessarily, all, the totality of actual totality. Variation or departure from the mean is the dynamic equilibrium among all its components. It is like the internal milieu, a homeostasis of an individual's physiology. In being generally applied to the special case TRU it is called Homation, the balanced processes of man and mankind. Eustasis applies to the functional equilibrium and mass action dynamics of TO and its sustained existence in TOF. The involved processes can be separated or fused and made total for SPALT and SPAT posits of the whole of TO. The dynamics and processes of TO, represented in OBT, SUT, ROT, MOT, SPALT and SPAT give a much better presentation. For a person, in TO, homeostasis is the optimal of health, fine-tuned to perfection, and is signified by, LETUN.

Mass energy undergoes many transformations in the transition, from its state in the physical universe through the combined ABSTO-RELTO in TO and into TOF and from the most absolute to most relative, generic to specific totality in the varieties of actual totality, TO. Mass energy exists in the most basic form and function of TO as a continuum, duality and dimension. For TO this is called the Á, or the prime mass energy dimension that is neither only that of absolute, nor relative totality. It exists in a state of expansion, explosion and acceleration to completion, ACOMEX. This explains the transformations and transit of TO into TOF, continued presence of their modes and the direction of TOF, especially by the speed and acceleration of higher cognitive powers of TO-TOV. These can be of various modes, qualities and rates, as well as various types of completion, as relative, absolute, toward completion, or by completion reversibly through maintaining optimal existing states and limits. Thereby it is the mass energy continuum or Á dimension that signifies much of the form and function of TO-TRU. How its O-S spectrum provides positioning in TRIT, SPALT and models, ACOMEX is a great concept that both reveals and shows how all comes together to prove, show the continuity of TO-TRU-TRIR by their power to reversibly adapt and alter TOF.   The levels and expansion of mass energy in TO-TRU are part of its basic form and function. The form and function of mass energy include characteristics of their role in the physical world, adapted by their actual pure essences in TO. The actuality of mass energy in TO and TOV-TRU-TIT-TRIR-TOF is most unique. It is the great convergent power of RELTO, that combines, unites and typifies TO, and TOV-TRU-TIT-TRIR-TOF, to be typical by the average sum of all the varieties in TO, and the effect of its own actual pure essences

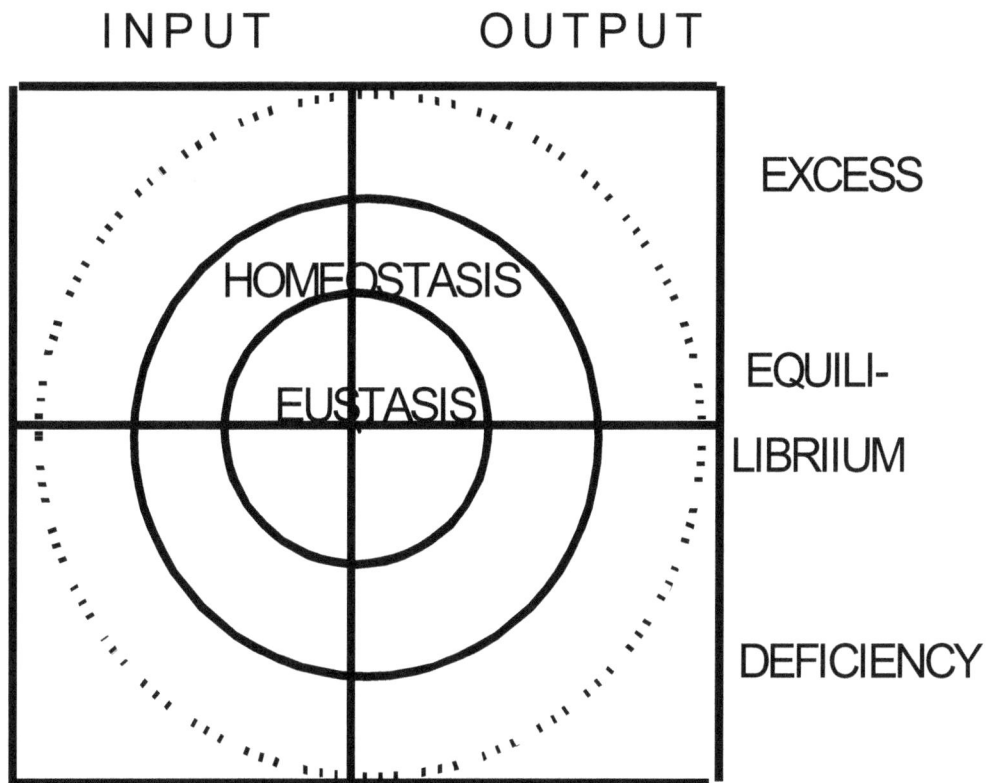

Fig 27-2 Eustasis, form and function of TO in TOT Future

Is totality moving toward the more special, or must the equilibrium of general to special be maintained? There can be no doubt that within TO, as evident from its kinds and varieties, there is much fluctuation and variation. TO is drawn by extreme forces that work internally and externally, toward extreme RELTO and extreme ABSTO. Those of RELTO have a consolidating and integrating effect, those of ABSTO a differentiating and diffusing effect. The more time approaches zero the more integral, the more absolute differentials. The resultant is an averaged sum and equilibrium that are optimal for ARTO and TO.

**Transitions, Transformations, Development and Determination toward
Completion of TO, TRU-TOF -**

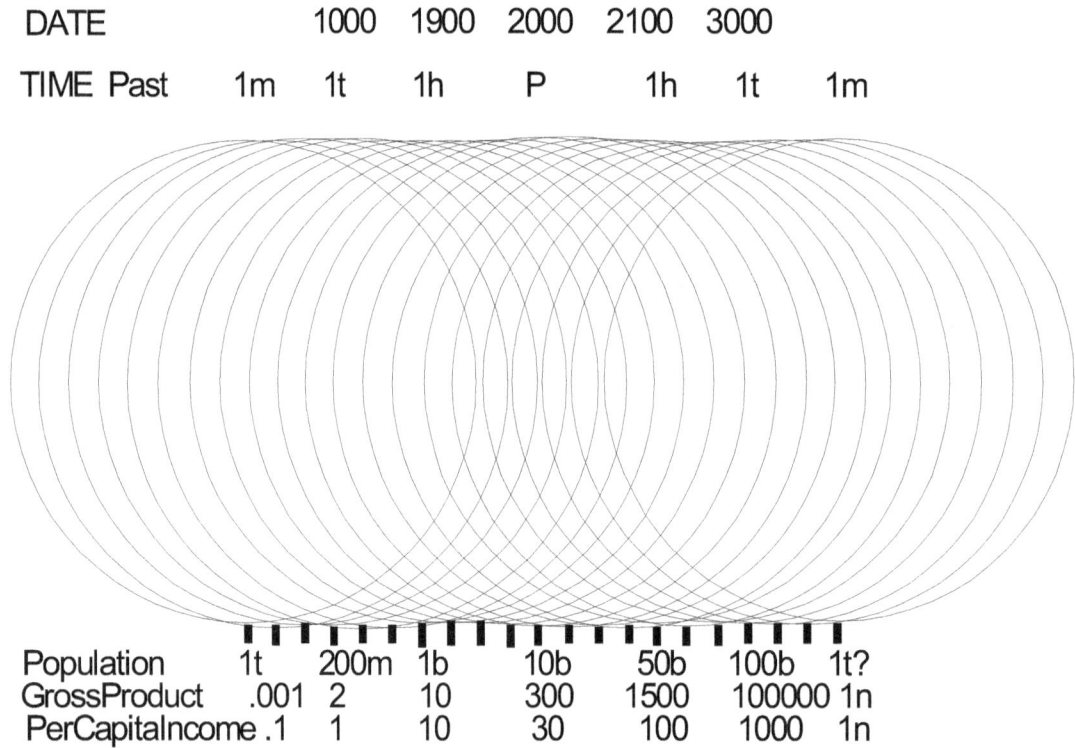

DATE			1000	1900	2000	2100	3000	
TIME Past	1m	1t	1h	P	1h	1t	1m	

Population	1t	200m	1b	10b	50b	100b	1t?	
GrossProduct	.001	2	10	300	1500	100000	1n	
PerCapitaIncome	.1	1	10	30	100	1000	1n	

Fig. 27-3 Serial Transitions of TOT-TO-TOT past, present and future, comparative valuations

Figure 27-2 shows serially the changing populations, GNP, PI and past, present, and future. Yet we must be careful not to equate overall worth, individual worth, or proportionate values by a set of monetary gross national products of all of the world. For if we would, the world in the next hundred or so years would outgrow it, not so much by population alone, but magnified many times by material, commodity and growth of the world economy. It shows how when populations expand rapidly, and the gross product does likewise, the supply necessary to support this population and gross product does not expand accordingly. This is immensely important by showing the greatness and necessity of REM in Action for and Control of TO, (ACC). REM, expressed in many ways, directly, indirectly, or inferred is a large part of the stability of TO and human life. This is seen in greatness, comparative bases for measure, ability to achieve, and ease of accomplishing desired ends and others.   How and how much passing time is involved, determines the form, existence and effect on the transitions and transformations of, and in, TO. Yet this, in itself, does not prove nor show how there is any overall movement toward the more special, as TO-TRU-TOT may suggest. However, we cannot assume the state of what is generic and specific in TO, are always the same. They are in transition and the media of their state, conditions, form and relations can be conceived as changing and fluctuating in various ways in reference to any set point in the progression of generic to specific through TO-varieties-TOT. This explains, certifies and proves the reversibility of transitory and transformation states in and of TO. It is the opportunity of this reversibility that is the hope and salvation of TO, TRU and best human life.   It is in this way TOV-TRU-TRIR can have its effect on TÓ, to provide the alteration of TOF necessary to allow the most advanced and highest progression of TO-TRU-TOF. It also supports the value and roles of combined proofs of TO-TRU. The improvement in modes, alteration and creation, of new and perfection of old essences, causes change, and differences in the form and function of the specific in TO. It exemplifies the role of TRU in TO, shows how the quality of the specific in TO changes, but does not change the equilibrium of generic and specific.

As human life and possessions have changed so have the values of that which might act as standards. Certainly they have changed. At some times it increases and at others decreases, in part depending, as in economics, on supply and demand. This is natural, for the less life we have, the dearer it becomes. We of the 21st century look upon those of the past times, who seemingly possessed so much less even in their own day, as worth so much less. This is measured by gross products and per-capita incomes. Yet per capita incomes, and gross national products, are only fair measures of value of the individual, or more totally, TÓ. To be exact, many other factors are needed in the equation.

Although TO and TRU, like TOT, exists internally as past present and future, it is by transformations in transition that distinguishes them. When transformed to become a new state, it creates new varieties, and even its modes and kinds change. Thus when transformed, as in a quantum leap to separate totality, or variety in the future, TO becomes a new TO-TOF. This depends in part on time, and degree TO is stable by its own actual pure essences, practical and functional values. In the future it will increasingly depend on how, and how much, humans manage and control their own destiny. Since TO is different from the physical universe alone, and consists of the entire spectrum of general to specific, specific modes, indefinite probabilities of what kind of totality TOF may be, and the needs and course of TRU produce, a new spectrum of outcomes results. TRU, as the core of RELTO, has and will probably with due human effort, increasingly affect these changes in TO and TOF. Many of these are unknown, yet knowable, some are most critical, and all are a part of the greatest of all givens, the most basic actuality, objectivity, subjectivity and reality of life, focused in totality proper. No matter whether it is the distant past or distant future, with many transformations, TO and its increasingly specific varieties, as TRU, consist of many universal continua and parameters that sustain continuity. How the positive transformations occur in the future, maintaining both the same and different of all, is the topic or theme of TO-TOF, and the direly needed concern of each and all persons. It is by these great transformations of actuality, whose series and gradients are revealed in the varieties of totality that we can most readily recognize, show and prove TO-TOV-TRIR-TOF.

Integration of TO, Convergent Totality in TOT - There is continual fusion and fission, union and separation, in the combined mix of totality from TÓ, to TOV to TRIR to TOF. This is and explains the problem of determination, to fill out, completely and continuously adjust TÓ, to conform to its actual totality. There is common ground in universals and universes whose superimposition collectively reveals and proves the whole, the totality of TRU-TO and adaption with TOF. The principle of viability, and selective forces of determination are proportional to TÓ, TIT and TOF, each in their own way, as evident in their actualities. Such selection is exemplified by the fact that a person can't do right in one department of life while occupied in doing wrong in another. Life, as TO and TRU, is one indivisible whole, its parts are like the notes in a great symphony. Never to much nor to little of its whole, and never to much or to little of its parts, all must be like a symphony. It is these forces by which transformation and transposition of mlt to ALT, to (ALT)3, confirms the total union of the coincidence of actuation, extension and intension, ACEXIN, from which is derived the specification and adaptation law of TRU-TO and TOF.

There are a myriad of different totalities and entities that exist outside as well as interacting with actual totality They are always both present and changing. The greater the time the greater their affect on actual totality. Some are potential to TÓ, some both in and out of TÓ, some opposing to TÓ, and many variant, infinitesimal, or null, relative to actual totality. They are important only to the extent they are positive to negative to TÓ, even though TO is practically and highly inclusive, and exclusive. Because it is permeable and interactive it is not free from the effects of all less, opposing and negative to itself. These effects vary with time, conditions, and the state, form, function, power and health of TO. It is the interactions between these and TO that are continuously at work in continuity and survival. The extraneous entities are important in the effects and understanding of the overall milieu in which TO exists. They are mostly not of TO, yet their interactions make them a part of the overall OITIO, in and out, state of TO, especially in transition. To prove actual totality, all that is a factor in its existence must be accounted for, to the right degree and way. This applies to actual totality as an exclusive whole, and to much that is less or variably interactive with TO. The end result, however is only temporary. Because our ancestors were the best there were, and adapted to their world, are much different from the immense human population today, and its extremely complex world. Unless they are controlled, and adapted to actual totality will be weak, deteriorate and lose its existence.

Even the origin and final end of TO and TOT can be more or less known. They can be made better known absolutely and relatively, for they can't be lost in the infinitesimal, nor singular unity of their existence. Since AA and RR are non existent in TOT and TO, TOP and TOF, except for their practical actual pure essence effect, we may assume the origin, and final end, will be progressive serial stages, forms, states, conditions and limits in which each has its own typical characteristics and properties. For TOP approaching AA, this will be largely absolute and physical, for TOF approaching RR this will tend to be largely relative and subjective. The kind of absolute-physical and relative-subjective will be less and less like that of TO, yet retain many of its actual pure essences, form, function and characteristics. These are some of the forces reversibility and variability continuously have in the progression of transitions and transformations whose ultimate equilibrium of absolute and relative is always what sustains the necessary existence of TO. The close relation, and importance, of absolute totality or the physical universe and RELTO, or universe of mind and man, is most apparent in the environment, how much, the environment affects RELTO and how much RELTO effects the environment, both positively and negatively. This is an all-time interaction, whose current status is now popularized. It is not well determined, for lack of TO, good judgement and follow-up action. It is how both mutually exist for the benefit of each other, and the whole of actual totality that is the ultimate solution to many problems.

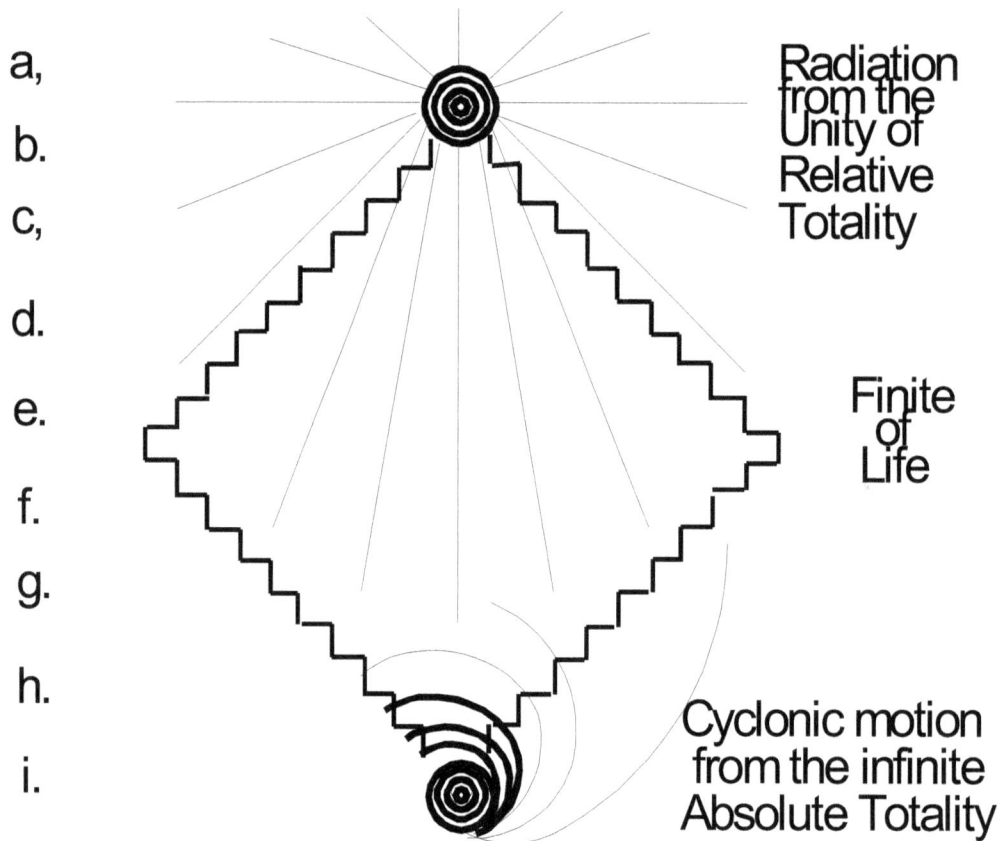

a,

b.

c,

d.

e.

f.

g.

h.

i.

Radiation from the Unity of Relative Totality

Finite of Life

Cyclonic motion from the infinite Absolute Totality

Fig 27-4 Levels from TÂ Unity to ÂÂ Infinity

This figure shows the relationship between the total shining forth light of TO and TOT from perfect unity of relative totality through the finite of life to the infinite state in absolute totality. It reveals the radiation and steps away from and toward the nothing and negative extremes. It shows the range of life, TRU and TO, between these two extremes, including,

a. Shining white light of relative/absolute, perfect solitary unity, absence of substance,

b. Initial divisions,

c. Gradations or levels,

d. Increasingly finite,

e. Optimal Finite, TROS of TO,

f. The hyper-optimal finite,

g. The infinitesimal differential,

h. The chaotic infinitesimal,

i. matter and anti matter,

j. The absence of light, black hole of absolute/relative.

The extremes of black and white holes are modified in TO-TOF by relativity and absolutivity. Absolutivity is necessary for RELTO and TRU, as relativity is necessary for ABSTO and the physical universe. This is the law of the moderation of extreme ABSTO and RELTO by relativity and absolutivity, LAMRA.   For TO, we are drawn in all directions, but maintain ourselves in the optimally differentiated and unified relative finite by TROS and ACEXIN, our equilibrium and homeostasis of all in TO. It is decreased by time, and loss of equilibrium and necessary existence.   Eschatology is the study of such final matters as death, half lives, mortality, divine judgement and the end of the world. This is a topic that has dramatic implications, and often erroneous interpretations. It is close to the hearts of those who are concerned with what is going to happen to them, theirs, and their world. Developing a rigorous, scientific, correct and complete view of the future comes with more information, knowledge of TO, TOT, their optimal fusion in transitions of time, and use of VAPO, SPALT and all that show and prove TO, and with TOT.

Although we would like to think of actual totality as completely stable, and a fixed set, it is not. This is why we speak of TOT, not only as potential to the actual TO, but as actual in its own right, with TO in time. It is also whey we divide it into two chapters, 27 and 28, whereby 27 treats TO and TOT together as they remain relatively unchanged in time, as TO-TOT. Whereas 28 treats them variably as more or less TO and TOT depending on the scenarios they assume in various foreseeable and less foreseeable time. It is the help that accrues from actual totality and its proofs, as well as the great expansion of existing accurate knowledge, that enables us to understand all, including the foreseeable future and its variations. This is the Law of TO-TOT states and variations with time, (LATOTSAV) or LATSAV.   Just as history tells us much about the future, so does the future tell us much about history. This is a part of the great integrative processes and reversibility that is present in all of TO, and much of TOT. It produces provides added values to the future transitions of time, and what TRA may be. Yet the further we get into the origins and ends of things the more they are not actual totality, but more or less illusions. This is both from being less and less actual, and being more and more particular. However transitions and what occurs outside of TO, as it is continuously interacting with TOT and the less existent, are aspects of TO we must always remember, ECRO. This is necessary for knowledge that departs from perfectly TO. Reversely we must also, and equally, always remember that the transitions, and TRA of TO must, to the degree and in the way that tbey are corrected, return to perfectly TO

CHAPTER 28 - Ultimate and Complete TO with TOT, The

Absolute Relative, TAR, to The Personal in TO

Future of Actual Totality, effects of accelerating relative totality, probabilities of potential outcomes, goals and purpose of life, equilibrium with absolute totality, extreme specific, eternal,providence, continuity, progressively greater TO,, transformations in transitions of TO-TOT

The duration and continuity of TO, and the direction TO with TOT will take, are vital questions that require the right answers if we are to prove and fulfill TO. This stems, in part, from the potential and opportunity of best human life. It is a potential for continued expansion and improvement of human capacity, mentality and all its accessories, as knowledge, technology and uses in life. This suggests a heavily weighted tendency to RELTO over ABSTO. Yet it will help if we know for sure just how and how well all is prevented, protected, cared for and works together in the future of TO.

The types of proofs featured in this chapter are those in which the great tendencies and powers of relative totality are increasingly dominant, and approach the absolutely relative, TAR. This is in contrast to the last chapter which also had proofs of and by the future of TO, but maintained its absolute-relative form, state, balance and dependence. The powers and tendencies of RELTO-TO approaching completions by TAR are reverse images of the beginnings and origins of TO. Its origins were of absolute totality, so that these ends are from relative totality and whose extreme is TAR. To understand and prove TO it is well to discover, explain and demonstrate its beginnings and its ends Since these are more or less open ended, depending on the power of TO, its constituents, time and external forces, we cannot with absolute certainty, give simple beginnings, presence and ends. Yet it can be developed, proven and given in a highly practical and effective way. Many of the same forces that have guided the progressive direction of TO in the past will more or less do the same for the future. From the obvious current existence of human life on earth its non self -assured state will depend on all the powers available in TO of the present and intentionally of the future to assure continued progression.

Since TO is in transition, albeit a relatively stable state, and RELTO are increasing the question is, "How and what does this lead to?" What changes and improvements occur in the special and generic absolute and relative in TO and TRU-TOT. Is totality moving toward the more special, or must its equilibrium of general to special be maintained? There can be no doubt that within TO as evident from its kinds and varieties that there is much fluctuation and variation. Yet this, in itself, does not prove nor show how there is any overall movement toward the more special, as TO by RELTO and TRU-TOT may suggest. However we cannot assume the state of what is generic and specific in TO, is always the same. They are in transition and the media of their state, conditions, form and relations can be conceived as changing and fluctuating in various ways in reference to any set point in the progression of generic to specific through TO-varieties-TOT. It is in this way TRU-TRIR can have its effect on TÓ, to provide the alteration of TOF, necessary to allow the most advanced and highest progression of TO-TRU-TOF. It also supports the value and roles of combined proofs of TO-TRU. The improvements in modes, alteration, creation of new essences and perfection of old, produce change and difference in the form and function of the specific in TO. It exemplifies the role of TRU in TO, shows how the quality of the specific in TO changes, but does not have a practical, major or permanent effect on generic-specific equilibrium.

Such an absolute relative, TAR will not hold if TO is in transition, or is limited to its viewing only. An absolute existence of TO, seems to hold by virtue of the fact that we must be. This is what our ALT and APE, and the APET of TO impart to assure and prove TO. For if we must be, our part in TO, TO itself must be, at least in some certain way. This is the enigma of life, our being, and each person. It is true for TO, if TO is considered autonomous, and relatively true to the extent TO-TOT maintains continuity. Such existence includes and ensures being that has sufficiency by the limits of universal constants as the speed of light, time zero, a center, periphery and form of TO. This is further proof of an absolute-relative continuum, and LANERA. We exist within this continuum of ABSTO-RELTO in TO, by the thread of life, which has the potential of approaching the absolute as this thread becomes all by binding it together in perpetually sufficient strength. As accurate as this is in the very long term future, to be practical, as demanded by actual pure essences of TO, we need to maintain a balanced perspective, of all as it is in totality, TO in the near, mid and potential extension into the far, and very far term.

The survival of human life is at stake. Is it lost, does it stay the same, or does it advance? Certainly, as a whole and its parts get better and worse variably, they depend on many factors, factors that are positive and negative on TO. As we discover, learn, know and produce more and more of TO, its basic constituents and their summation reveal the massive dynamics of TO, how TO is relatively exclusive, yet in transition with transformations that minimally to maximally change its existence. This is how TO, its varieties TOV, and especially reversibility in its most special varieties, affect and develop with potential TO, TOT, to form better and greater TO of the future. Among these new forms of TO altered by TRIR, TIT and TRU is TAR, TO as the Absolutely Relative. It varies from a fully combined part and role in TO, to an extreme separate totality in itself. By describing the potential courses these transformations can take, and finding their best progression and ultimate unified totality helps to show and prove TO and guide our lives in continuity.

The next question is, if TO tends to progress toward the more and most specific, how far can this go, what is likely and what is optimal. Can there be an end, or what kind of completion? The answer and proofs are, as it has always been and will be, generic to specific only by their quality, form and function, the power of continuity by reversibility over change. Since TO is all, and by its APE necessarily set, the definition of generic and specific are not independent, but, like ATO, within TO? Yet since we know TO changes and is less or more transformed with time we must conclude that its APE changes. This is evident in the potentiality of TRU altering, if only by slight degrees, TOF. Herein, as in the modes of TO, are certain proofs of TO, TRU and TOF, as they are in combined relative transition, unity and equilibrium. It is by these great transformations of totality, actuality, objectivity, subjectivity and reality, whose series and gradients are revealed in the kinds and varieties of Totality that we can most readily recognize and prove TO-TRU-TOF. Since TO, for the near and mid term, is practically stable, and in equilibrium any absolute final answer to a course toward an absolutely relative TO is not as important as its maintenance and continuity. Certainly there will always be some kind of ABSTO in TO, so that RELTO to progress toward completion, and a kind of TAR, will necessitate some kind of simultaneous progression of both ABSTO, RELTO and TO.

From its non existence and incipiency to its potential completion in continuity the enigma of TO is described and clarified. Neither non existence nor any completion exists, except by their recognition, explanation and final absolute of the absolutely relative. The duality and spectrum they form are the positive imperative that makes, saves, is, shows, and enables us to live by this TO-TOT-TO kind of totality. All who do not follow this imperative power of TO driven by its special varieties are to that extent relegated to a decaying life of non existence. What is negative to TO-TIT-TOT will, sooner or later, be selectively rejected or eliminated, by natural selection or more rational means. They, by the lack of actual pure essences necessary for existence, TO, are already irrelevant, only physical, thereby extinct. The imperative of an absolutely relative existence is actuality, the most fundamental of all modes of TO. This is evident in the fulfillment and equilibrium of the modes of, actuality, subjectivity, objectivity and reality, and all that constitutes TO-TIT-TOT. Why this enigma exists is explainable by the form, function, and inclusions of much that seems to refute TO-TRU-TIT-TOT. All that proves TO and TRU helps to prove this enigma. The absolute is not all, by itself it is non existent, it is the absolute-relative that transcends all, and as ARTO in TO.   As noted in the beginning of this chapter on proofs by TAR, the immense tendencies and powers of relative totality to be dominant and expand cannot progress to an absolute end. Since an absolute end, like an absolute beginning is not possible, some, if not a balanced absolute-relative existence in TO-TOT is necessary for its existence, or actuality.

The Principle of Viability, POV, by relative totality or RELTO, is largely proportional to TRU in TO. It is the collective and cooperative essence of TIT-RELTO-TOV-TO. How we live, work, and achieve TRU-TO dynamically and interdependently all our lives. The Principle and Proof of Continuity and Duration, PROCAD arises from the fact that humans are primarily a mixture, often more separate from TO than TIS, TES and TRU in TO. To overcome these, less than actual and essential being, and in and out, OITIO, actions of the negative, is a major purpose of best human life, TRU in TO. Humans can coexist only in so far as there are a great TRU-TO-TOT past and future. In TO, its past, TRIP, and future, TRIF, are as much of TO as its present. Past, present and future in TO are roughly equal and define its time. This is an averaged sum for TO, not so for each TOV. So is it for TO-TOF, which explains the great power and role of TRU and TIT in TO-TOT, and v.v.

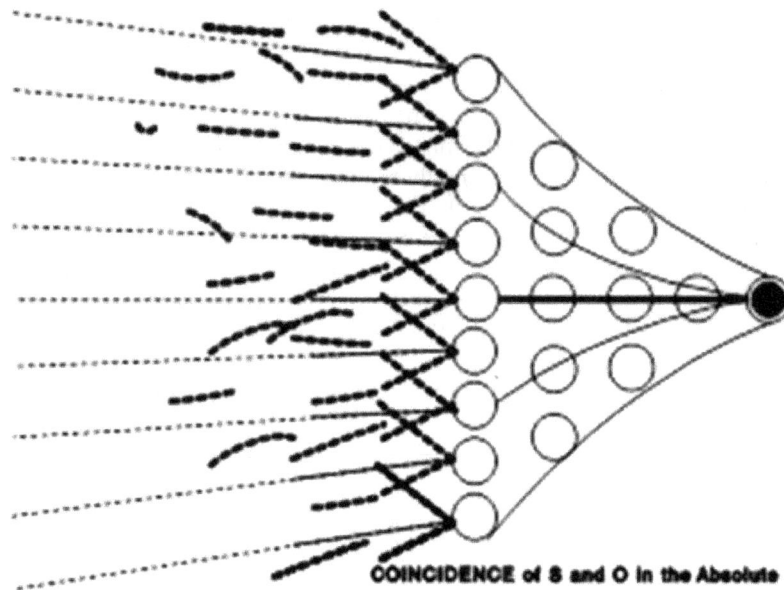

Fig 28-1 Coincidence of S and O in the Absolute

The Coincidence or identity of SO in the absolute was noted by Schelling and is signified in TO by, COSÂ. It shows how the absolute is derived from the union of many elements of S and O in TO achieve, in higher states of actuality, an Á that approaches completion of TO with TOT in the relative absolute optimal of TAR.   The reality, produced by S-O and its applications, generates the great powers that propel and prove TO, and must assure its continuity into the absolute, TAR. The absolute is realized, through intuito, the integration and summation of all that comes together in TO, especially as produced by self knowledge and COGNOS, This absolute arises from being the supreme power, as God, Brahma or the world soul. The union of all in this absolute, by COSA, although showing the mechanics by which TU and TÂ are derived, is not the whole mechanism. This coincidence and union are attained by recognizing the absolute in its overall form, mode, aspect and the totality of conditions or states, from which it is produced. These include, as well as Á or SO, COSA, length, time, world ground, divine will, absolute ego, an organic whole or real, and perfect being.

Are the tendency, inclination, goals, purpose of TAR, the best TO is, or the best TO can be, or the best relative totality can be? By religion and the teleological argument, God does not change basically. For TIT it is by it's autonomous at once being, ATO, whose derivation, development, determination and existence are associated with propelling and compelling forces that continually operate to sustain, maintain and assure its survival and fulfillment of its design, form and purpose. However for much of TO and TOT it is greater and more complex, to assure the optimal adaptations and alterations of TO with TOT that maximally sustain TO. The difference between these three potential courses depends on whether they are of the near, mid and far term, the actual pure essences, strength and stability of TO, the external forces, and how well continuing TO can adapt to and alter TOT, including reversibility returning into TO.   The problem, and its solution, is, how absolute totality, absolute-relative totality or ARTO and relative totality, are the same and different? How is the extreme of relative totality, or TAR, the same and different from TO-TOT. TAR, or the absolutely relative, is much different, even opposite from, the absolute-relative, ARTO. That absolute totality, relative totality, and TAR are opposites and extreme limits is the resolution. Relative totality and TAR are different by transition in time, present, or far term future.

TAR exists and grows only as TO does, Only, in a highly developed and advanced TO, does TAR become most prominent. This is more or less prominent in never ending cycles and periods of extreme forces of generic and special varieties that always tend to return to equilibrium within TO.   It is TÂR, the absolute of the absolute relative, that explains the enigma of the existence of TO in future time. How the positive imperative that makes and saves our being, reveals what it is, is shown, and how to live. All who do not follow this imperative are to that extent negative to TO and TIT-TOT, and will sooner or later be, or, cause selective rejection or be eliminated, by natural selection. They, by the lack of actual pure essences necessary for existence, TO, are already irrelevant, only physical, thereby less, to without existence. The imperative of an absolute relative existence is actuality, the most fundamental of all modes of TRU. This is evident in the fulfillment and equilibrium of the modes of, actuality, subjectivity, objectivity and reality, and all that constitutes TO and TRIR-TOT. Why this enigma exists is explainable by the form, function, and inclusions of much that seems to refute TO-TRU-TIT-TOT. All that proves TO and most that proves TRU helps to prove this enigma. The absolute is not all, by itself it is non existent, it is the absolute-relative that transcends all, all in TO, and all they become with time, in TOP-TOT-TOF.

Although TAR may be thought of as completely relative or RR, of or separate from TO, it will always have a balance of absolute and relative totality, as ARTO. The resultant product of continued TOF will more or less and always consist of a large amount existing TO. Much of this retention will be by TRIR, and human will. This may not be the same as in present TO, and may even look, in the far term, quite different. As the relative totality in TOF expands so will the absolute totality, and have features that are more or less alien to TO. Yet much, of absolute totality and relative totality that are now in TO, will remain. This is like how the general varieties of TO have existed through much of TOT. Some departures from TO in TOF, is so far in the future that practically, most of TO will have optimal equilibrium and reversibility. Thereby it will remain the same, except for that associated with the revolution now occurring in TO-TOT. To predict the future it is a great help to know TO, how TO is affected by time, and how it interacts with TOT. It also helps to use what we commonly know, extends into the future, and apply it, as its actual pure essences become their part in TO and TOF. When all the ways the future can be known are collected and brought together, corrected and united in TO and TOF we can better understand, from where we came, what we are, where we are going, and how TO can be perfected, proven and assured success in the future.

The ultimate goal of actual totality and its pace setting special variety, best human life, is only in part the absolutely relative, for this to be complete, like the absolutely absolute is highly improbable. The ultimate goal is to continually exist and expand at an optimal pace for both absolute and relative totality. This is a relative totality in TO and TOT whose production provides the optimal state and conditions for best human life. This is evident in the reversibility of TIT, TRIR, in its interaction with TOF. It has been, and is, the purpose and result of this work to provide and use knowledge and proofs to further TO. The actual pure essences of the major components, continua, universals, previously unknown and unnamed new concepts, and their descriptions help to fill the voids in knowledge. We can through operations in SPALT, SPAT, and more advanced total models and representations bring out a much greater knowledge of the future. In the closed set TO-TOV-TRU, when infused by TRIR in ultimate totality, TO-TOF, we can discover the most profound and subtle form, function and facts of all, TO-TOT. The result of this is both a much better guide for management and control of the future, it also helps to prove TO, TOT, and their constituents. It fulfills the positive imperative of being and adds tremendous incentive and motivation for each and all in their part and role to know and be with all that enables best human life, in the permanence of TO-TOT. It substantiates the immortality quotient of each and all persons in the common immortality or permanent survival of the relative universe of best human life, TRU, in the ultimate unified totality of TO and TOF.

Humans have been, and are out of touch with their actual pure essences, both are out of touch with TRU. TRU is out of touch with TO, and TO is out of touch with TO-TOF. This explains why there are so much trouble and suffering in the world, and much worse to come without the identity of all with TO-TOF. For every person and their world to have the reversibility and reconcile with all necessary for TO-TO, it will require a great amount of capacity, experience, training, education, interest, love, will and endeavor by and for TO. One of the greatest problems of today and accelerating in the future is to retain our natural settings, conditions, capacity and ability in a rapidly changing and variant world of increasing technology, scientific advancement and population. The benefits of applied science are immense, yet their down side is the degree to which they lead us to extremism, alter our lives and habitat to the extent we lose the struggle to prevent, protect, maintain and advance best human life and TO. The greater human population is, the more resources are permanently lost, and the less we comprehend these and all great problems affecting the totality of which we are a part, the sooner and worse their effects will be. They can only be prevented, managed and controlled by understanding, learning, remembering, living by, and benefitting from all, TO-TRU-TOF, their varieties, future and their ideal ends, the relative absolute, or TRÂ. Accounting and reconciliation of assets and liabilities of TO-TRU-TOF through time require spreadsheets and are found in the addenda and associated works.

In order to show the use of TO-TOV-TRU-TOF and some of the major needs, questions and problems at various times in the future, analysis, accounting and reconciliation must be made. With TO-TRU-TRIR- TOT and their models as background detail, precise knowledge, representation, positioning, valuation and well accounted interpreted and applied spreadsheets should enable an effective guide for best human life in TO with TOF. They should help to prevent, protect, manage and control those negative factors and forces that are so numerous, inevitable and likely to cause loss and failure. This is a perpetual long term need and endeavor. It only scratches the surface of what can and must be done for all TO-TOF. Yet it is imperative that all be accounted for that will or will potentially require attention and action.

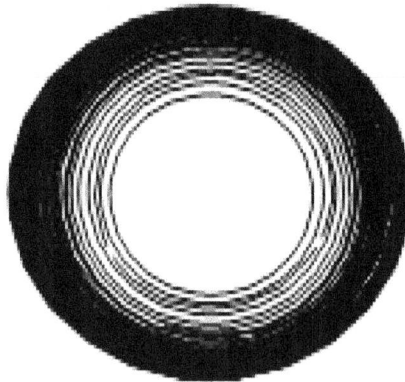

Fig 28-2 Reversible Densities, transitions and transformations of the absolute and relative of TO, TOT, and all.

Throughout the physical universe and much of TOT and TO there is a great variation of density. Matter is distributed from a near vacuum or zero to an absolute limit. This is evident in the astronomical content of the universe, to the different states of matter, and to its role in actual totality. Astronomically it is most dramatic in the big bang, black and white holes, and the distribution of dark matter versus dark energy. In this figure we show a white hole. This is a reverse image of a black hole. Instead of being the vanishing point of the observer and relativity, it is the continuous beginning. The white hole arises from how the nature of a thing is in passing through. It will not have complete form, identity or finality in any specified position or way. It may have a predominance, its fundamental nature may be determined or even mostly made up at a certain level but never complete until it has been reduced. This is the continuation of the form, and is, Continuation of Form of an entity, thing, sense, and TO. CONA, is similar and related to universals and TOT, only in different ways. CONA is more limited to the entity, thing, sense or universe in which it applies and operates. The continuation may have sharp, acute or obtuse asymptotes, depending upon its relationship to the axis of its transit. CAORT is Continuation of Form Relative to Total Transit, and ultimately by TO.

Black and white holes are mirror images of each other, the resulting processes of origin and final states of existence. Such existence and all that they consist of in between are like a continuum or duality. One in which their consistency becomes more and more like that of TO with advancing RELTO on ABSTO in TO. A white hole is the beginning of absolute assurance of being, of TO. A black hole is the opposite, the beginning of absolute assurance of non existence. Such a figure would be one in which all vanishes through its center. The differences are opposite and together they make up a polarity of the existent, ATO. It is, as any polarity, in a reversible position, whereby extreme density and extreme lack of density reverse themselves. This polarity is what is between the complete absolute and relative, or the world, as our own, existence. Toward the vanishing point there is extinction. Toward the continuously beginning point there is perpetual and expanding survival and life. Existence as it contrasts between a black hole and a white hole is dramatic. Many kinds of RELTO participate in this dramatic state. This is a most important mechanism, concept, principle or law. It provides major proofs of TO. A good example is the great power of the future in TO, especially that of human mentality and all derived form it. It is by the power of the deity, personal identity, subjective-objective and reality in TO. This power is of present TO, and all TO-TOF. Their differences increase, at first largely quantitative, then qualitative in the far term. This is the power of future mentality and COGNOS in TO, In their extremes, neither of these two are in TO. Like much of life, nature in general, and RELTO, it seeks balance. This, in the breadth of necessary existence, is the thread of life, the enigma we describe in Chapter 37. In prior chapters we show these relationships, as they pertain to their vantages. These are extensions, as by reversibility of TO, neither a black hole nor white hole. It is a stable state whose vast extension of time provides the potential for eternal life. There are many ideas and topics which can be derived from such a view, including:

1. Relationship between light, life, TOV and TO.
2. Light, relativity, physical and relative universes,
3. Density and lack of density, general theory of,
4. Density and lack of density in TO
5. Reversal of density and lack of density of TO
6. Relations between density, its lack and REM,
7. Their comparisons by generic, specific, ARTO and TO
8. The dynamics of the continuum formed by the polarity between black and white holes in TOT-TOT-TOV
9. Relationships by positive and negative forces
10. Absolute and Relative, ÂÂ, ÂR, RÂ, RR and TÂ
11. Representation comparison with black and whites holes, or negatives of photography
12. Law of Individual, RU, TES and TRU, Immortality, or LIRM, largely from TO, TES and the individual
13. Roles of many great concepts and laws, interpreted on the basis of extremes of density and TO, ATO, TRIA, and TÂ.

The summation of black and white holes are in balance in time, with TO at the center, time zero. This may vary for which adaptation and compensation must be made. The degree of continuity in balance tends to permanence of TO. A white hole tends toward TAR and TÂS, which must be controlled if TO-TOT is to remain permanent. It is thereby self-evident, yet so often needing remembering, TO-TOF tends to be "is and is not." This depends on how well its FAF and SCARLS adapt and find equilibrium in the transitions between non existence and TAR.

BLACK HOLE	WHITE HOLE
VANISHES	ETERNAL EXISTENCE
LIGHT EXTINGUISHING	LIGHT BRIGHTENING
EXTREME DENSITY	EXTREME LACK OF DENSITY
DECREASING	INCREASING
ACTUAL to NOTHING	ACTUAL to EVERYTHING
LIMITING	UNLIMITED
CONTRACTING	EXPANDING
NEGATIVE	POSITIVE
Overwhelmed Externally	Eternal Internal Predominance

CHAPTER 29 - The FIRST PERSON, I, in and of TO to Knowledge

Self Focus, Actual Pure Essence, Vehicle of TO and TRU-TIT-TRIR-TOF, Subjective development, divinity, TES-TÍS, Soul, Hope, Will, Achievement, Fulfillment

I, the first person along with here, now and mass energies that exist in TO, and project into the future, TOT, are the convergent center and great power that make TO, and determine much of its future, TOF. A great problem that exists in all time, especially today is the fact that a person only exists more or less in TO. Also here and now and mass energies that are of TO are not what appears or we commonly believe and follow. Thus the first step is to correct one self, and the part here, now and mass energies have in TO. The latter were done in prior chapters, cf. 16.Each and all persons, and groups, do not only have their own perspectives, they are more or less set, in their own beliefs. Another person or group also has their own perspective, and more or less set of their own beliefs. It is this world we live in, which consists of separate persons, groups, perspectives and beliefs, that are often stubbornly held, additionally set and frequently opposing. This and the associated feelings of self preservation, narrow interests and greed, explain many to most, human conflicts, despair and failure. Generally the cure is simple, to bring each and all perspectives and beliefs to their least common denominators in the ultimate unified totality, TO. More specifically it is most subtle, profound and complex. It requires not only the right actual totality. It requires its full acceptance, and identity by each and all persons. Lives that not only identify with all that is TO, but dedicate themselves to know and do all need to fulfill it. This is by personal universal knowledge that comes from, and guides all other knowledge, for ones life in TO, TES.

Aside from being the great carrier and center of TO is the unending problem of why and the extent to which we are not our part and role TO. Why and how we have failed to discover the spectrum of correct existence in TO. And adversely why TO and less than TO is as it is in persons groups and all people. The great and complex world of peoples, languages, groups and individuals produced by immense increase in population, knowledge, technology and transportation in modern life has not been accompanied by the necessary existence of union in TO.It is the principle of individuation and regeneration in dependent union by the person and from the present in TO that permeates and proves their actuality, relations, universality and thereby TO, and oneself in it, TÍS. The ultimate identity of the special partial, ones soul, with the absolute relative of TRU in the absolute-relative of TO, gives differential and integral immortality to both the part, as the person when TÍS, any person when TES, and when together, all in TO-TOV-TRU-TOF.

In everyday life each person is more or less much the same in the core of actual totality, totality proper, TP. They are very familiar with that close to them in size and time. Yet even here their attention focuses on differences and misleads them into believing whatever sense or story they are told. What we sense and are told are largely far from the actual pure essences and reality of actual totality. Each person is rarely aware of all totality beyond its core of internal and external dynamics. When aware of right reason, it is highly partial, fleeting and lacks valid and proportional perspectives of the whole. This is why it is so important that all persons be raised to their part and roles in TO, learn and develop the most accurate and full identity.

There is a dependent union of the divine, sacred, important of individuation and regeneration by each person in TO. It also includes the necessary respect for one's own, others and the life of all that enables TO. These include, the first person, thought, knowledge, wisdom, others, love of all that is great and practical, in TO. The predominance of human life and mind, is a great proof of human ascendency in RELTO of TO, especially the reversible effect of TRIR, TP, and the best in the universe of human life, with the best in accuracy of its proof in TO. The deciding factor in each person's life is, 'am I equal to my counterpart in actual totality'? Am I my soul, and are my soul that of its role in God, as both are their equals in actual totality. The absolute certainty of me, myself, or of ones own being is the common and obvious belief of all. Yet I am dashed to pieces by many painful experiences in life. Their resolution is especially realized when qualified by its actual pure essence, acceptance, reciprocity and role in TO.

This is the subjective focus corrected by the objective providing reality, especially actual pure reality of oneself in TO. Asceticism is not only physical, it is more a frame of mind in which one and all allow only what is of all actual totality into their lives. This like the spirit of Christianity and other religions, is the meditation and practice of living by the proof and paths actual totality demonstrates.

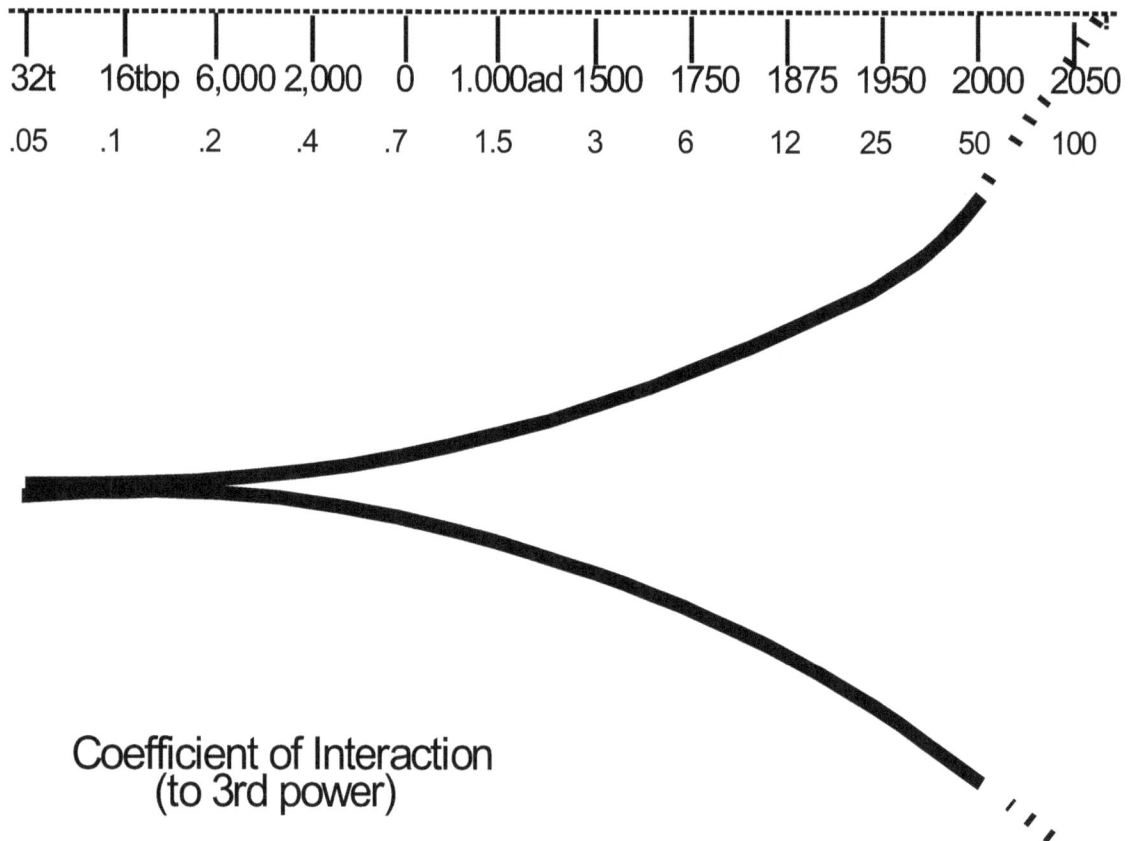

32t	16tbp	6,000	2,000	0	1.000ad	1500	1750	1875	1950	2000	2050
.05	.1	.2	.4	.7	1.5	3	6	12	25	50	100

**Coefficient of Interaction
(to 3rd power)**

Fig 29-1 Rapid Expansion of Human Interaction, lack of total actual identity in TO-TOF .

The great increase in interaction by mobility, communication and the electronic media has not been accompanied by a similar increase in the capacity and best of basic human nature. Poorly does one identify with the actual pure essences in themselves or in others. How much do doctors identify with patients, or politicians with their constituents, or the public with the actual world, TO. Both by nature and nurture humans have not kept pace with their rapidly expanding world. It is a world that is highly and increasingly at odds, with much of TO-TOT. The greater the interaction, mass of existence and TO are, the harder it is for a person. Identity is a total, as well as a very fine and critical function, of the mass energy continuum. Its failure can be discovered when we analyze its elements, how little the first person, second person, doctor, patient, public, subject, object, or entities, have either in common with each other, or how much they are less, not or negative in TO. This is the condition of partiality and non identity the world finds itself in today. This is most evident in focus on money, power, their greed and misuse. Only superficially in interpersonal relations, routinely in work, or fleetingly in principle do we touch the kind of good will and identity all must have in TO. The great increase in human interaction only aggravates this basic weakness. All must learn or be trained to experience a deep love, interest and dedication to TO-TOV and its convergence with TOT. Identity can only come with the greatest wisdom and toleration this best of worlds, TRU-TOV in TO, can provide. This, when well proven, appreciated and popularized, will help all be immortal and assure the continuity and permanence TO and RELTO must have. The importance and extra attention needed to compensate for the growth of special TOV, RELTO and TO and their role in TOF is evidenced by the rapid expansion of human interactions. Since

184

these are accelerating at such great rates each and all persons will succeed only by learning them, their part in TO and their most effective management and control in TOF.

By achievement, and providing any and all that is needed by TO, is the role, test and proof of every person. What works they have done, what benefits they have given and what reality they have added to TÓ. This is not achieved without a strong and continuous learning and effort in life. It requires the best possible genetics and health from conception through old age and death in all stages along the way. Especially important are training in early life, well-guided experiences, education and cultivation of the best interests and skills. It is from knowledge of TO and ts actual pure essences that all can be best ordered and proven. If one would doubt the great and vital role of actual pure essences let them discover the lessons learned from people who have experienced great and deadly trials and lived to relate their convictions. From Abelard there was an ascension from negative and opposing worlds, and assumption to ever greater levels of one's actual pure essence part in TO. 'Sic and non', yes and no, right and wrong, is not by will, intentions, acts, etc., so much as by necessary identity of one's special part, TES in TRU of TO. The individual, as a thing in itself, must come to understand, act and live by actual pure essences as special parts proportionately, properly and fully to become TO.

As great as an individual's consciousness, attention, and self centered importance are taken the experiences of maturation in life teach us otherwise. Mind is no more the ultimate totality than matter, our persona is no more real than some objects in the physical universe. This must not deny the greatness of one self and mind when their actual pure essences equal their part in TO. It proves that TO does not consist of its extremes only, by ABSTO-RELTO and all TOV. TO is a qualified combination of all that gives it actual, pure and essential existence. This is revealed and shown by TO and its varieties, modes, actuality, objectivity, reality and subjectivity, and the power, potential changes and transformations in TOT future expansion. This power is what gives TO like the big bang and potential of human mentality, much of its form, function, power and ability to reversibly maintain itself and alter TOT to assure TO with TOF.

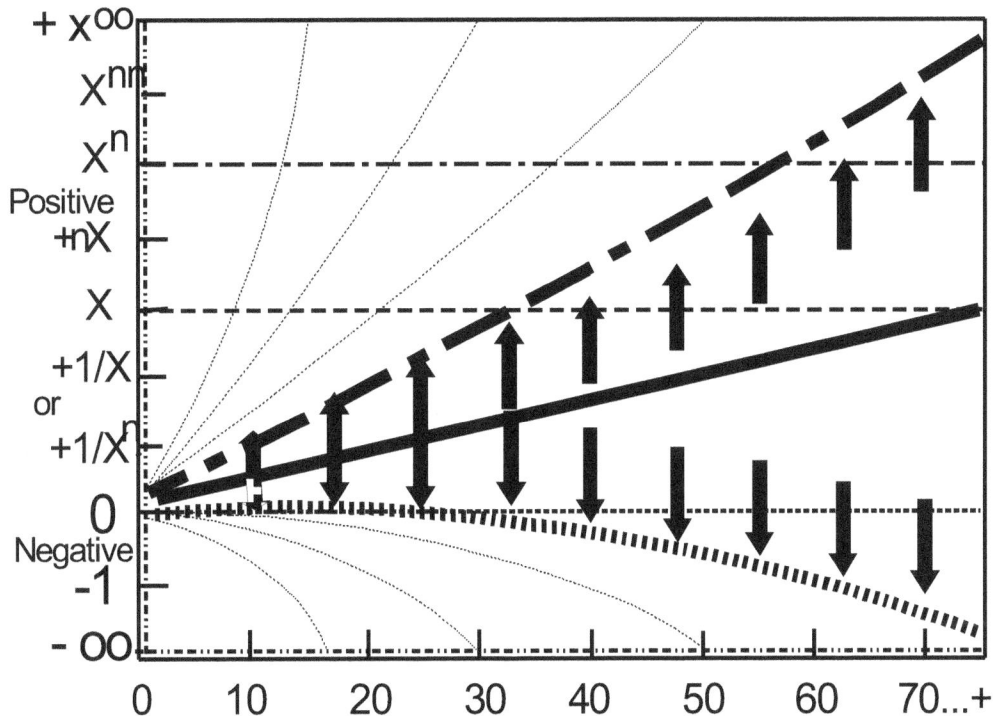

Fig 29-2

Life's Positive and Negative Dynamics, a person in TO, Í-TÍS-TES-TRU-TOV-TO

All people, from one, to groups to the total world population, are in the same state, their part and role in human well being and survival. Individually they do not start the same, only on average for all.

This is more or less positive or negative. If positive and above par, x, in their role, it is known as immortality. The degree this is positive is their Immortality Quotient, ImQ. Everyone should be cognizant of this quotient, for it is the most direct path to making life meaningful, as well as survival, TIS-TES in TO. All that adds to the immortality quotient, which generally is the essence of all that benefits the best of worlds, must be the principle, standard, and test of ones life. The futures in TO, TOF, reflect this progressive intention, positive imperative and continuity of action, in which the more special varieties of TO, driven by best human life, TRU, have an increasing role.

It is the development of identity in life and TO, the right kind of love, feeling, knowledge, gratitude and joy of achievement and fulfillment that must be at the heart of all, persons. This is what enables one to love actuality, subjectivity, objectivity and reality and all that is the greatest of TO in all its glory, and how to prove, save and advance TO and TOV-TOF in eternal progression and continuity. Although it shouldn't be necessary that rigorous training produces the best thinking, knowledge and actions in a person, they can more readily be acquired by native inclinations and experiences in life that stimulate one to be their better self. This comes from experience in those angels that have directed one to better thoughts and deeds, and a growing appreciation and love of ultimate totality, Such were the daemons of Socrates that led him to search for those ideas and essences that are the basis for actual totality, (cf. 20, 30) Among the training routines for developing the best knowledge, beliefs, habits and behavior in all generations, ages, sexes and kinds of people one of the most important and critical for resolving many of the social problems that we suffer from is to identify psychological differences and problems as early as possible. This is especially needed when extreme or likely cause of loss in TO. It is by perfecting techniques for complete management and control. Such abnormal differences and problems include,

List 29-1 Attitudes, Beliefs and Training,Problems and Solutions

Lack of Universal knowledge	Life long learning of TO
Lack of TO or equivalent.	Life long learning of TO
Self Centeredness	Early experience and training
Stubbornness, Intolerance	Early experience and training"
Beliefs, Partial, Wrong	Early experience and training
Impulsiveness, irrational,	Early experience and training
Habits, bad, destructive	Early experience and training"
Behavior, aberrant, asocial	Early experience and training
Psychological, Social	Early experience and training.
Abnormalities, Physical,	Genetic through lifelong care
Criminal tendencies, Misfits	Genetic through lifelong care
Psychopaths, Psychotics	Genetic through lifelong care
Special differences, problems	Target each through TO
All not or less than TO	Learn and Live by TO

To understand oneself, as to understand all, the means and extremes by exclusion and inclusion in

TO, must be known, realized and compensated for. The individual, as by ones soul, must come to understand, act and live by the actual pure essences as special parts properly and proportionated in all TO. Without knowing this one can never adapt nor have a successful way of life. Although the individual may assume independence, or even opposition, they are inclined to TÓ, by the convergence of their own form and function and that of TO, Yet TO and TOV-TRU-TIT-TES only accept and contain what is within, an essential inclusion of that needed for fulfillment and equilibrium. This explains the great divide between any person, the person, TÍS, or oneself in TO. It is at the crux of being human, differential determination that proves how all that enables TO and its relative universe of best human life, TRU, exists in the hierarchy of absolute-relative of TO.

Religion has been the greatest in its contribution to understanding and making vivid the relation between an individual and God, or TO. From the bond or covenant between man and God, to faith, commitment and devotion, the seed planted with expectation and prayer, each person must in their momentary and daily lives continually identify with God, that God which is equivalent to TÓ.

In terms of a person in the world and in ascendence with actual totality, TO, it is simply to,

1. Learn and know what this world is,

2. Learn and know your place in this world,

3. Learn and know what is needed to be in TO,

4. Live by the first three and help others do also.

This is, the most simple basis, keys to, and proof of TO, and to a person's life. When supplemented by all the essentials of equivalent knowledge much of TO is readily available for all.

The world, TO, individuals and TES are pervious to all around them. Each must continually sustain give and take, positive and negative existence. The larger dotted circle is TO, lower right smaller one, TO in decay. Individuals intend to do better, but often tend to do worse. What is needed is to have had, be, and endeavor to become a component of TO that has great physical and mental strength, and enhance all necessary for existence in, and that of, TO. The proofs of TO, and much of the detail of this book, provide the facts that produce such a person.

It is because so much we are familiar with, are in interaction with and follow, is external to, not and opposing to TÓ, that humans have, are and will suffer interminably. This is especially so for the individual, groups, societies, cultures, civilizations and species. They fail, disappear, are ruined and become extinct for many causes, external and internal, mostly because of attrition and failure to sustain life. This is increasingly so internally with human life now and in the future. It is partly due to a loss of optimal natural selection, built up abnormal deleterious forces and failure to adapt to the environment and complexities of modern life and technology. It is the rapid expanse of quantity over quality, of which there must always be an optimal balance. Yet the greatest of all is human against themselves due to failure to discover, experience, be trained, educated and mutually live by what is best in life, or TRU in actual totality, TO. Each and all persons must be their part and fulfill their roles in TO. Without cohesion, unity in totality, tolerance, appreciation, and dedication to all that enables best human life in TO, all generations will continue to repeat the same suffering and misery of history. At worst all will gradually or more rapidly die, fail, disappear, be ruined and become extinct. Each and all persons are no more nor less that their part and role in the whole, TO. This is the opportunity, duty and glory of life for all who make it so in TO. It is from failure to utilize our naturally given capacities, abilities and talents, complacency and lack of right ambition, that has, is being, and will lead us to loss, failure, misery, death and extinction. It is easy to think it can't happen to me, or doesn't apply to me, or I will avoid or escape from disaster, even when all around is falling apart. Human life is at war, whose battles must be continually fought if all that threatens is to be prevented and the thread of life is to be protected, prepared for, planned, managed, controlled, acted upon and properly and optimally lived by.

The greatest problem of all is, the fact that a person doesn't understand, actual totality. This is largely because lack of opportunity, limits of the mind, being poorly known, complex and resisted even when well vertically integrated. The inability to understand TO, is also due to all those factors that keep a person from being their part and fulfilling their roles in actual totality. It is also due to preoccupation with the everyday demands of life, and making a living. Not the least of causes for failure are those media, cultural, sectarian and partial influences that disable total understanding, or are used as vehicle for self and partial gain, or as an excuse in denial.Some of this great problem of modern times, is the fact that each individual and generation, is or becomes lost in the mass of differentia, trivia, figments, illusions and fiction they are born into or exposed to. Without the best of education, and an open discriminating mind, this is a loss they stubbornly cling to, in the belief it is the world, universe or deity that is all. Only until they care, wake up, discover, learn, know, and live by what is right, proportionate and proven of actual totality by a united absolute and relative totality can there be peace on earth and good will for all.

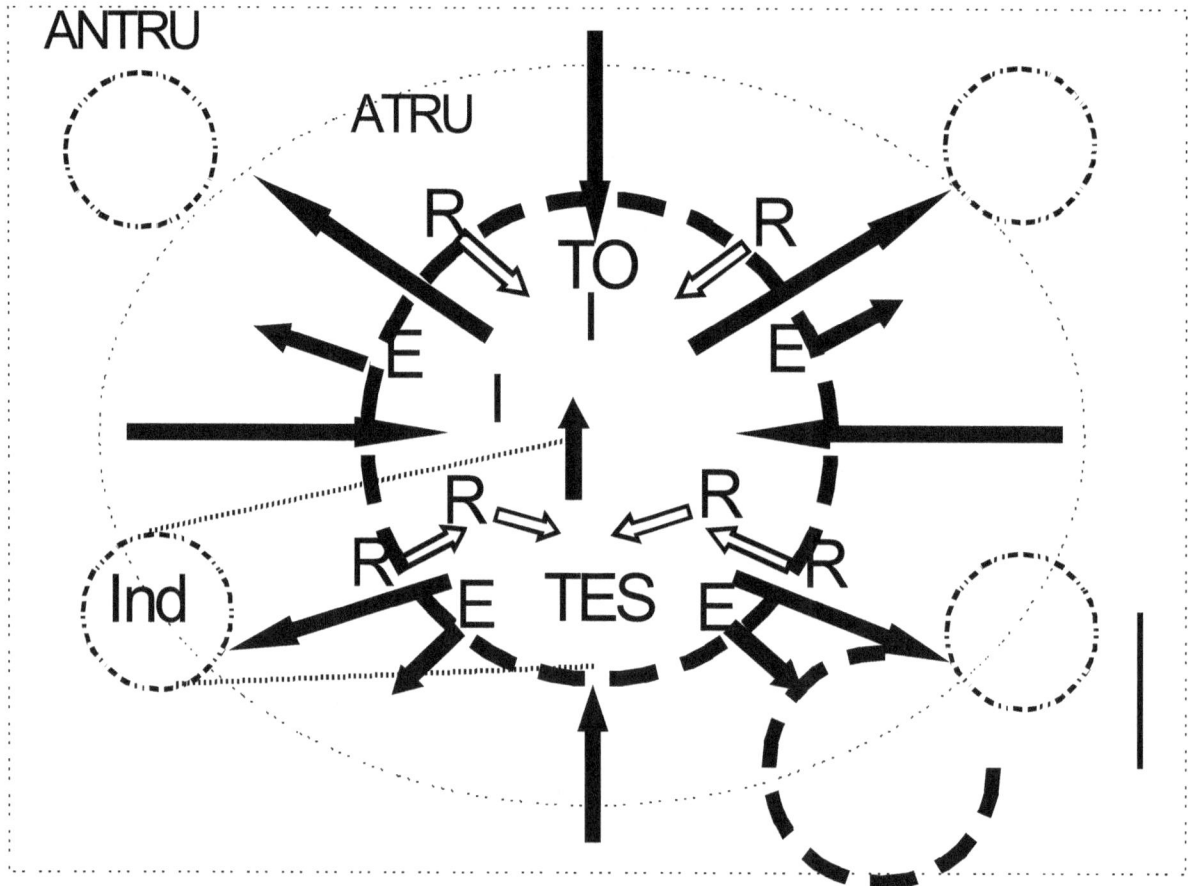

Fig 29-3 Positive, negative action-reaction, TES in TO

Much in actual totality, is the same entity, only of different quality, form and function when separate, and not TO. This Law of Existent and Non Existent Parts, LENEP, is a most fundamental law of parts. In contrast to LOPAN, the Law of Partial Necessity, LENEP signifies that most basic difference between the same entity, how it can be both internally existent in TO, yet also external. This applies to all entities, and is especially notable, and personal, in the lives of each individual. The reason for the state of existence that LENEP, makes imperative, is the fact that to be a part of TO one must be in a way that is typical of TO, i.e., with actual pure essences that meet the requirements of the part in TO. When non existent, even with actual pure essence, the same entity lacks the actuality sufficient for TO. It is in this world of LENEP humans exist, most non existent, more, hopefully, becoming like LOPAN, their part and roles in TO. LOPAN is internal. LENEP is external and internal of TO. The law has wide significance and is similar to the tradition of individual, soul, God, or Christ or Mohammed as the messenger between soul and God. It is the most critical step in the rebirth of a person, their identity, trust and faith in their actual pure essences and its adaptation to those of its part in actual totality. This is why learning, knowing, living by, and applying all actual totality as a guide to ones daily, and entire, life is the first and greatest task of being. Each step in the three stages from non TO, intermediary, and TO, must be understood, especially what the part is in TO, and the intermediary step that led to TÓ.

By the first person and actual totality in climax ones special partial identity is enabled and enhanced. By the absolute love, bond, fervent devotion, actual pure essences and identity of each with others in TO all is secured. This is one's soul, in the divine being, the progression of TÍS to TRU-TOV inTO and by TRIR with TOT, evident from Sanskrit Theology as bhakti. It is the sympathy and appreciation for our special partial that was shown by the great bond between each person in Christ, God and the Holy Spirit. It is this principle of individuation and regeneration in dependent union by the person and TO that permeates and proves their actuality, reality, relations, universality and thereby TO and TÍS. The ultimate identity of the special partial, as ones soul, with the absolute relative of TRU in the absolute-relative of TO, gives differential and integral immortality to both the part, as the person when TÍS, any person when TES, and when together all in TO, and TRU with TOF. A person's duty is signified by LIDAR, Law of Individual Identity and Responsibility. It is the most specific form of LOPAN.

Absolute Love of their Partial Bond and Partial Kind of Identity within the whole, is TÍS-TES-TO, (ALET). This is the ultimate definition of love, the higher kind of relation and bond that the subjective in all its power produce throughout TO. It is a definition that raises much that exists in any way or by any value in a person to their full role and proper value in TO. A lack of knowledge, appreciation and existence in TO produce separation and alienation, a narrow mindedness that not only deprives a person of the benefits of understanding TO and TRU, they continually suffer from this lost of actuality. They retreat into denial, neglect, and opposition to, what is right. All is changed by actual pure essences to have actual existence. This is by their typical form, function, parts and processes in TO. For it is by the actual pure essences in all that when becoming its part in TO, makes the critical difference between any and the, any or non existence, and the or necessary existence of all, only as it is in actual totality. This separates and proves both, all things, senses, entities and TO. This is the existence in, and the bonds that exist among, all that is in TO, especially between TO and its parts. It signifies how love, both generally and specifically, permeate and largely dominates much in TO-TRU. Love being that special kind of bond of dependence that increasing RELTO and best human life add to TÓ, and TÓ reciprocates in turn. The many kinds of love, when of their correct proportions, correct many current disbeliefs, and departure from TO and TO with TOT-TOF.

It is by thought and thinking that humans make a very complex world more meaningful. Senses become ideas, and ideas become words and concepts which when thought about in the right way reveal the great subjective powers in the proof of actual totality. There are hardly any abilities and properties of human life that are better and more productive of quality and success than the right and best habits of thinking. These are not spontaneous. They come from intelligence, interests, training, education and a most stimulating intellectual environment. It is also enhanced by beliefs based on the highest forms of knowledge that is organized, ordered and coordinated by the vertical integration and unity of actual totality. Thinking processes like the composition of great books are the means by which all can be expanded and focused on what is most contributory, actual totality and its use in life.

By the law of partial necessity, LOPAN, all parts in TO are more or less needed, some greatly, acutely, or both. This is no where more evident than in the human body, where some parts cannot be done without even in a very short time, e.g., minutes. It is the same for the world we live in, TO, all its parts are more or less needed. The greater they are as actual pure essences in its proportional parts, the more vital they are for survival over extinction. This is closely associated with much in the dynamics of TO, e.g., proof of integrative processes, PIP, and proof of integrative negation, PIN. Not only is a part often vital. When combined they exponentially compound their necessity for survival and will often combine negatively to weaken and lose the vitality needed for life over death. Personally these are the most direct concepts and guides to life, self improvement TIS and TES. We must be grateful for, all that has made us, make the best use of all we are, and live to enable and assure all we, and TO, can be. This is a lesson of life each generation must learn, and all throughout their lives must follow if all are to prosper, be happy, and give to posterity all it will need. The total and all exclusive value of TO, OV, are what makes it deserve all. We as vital partials by LOPAN, must return to posterity the actual totality it deserves.

CHAPTER 30 - KNOWLEDGE and Certainty, to Higher Forms in Actual Totality

Personal Ignorance and doubt to Knowledge, Knowledge to Certainty, and Actual Certainty in the Knowledge of TO, Truth and Kinds of Knowledge, Organization of Knowledge

The active knowledge that is within each person and even more massive and beneficial knowledge in all persons and civilizations are much of the form, function and substance that make TO what it is and provide its certain proof. No matter how great some entity, object, topic, field, system, deity or universe is in itself or in actual totality you can't approach, hold in perspective or properly know and use it as it exists separately in itself. All things exist only as they are their counterparts in the ultimate unified totality that is actual totality, TO. Their actual state, conditions, limits, relations and necessary existence are determined by their identity, conformity, and dependence in actual totality. It is by learning, knowing and applying this difference, as much as it appears or is held otherwise, that we discover, learn, known and prove actual totality. This is law of different and same discovery, knowledge and proof of actual totality, LADKAST. If it is great in common knowledge, it will be more or less potentially great in actual totality, only in different, and often more essential ways.

Knowledge has been the continual accumulation of experiences throughout human existence, especially since written records enabled more extended and reliable means of transmission. Yet written records were not the only method, nor means by which knowledge was cherished, and love of wisdom flourished. This contrast was even known to Socrates whose method he compared to that of Anaxagoras, e.g., didactic versus written. Plato with essences of ideas, and Aristotle with a more scientific approach expanded on the tradition of Presocratic Philosophy. This they did by expanding and ordering knowledge of politics, ethics, social behavior, personal betterment, and other topics central to proof of TO. Most of this knowledge is known today, but lacking in much of the public. Because it lacks essence, integration, order and the spirit of its originators, actual totality and its proofs remain unknown. With validity, proportionality, actuality, totality and all the types of proofs given, convergence, knowledge and use of actual totality are obtainable. Because they are necessary to exist the actuality of all leads to the summation of the extended actual totality, TO. As Socrates sought to exfoliate the essences of the great ideals of human life, so do we seek to exfoliate the actual pure essences of all the great aspects, features, modes and major components of actual totality, TO. Such is the Law of Exfoliation of APET, LEXAPET. In this way the most profound and subtle knowledge of TO, and its greater proofs can be found. The coincidence of these approaches, their answers, results and effects is further proof of TO.

Although poorly and highly variably held, the beginnings of knowledge of actual totality have been slow. It is largely a product of the development of humans, their brains in relation to the world around them. There has always existed, in the past, present and hopefully less in the future, many who know neither the world they live in nor themselves. Some American Indians, a hold over from paleolithic times, believed in a great spirit, how nature and their lives worked together in unity. Ancient Greek Philosophers with the benefit of writing, better language and a religion and culture that championed wisdom, strength, and refined artistry, held a series of beliefs that culminated in a philosophic approach to life and its many skills, fields and sciences. Yet even today, as in those days, the majority of people have little to no accurate knowledge of actual totality or its equivalent. The strength and survival of cultures have largely been a function of how well it is ordered by its most fundamental knowledge. Many of these basics have become a part of knowledge, yet are not known or ignored by most people. We cannot be misled by lesser principles, priorities and actions. We must be guided by the accurate and great knowledge that has been given us and bring it together, as it is by, of and in actual totality. We cannot find our purpose in life randomly, with denial or with little effort. Knowledge of actual totality is the most serious and greatest hope for mankind, now and in all future. It must be correct, available to all, made palatable, with the best training and education possible. Whether it is called by the name, or acronyms, of actual totality is much less important than its equivalent that exists within much the same approach, constituents, form and function. Like absolute totality largely exists in actual totality as the physical universe, so does relative totality largely exist in actual totality as the relative universe. This is somewhat different for God, which is unified and total, only lacking both knowledge and interest in developing its basics. The different kinds of actual totality, and God, are much the same both by virtue of their purpose in best human life, and the rule of mind and identity. They are different in several respects, mostly by the fact that God, as a major component in TO, is much different from how God is known in various religions. Yet God as a part of TO, has, and continues to add much to TÓ, both by what are the same as relative universe and what is different. It is the actual pure essences of all, modified in TO, to be their part and role, that is key, and proof, of TO. If they add the same ingredients to TÓ, the credit goes to both, depending on how well they exist in, and support TO. How much it, and its essences, has been a part of actual totality in the past, whether known by that name or its own.

Currently humans have collected an immense quantity of knowledge. It largely covers their world, more or less needed for their existence. There is much detail of poor quality especially when used negatively, lost in trivial detail, lacking vertical integration and without organization. The state of knowledge in the modern world is both highly beneficial and harmful. All, more or less, partake of the one right totality in a very common, unqualified and disordered way. This cannot be the ultimate unified totality without the highest level of validity, actuality, proportionality and totality. Their most fundamental doubts have led to false beliefs, confusion and turmoil in our ideas, words, purposes and lives. It has made consensus and proof impossible. If the physical universe, God, Allah, or humans are perfect, or even mostly right, actual totality would be unneeded. However, they are not well based, or complete, in fundamental ways. To this extent, the needed actuality of TO, increases proportionately. Questions and errors in knowledge are often found in what they leave out. Answers and corrections are often found in, and are proofs of, TO. There have been consensuses in various ideas of one God, one world or one physical universe. Yet these have been of varying quality and probability. Their needs have been expressed, but a consensus in the unity of ultimate unified totality has failed. They have not met all the requirements and proofs of TO. As long as an animal, humans, knowledge, TRU and TO are uncertain, they are suspect, in doubt, suspicious and likely to fail. To acquire certainty is the great driving power of higher animals and when fully rational and comprehending the great characteristic proof of TO and TRIR-TRIS-TIT-TRU-TOV. By relation their purpose and goal are to thrive, survive, perpetuate and seek permanence through bonding with, or a predominance over that not, less than or in opposition to TÓ. It is the fruition of actual pure essences in TO and TRU that the critical and absolute necessity of humans in all that is of absolute to relative totality occurs. This is especially so if they are to maintain and sustain best human life. How imperative certainty overcomes uncertainty is what we must have and live by if we are to achieve what is most precisely and fully right, in TO, its TOV and their validity, actuality, proportionality, and totality, (VAPO).

Failure to achieve the benefits of knowledge does not only come from ignorance or doubt, much comes from our lack of adopting and adapting all the great knowledge we have to its part, role, use and benefit in TO. When we rise to understand these, we will show how the level of knowledge in modern fields and sciences is not well used by the general public, nor appreciated and lived by without a like high level of knowledge of TO. It emphasizes the disparity between us, best of modern knowledge and knowledge of TO. This is the disparity of separation of non existence, poorly existent, and its level of opposition to, TÓ. (DOSTÓP). This is the kind of world we now live in. To seriously understand this is the beginning of a good realization of the proofs of TO and the more effective transformation of all to TÓ. Knowledge and all that are major proportions, or divine, in TO, are necessary, APET, and large parts and proofs of TO. This seemingly obvious fact is also most subtle and profound, for without realizing the great components, and TO itself, all is less, lost and the sooner all becomes extinct.

Detail, like differentiation is a great part of actual totality. Like vertical integration to the unity of actual totality, differentiation is equally important to provide the necessary equilibrium in this dimension. Yet neither integration nor differentiations stand alone. They are highly interdependent and combined in the unity of TO. They consist of numerous kinds. To understand TO, these kinds need to be known, and to be known they, needed to be ordered, as all in TO, by its categories, modes, kinds, varieties, form and function. By such criteria, differentiation and detail are measured. Detail, unlike unified integration, presents more diverse problems. The greatest of these problems are the limits differentiation must have to be practical, functional, and meet the requirements of actual pure essences in TO. The solution to this problem, as many problems, is to look at, analyze and develop it as a continuum, from the simplest integral-differential to the most complex. Then develop knowledge that best suits this level, but never being satisfied, and continuously perfecting and expanding it for improvement and successful use. Each occurrence will need to adapt to the use and level that best suits the circumstances in life.

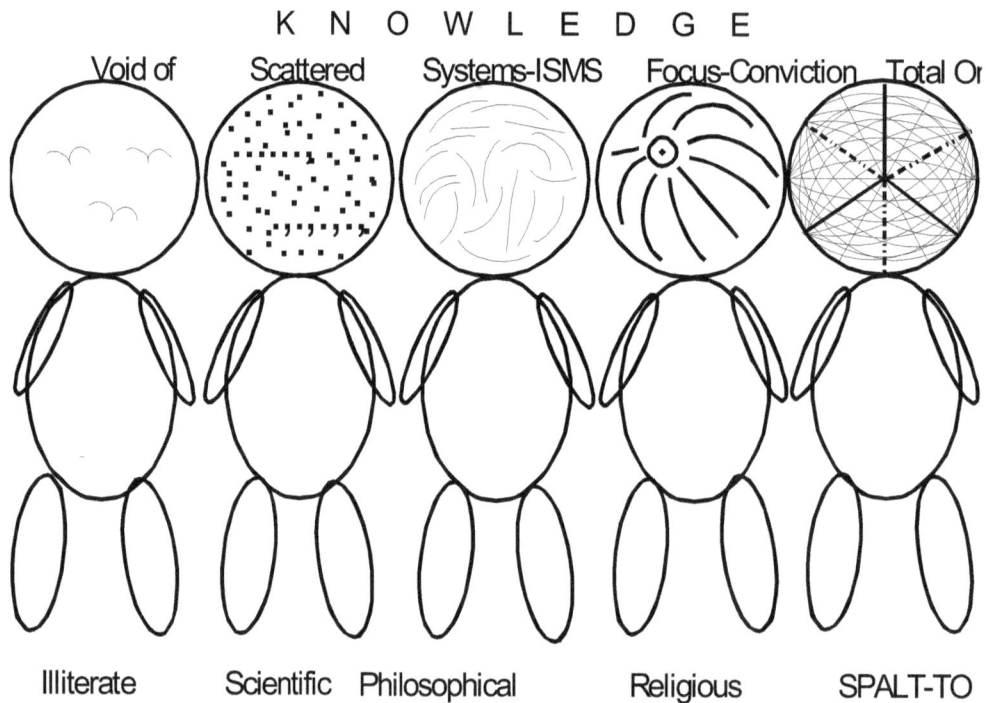

Fig 30-1 Kinds of Knowledge, Comparative Mentalities and World Views

This figure is not intended to cause any lack of honor or appreciation of the great contributions and, experiences of human life, science, philosophy and religion have, are and will, provide. They are large parts of TO. In order to get the most out of actual totality as a topic, it must be shown both in contrast to what is not, less, or not directly of actual totality. The great purpose of the text - as much of TO - is to follow the best available knowledge and how it can be interpreted and presented so that each person may gain optimal TES, TOV and TO. If metaphysics and epistemics are ever to be concrete philosophies and sciences, they can't be limited to either bits, pieces, or the polarities in which they exist. It can do so only as it includes all, and these by VAPO. Even when we extend physicalism to realism, we are still operating in a limited field. This is the Law of Limits of Isms, Oppositions, Polarities and Operations in Knowledge, LIPO. Overcoming these limits is what enables us to rise above and break the bonds and confines of little minds and bring them to their full role in and of TO. Isms tend to be carried to the extreme, become partial and reactionary. They deceive us, can never be correct, will always be defective, and harmful. Isms seek to champion, adapt, and apply what is disproportionate of TO. Metaphysics, is possible only when TO, or its equivalent, TREQ. Perspectives of each and all, are no more nor less than their form and function, parts and processes or roles in TO. This is why a good knowledge of TO, its models and representation are so essential to the right way in life. Actual totality, TO, is a product of developing knowledge, including the anthropological experiences of the illiterate, the many great observations, experiments and discoveries of science, the profound wisdom of philosophy, and the faith oriented centering of totality by religion. This diagram is not intended to lessen their roles, it only seeks to stress the vital need, features, form, and use of TO.   Each case must be interpreted by its own merits, and all must be reconciled in the unity of TO, and TO future, TOF. They often define their differences and thereby are key to explaining the DAOST or different aspects of the same thing in all. Laws arise from the need for limits and the inequalities of opposing differences. That laws arise from the need for limits imposed by differences are most basic, the law and proof of TO. LANLID, It is in this way, the limits imposed on, and imposed by TO can be discovered, known and proven.

Truth, reality, and the commonly used word, 'is,' have much in common. They depend on, and to the degree they depend on, identity of the observer with the observed. This is relativity, without which the physical universe has a different, incorrect and incomplete meaning. It is the meaning relativity gives to TÓ that proves its existence and difference from all other worlds. This is the Law of TO from the observer observed meaning of relativity, LATORM. Truth and validity are nearly synonymous, only validity is more scientific, emphasizing evidence and proof. It is how to bring truth to validity, and validity to actual pure essences and their part and role in TO that ultimate knowledge can mirror the ultimate unified totality, TO-TOT. Both truth and validity increasingly identify with what is necessary for existence, its actuality. What is positive over negative and only if it is actually relative to the observer. This provides a truth, whose validity, proportionality, actuality and totality can be known when we realize its part and role in TO. Knowledge cannot assume any being, truth or even validity. It must rise to its actual pure essences level in TO and includes as much different knowledge, as TOV-TRU-TIT that is practical within the correct limits of the observer when TES in TO. Truth arises from the depths of existence, necessity and actual pure essences in actual totality.

The modes necessary for knowledge and proof help form, organize and advance knowledge in and of TO. All actuality, objectivity, subjectivity and reality prove existence, both of TRU, and how TRU exists in and proves TO. The greater the subjectivity or observer the greater the reality, the greater the reality the better the knowledge, and greater the knowledge the better and greater the proof and existence of TO. This is the great role of the modes of actuality, objectivity, subjectivity, and reality in TO and TRU. By their actual pure essences or actuality, TO and TOV-TRU, maintain sufficient existence to sustain the highest level of certainty necessary for adaptation and alteration in time. This is why it is critical that all learn, be their part and fulfill their role in TO.

Knowledge sufficient to bring certainty to actual totality is not simple. It is much more than the use of existing ideas of wholes, totality, unity, actuality, existence, necessity, same and other, subjectivity, objectivity and reality. Any use only becomes a vicious circle released from aleatory tendencies of the uncertainty of differential oblivion. This is what any events, contingencies, relations of good and bad, profit and losses produce when not vertically integrated properly. Much remains unknown and thereby uncertain, and much more will become known and made certain. Not only are all the pieces of the jig saw puzzle of TO incomplete, those of TO with TOT will require continuous completion and certainty. Yet when we build on all that is actual totality, its modes, continua, kinds, varieties, and fuse absolute and relative totalities together their proofs converge in TO. It is this that yields a high level of certainty, to provide, show and guide our lives in actual totality. The actualities and operations of all 1,000 parts in SPALT of TO give the precision and order that clarify knowledge, its basic concepts, principles and vertical integration to differentiation, and prove TO. Such an objectively accurate approach and representation help to orient all to the unity of TO, although it is limited to the practical present and form of the averaged sum of TO.

Though universal knowledge by vertical integration to unity and exclusive oneness of TO is the keystone of both TO and human life, it is not all. The part and role of each and all persons must always be in TO, with limits that necessitate knowledge that fits their capacity, ability, talents, inclinations, interests, experiences and best contribution to TÓ. This means that the quality and quantity of knowledge of TO will vary with the universal, partial, special and essential knowledge to provide optimal effectiveness and efficiency for equilibrium in TO. This and a person's epiphany, devotion and continuous transformation into TO are the vehicles by which humans sustain and advance TO and all it consists of. It is the major theme and story of everyone's entire life, their guide and path to quality, union and immortality in TO.

Degree of confidence is a statistical expression of the probability that the interval-random contains the parameter fixed. This is equivalent to saying the subject contains its object, or the subjective half of the truth equation. Proof of TO, depends on its basic APET truths. Truths depend on correct meanings, and the reality agreement of subject and object. Yet, subject and object, are limited to their adaptation, alteration and warping of the spectrum of the mass energy dimension they are in by TO and TIT. This is produced by the correction supplied by the actual pure essences of all that enables TO. Thereby degree of confidence, truth and proofs are convergent on their actual validity in TO and TIT.

Of all the factors, forces and causes that maintain and assure actual totality, some are of basic mechanisms, some are major, minor and of less significance. Thereby the maintenance and assurance of actual totality are both profound and subtle. Yet it is one, if not the greatest, of all features of TO. To be sufficiently accurate many of them need to be given, depending on value and need of the determination. To meet this most that is available are described through the text. By value none are greater than the mechanisms of natural selection, self determination and powers of higher knowledge and achievement. Although natural selection and the external internal inter actions of the absolute and relative have been the major factor and force that maintained and assured TO, they and self determination, alone will not be adequate. The future, and increasingly with time, the maintenance and assurance of TO will, and must, depend on the highest factors, forces and powers of knowledge and achievement for best human life. This is much of all that RELTO and TRU can and will provide for TO. Even all that is unknown, or incompletely known can, is, or will be, a major proof of TO. This is why it is imperative that the searches for TO, and unknowns yet to be discovered, are a paramount concern for each and all persons, at all time. To be sure, our knowledge of TOT, TO and TP is the best it can be.

The Proof of TRU, PRUT, is obtained with the knowledge of TO, KOT, and especially by understanding the relationships of TO, as seen in the interactions of its concepts, laws and proofs in summarizing representations. Each person alters their position in all that is less than, not, or negative to TÓ, OAT, to best return to TO-TU. These together bring one and all to TÓ. How they revert to TÓ is, the Version, Reversion to and Verification in TO, VERTRU. In a sense AU, RU, and even that less than all TO, is superfluous. For all that is other than TO, or less than its totality, is to that degree, superfluous. This is seen in the differences in TOV and RELTO, even entirely within TO. Yet, as knowledge of all is, more or less, necessary for the individual relation to TÓ, so does knowledge take some precedence, seen in epistemics of TO, EPI. This is a paradox and fallacy that, if not properly held in mind with others that qualify TO and its observation, are less able to discover, know and show TO. This is, the Law or Fallacy of Perspective Precedence. LOPRE or FOPRE. When of a fallacy it is a part of the loss and limits of TO, UPTRU, and other LOLS, Laws of Limits.

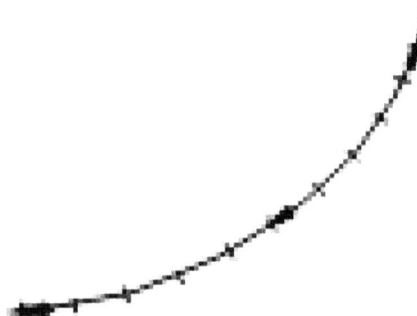

Fig 30-2 Accelerative Problem Solving, with increasing modes and kinds of TO

This figure serves to initiate the enablement and enhancement of understanding the discoveries and solutions in TO-TRU-TOF. How in the whole course of human development there has been accelerated expansion of our ability to solve problems. This has been the result of many factors, as evident in the modes and kinds of TO, or increases in our mass energy capacity and ability. All that has contributed to human mentality has participated, including perception, apperception, attention, anticipation, ideas, language, thought, reason, planning, discovery and comprehension of the world. This has been supplemented by art, technology, communication, industry and business in an ever increasing variety, depth, use and effectiveness. So that, even within the limits of individual ability, ever greater solutions are and can be provided. This makes up for some, but does not eliminate many human shortcomings and threats. These also must be solved. The great acceleration of problem solving has not only been by individual ability and technical knowledge. It has been by the imagination, ingenuity, determination, will and far reaching purposes and goals of life, seen in TO.

Cultural perspectives both worldwide and in various regions at different times in history, present and future, are very important evidence, indicators of the background and proof of actual totality. Their great changes have been most amazing in the variety and development of knowledge. Although it has more or less occurred in all languages and cultures it is in those that are better documented that we can find their part and role in the development of actual totality. This has been no where more evident than in the Ancient Greek Religion and especially Philosophy from the second millennium BC until into the Roman Period. Indo-European Religion developed an eclectic system of deities that increasingly honored knowledge and wisdom, e.g., Apollo and Athena. The Ionian and PreSocratic Philosophers left a developing trend of open questioning all, to steps in focusing various approaches, arguments and methods for certainty. Beginning with Thales and Anaximander this was expanded on by Heraclitus, Democritus and Anaxagoras. They and the associated cultural milieu of knowledge and writings in knowledge bore the great fruit. This culminated in better documented works of Socrates, Plato, Aristotle and others. Knowledge since has increasingly expanded in many fields, sciences, topics and skills. Yet this hasn't solved the need for knowledge of actual totality.

Current knowledge is vast, most valuable and the driving power of humanity. Yet it is poorly qualified, ordered and highly abused. It is how to settle the arguments that inhibit knowledge of whether and what the whole is that is the challenge of today and the near future. That there is a unified whole is a categorical imperative demanded by all that makes up actual totality. It is also logical that the diverse state of beliefs and knowledge in the physical universe, science, philosophy and religion lack both natural and normal unity. They defy the unified purpose in human life, its quality and survival. This is why it is of the highest importance to both compare and correct current knowledge with what it can be, as we seek to find knowledge of actual totality.

We can never know from the bottom up, the top, one side, another, or in any way, what is this wonderful one and only-world we are parts of, until we discover, develop, learn, perfect, order and proportionately best apply all ideas, concepts, laws, principles, and proofs of actual totality. It can only be done by knowing, proportioning and ordering actual totality when reversing vertical integration from the top. This is the only way the right kind of ultimate unified totality can be achieved.

CHAPTER 31 - Higher Powers of KNOWLEDGE, to Survival

Knowledge of TO, truth, future TOT, purity, reality, actuality, Pureaclity, TIS-TES-TO, knowing one knows, Cognos, kinds of knowledge, KUPS, benefits, exponential sufficiency and completion, purposes, providence

It is an infinite spectrum between the singularity of instantaneous attention or thought and the whole of knowledge, as well as that of actual totality. For one or all persons to master this spectrum in a way that is an accurate unit for all time use is an end to which knowledge, the human brain and actual totality converge. Such personal powers of knowledge, thought and their role in life is not out of reach. For the human brain and mind have almost unlimited powers of comprehension when cultivated. This potential is what a person can achieve, when their part in actual totality has been given, and when well conditioned. Since to know, and to know correctly, practically and most successfully are the hallmarks of being human, this greatest of assets of relative totality is only in the middle of its great transformation that assures TO. This is largely the duty and work of its higher powers whose growth has, is and will change human life in ways many don't appreciate and only actually realized by a few. Such higher powers include discovery, reasoning, logic and right judgement based on proof by the most valid and proportional world view, TO.

The proof of anything and everything, is in how well it works, does it serve and fulfill its purpose? Are these precisely as they exist in TO? This is not only a basic principle of functional value and actual pure essences. It is of all actual totality. Nor is it simply of actual totality, it is how all in and of TO must be formed and functions. And most directly, for each and all persons, it is the degree to which they work well, serve their purpose, are their actual pure essences in their part and contribution, as well as rewards in the survival, success and well being of TO. Since truth isn't absolute by LACERA and LARES, (Ch 12), the problem is how are higher levels of truth and knowledge achieved. It was Protagoras, who held that all things, are only relatively true. The flaw in this was quickly picked up by Plato who held that the proposition made itself false. Yet the greater the truth the more complex it is. Simple truths, like sound bites are often, more illusion than truth. There was much truth in Protagoras' thesis, and error in Plato and many since. They tended to dwell in simple absolutes and failed to realize the complexity, limits and absolute relative spectrum of truth. Higher levels of truth and knowledge are not achieved by simple words nor propositions alone, even when suggestive of great truths. They require much that is given in this book, most notably the unity, multiplicity, diversity, proportionality and order of the categories, modes, continua, kinds and varieties of actual totality.

It is the overwhelming form and function of what is positive and right in existence and necessity that most identifies with TO. Although individuals come and go, and usually leave little marks on, or contribution to the world, this is attained by the sum of all, and what is positive and right. Right is carried from generation to generation, the history and entire development of human life in time to become the great power of its mass energy, actuality, knowledge and reality in TO and TRU. By identifying with what is right by TO-TRU-TOF, each person can escape from oblivion, and greatly achieve more eternal, life. By this we find consolation and joy, not in any social praise or condemnation, material gain or loss, but by the ultimate exercise of our total, special, partial, essential and combined knowledge as it is actively lived by, in, of and for TO, and TO-TOF. The power and force of all that is right and positive not only reveal many of their mechanisms, processes and interactions, they help to substantially prove TO, and guide life in TRU and TOF.

The absolute value and divinity of the mind, a part of the mass energy form of TO, were unknown before humans. Even in humans the mind and knowledge are variably treated, inadequately regarded and often neglected. The human mind, its power and most special and unique form and function do not exist in the astrophysical universe, at least in the nearer reaches of space and time, nor without relativity or quantum mechanics. Because of its value and divinity the mind must be raised to its highest levels if TO is to survive and flourish. When actuality purity and reality combine it is called PUREALITY, and when they bond with the VAPO, SCARLS, essences, qualities and other features they provide the knowledge needed to reveal the most important form, function and proof of TO. When the greatest features and forms of TO are joined by other vital aspects, divisions, parts and processes and integrated into the unified whole, we can begin to realize and identify the intricacy, beauty, omniscience of all that is actual totality.

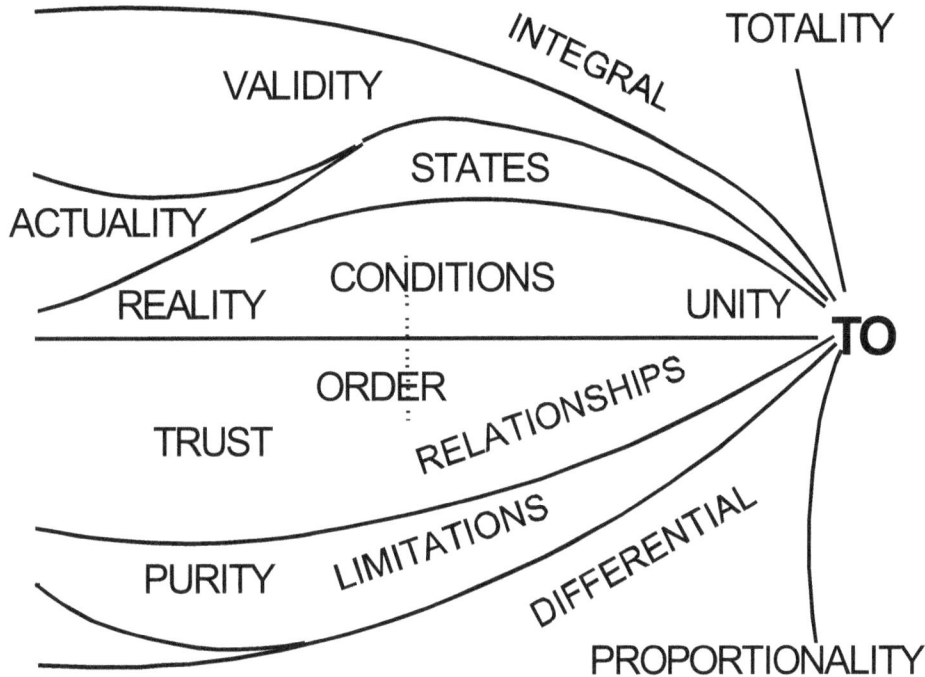

Fig 31-1 Pureality, VAPO, SCARLS, and qualities that enable TO.

The Law of Actual Pure Reality, LAPRIC, signifies how they exist and are formed together in Completion. It is neither separate existence of basic qualities, nor their separated existence within TO that give them their greatest value, strength and power. It is how they are formed, exist and function together in a variety of ways that bring them, and TO, to fulfillment. Pureality, although chiefly focused on a person's quality in life, pertains to all components of TO. For each person this means that if we are to overcome and overwhelm all adversaries, avoid decay and extinction, we must have the purity, reality and actuality that are combined in TO. The capacity, means and will to coordinate, order and command individually, by groups, all mankind, and all that is and is apt to affect TO. This is the heart of enabling the survival of TO. The Law of Order, (LO) is not of any order or law, it is the order and law of the unified whole, shown by the optimal form and proof by major qualities of TO. Only with a thorough, complete, and ordered knowledge can the evidence and information be provided to show what is negative, trivial and in flux, and what is most positive, proportional, permanent and total, TO. The pureality, or purity, reality and actuality that are necessary for each and all persons to aim for, have as a goal, and achieve as much as possible, is equal to the quality, subjectivity-objectivity and necessary existence of TO and all its components. This equality reflects the reversible bond that exists between the proven TO and its components, how their basic qualitites and modes of necessary existence are identical, as much in TO is the same. It is how ones soul, TIS and TES, exist and function in TO.

The kinds of knowledge, required by each and all people, are major determinants of their survival, success, soul, state of Pureality, TIS, TES and permanence of TO. It is signified by, KUPS, or 1. universal knowledge, 2. essential knowledge or universal becoming actualized, proving TO, 3. particular knowledge or increasing knowledge of the differential and detail of TO, 4. special knowledge or increasing knowledge and practice of ones skill and work in life, 5. partial knowledge or how to be a good handyman, active citizen, member of society, husband, wife, father, mother, child or relative, 6. Practical knowledge or how to elevate all by a positive, hopeful and cheerful use of actual pure essences whose interactions bond appreciation with principle and the joy and success life is provided with, by TO. The sum of these kinds of knowledge, KUPS, is exemplified by common sense. It is the sensible knowledge individuals and cultures gain by experiences and thought, when combined with great meaning that serves them best. It is like using ones ideas and mind effectively and essentially, e.g., Socrates and Plato. The most successful cultures of the past had populations that had superior sensible knowledge, i.e., inclined to, TÓ. Together KUPS provides a mastery available to everyone, a simple and most effective means of action in pursuing TO and their own well being and continuity. It is by KUPS that one can most readily achieve pure-reality. The essential nature of different kinds of knowledge, both separately and together, were known in Presocratic Time. Pedagogy, or art and science of education, was the needed supplement to experience in life, the solution to an ever more massive, dynamic and complexly opposing worlds. The flowering of Ancient Greek culture in the fifth and fourth centuries B.C. shown by the great perfection of many skills proved for all time the direction and benefit that different kinds of knowledge can provide. This was especially so when they could be guided by the unity of universal knowledge, nous, a cosmos, universe or totality that necessarily exists, as in actual totality.

Of the kinds of knowledge each and all persons must have universal knowledge is prime. This is the Law of Universal Knowledge from One, or Unity is Prime, LUKOP. Knowledge can be based on neither the infinite, infinitesimal, nor a singular unity. All produce diminishing returns, loss of orientation and confusion. It must come from the continuum, infinite to infinitesimal, or absolute to zero united in its own typical medium, cf., Figs 21 1-2. The initial view and coordination must come from one, the unity of the totality that most effectively, efficiently, practically, functionally and correctly is actual totality, TO. Universal knowledge based on LUKOP is a major proof of TO. Without it, like a vague deity, we lack the right knowledge and appreciation for the whole, TO, and are forever lost. Like lost sheep the public knows not, nor desires TO, or its equivalent, TREQ. The right knowledge is all that is of TO, which when perfected is to know, and know that we know, by COGNOS.

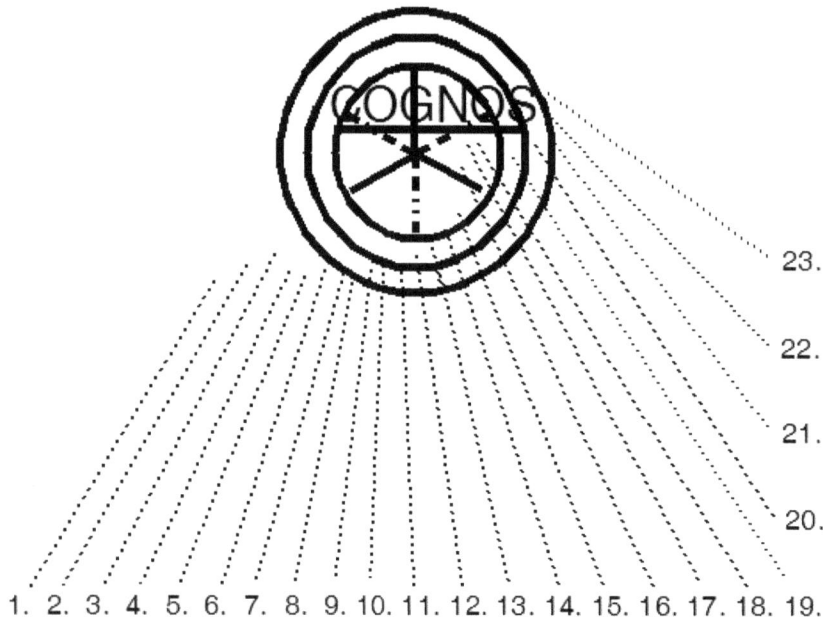

Fig 31-2 Components of COGNOS, from unknowns, to knowns, essences, actual pure essences and highest total states of organized knowledge of TO

The Ultimate Objective, (TUO), and purpose in life, is not a narrow and external exploitation of the material or objective universe, nor a narrow belief and adherence in a deity, religion, field of science, system of philosophy, economics, or politics. It is an internally subjective and objective bond in actual reality, an approach to the mass energy form and function of all TO, demonstrated by its knowledge of TO and certain constituents, TOV, SPALT and COGNOS. COGNOS is the highest stage of knowledge for a, any, and all, persons. It is the sufficiency in knowing that 'we know' and knowing with certainty that what we know is right. It is the ultimate polarity from no knowledge to perfected knowledge in TO. COGNOS is, in and out focusing, with infinite subjective power, of and on all the multiple faces of actual totality. It is total cognitive powers of absolutivity fusing with TO. It is the most special proof of TO and a large target of our immortality. The convergent greatness and power of COGNOS and its proof in and for TO, are better appreciated when we realize its need in the future, as well as all those steps that have caused its formation, including, Unknown, never known, 2. Unknown, not likely known or to be known, 3. Unknown, possible to be known later, 4. Unknown past or present, probably known later, 5. Unknown, not known in past, more later, 6. Unknown now, but known past and future, 7. Known slightly in the present, 8. Known moderately at the present time, 9. Known simply and separately, 10. Known, much of mixed AU-TO, undifferentiated, 11. Known much of mixed AU-TO slightly differentiated, 12. Known much of AU-TO, moderately differentiated, 13. Known much of AU-TO, well differentiated, 14. Known of AU-TOT-TO and well differentiated, 15. Known of TOT--TREQ and well differentiated, 16. Known of TREQ-TO and well differentiated, 17. Known of TO, KOT, 18. Known of TO extensively, OM, 19. Known with actual pure essences identity in TO, 20. Identity in everlasting life by instantaneous, coordinate, massive attention, ELTICMA, 21. Known by COGNOS, fully to infinite, 22. Known by COGNOS to TO completion, and 23. Known by COGNOS to TÂ completion.

The likeness of the radiant effects of COGNOS on knowledge to that of the hands of past deities, reveals their common base of TREQ - TO. From this demonstration knowledge is not a simple state. It is of gradients toward fulfillment and perfection in COGNOS. It can provide great advantages, force and power for one and all in TO. It approaches the ultimate means from epistemics in TO, accelerating KOT exponentially to TÓ and TOF. This is how COGNOS, as THESÂ, advances through THESOÂ toward SÂ and TÂ. The most profound meaning and reality of COGNOS are its actual pure essences in TO, the identity of the knowledge of each and all persons with TO. COGNOS is the cognitive power that highest subjectivity-objectivity realizes its identity with itself and all in TO. By the subjectivity-objectivity and relativity bond evident in the neuromuscular the convergent fusion of mind and matter in TO is shown and proven. It is how cognition and conation, idealism and materialism and thought and action are not separate, but in TO, two different aspects of the same thing, the mass energy S-O state in TO. Conation and action of plans, prevention, protection, maintenance, control and advance of humans in TO-TOF achieve the predominance and mastery necessary for survival. The mechanisms that largely provide this bond are the actual pure essences of the means that bring separate objects and subjects to their mass energy, and the essential, functional and practical effect in the bonded actuality of TO.

We receive, and must in turn, give many benefits and blessings in the highly interdependent life of actual totality. In order to do this, by the positive imperative of our essential existence in TO, our purposes and goals must be right, and include,

1. The exclusive sanctity of best human life in the special absolute-relative variety of actual totality
2. Honor and uphold all the greatest and most basic aspects, features, modes, form and function of TO
3. Take heart and joy in the sacrifices you make that assure goodness, greatness and permanence of TO and yourself.
4. We must understand the reasons and proof of why all that is TO, and not any universe, totality, or other world.
5. Always remember and live by the greatest of TO, e.g., its differential-integral form and practical limits.
6. Honor all equivalents, as the physical universe, God, philosophical systems and other worlds which to the degree they are in TO, is their justification.

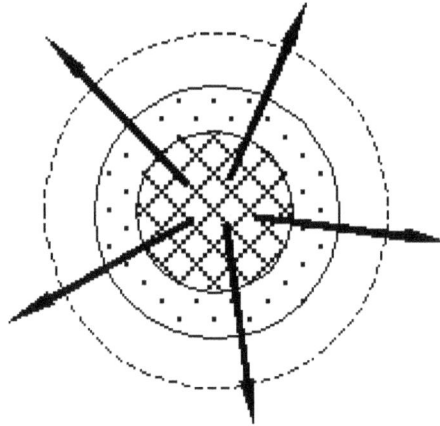

Fig 31-3 Exponential Sufficiency and Completion of TO

LEXS is the Law of Exponential Sufficiency. CEX is the Completion of the Expansion of TO. The central meaning of COGNOS lies in the advance of S-O reality and actuality of/in TO. It is a filling up to completion. The sufficiency of TO is fulfilled when highest knowledge and all, needing to be known are known, as VAPO induces COGNOS to equal TO. In this way "we know," and "We know that we know, and know ad infinitum." Thus we know what to do, what is not necessary, what not to do, and all exponentially sufficient unto permanence, or the infinite, TI, and TA. TO as TÂ is exponentially sufficient, law of exponential sufficiency to TA, or LEXS. Yet this is out of itself not a decreasing to zero effect. It is one of exponential perpetuity. It is a symphonic and harmonious expansion of each and all of TO. All polarities, including the objective, are in harmonious expansion in the tremendous success of TO. As S-O, or Á, fills up there is simultaneous broadening. There is a constant equilibrium of this filling up and the broadening of S-O and all dimensional formulation and ACEXIN of TO. By systematic unity and equilibrium filling up is by the concretion and construct of TO. To a large extent the equilibrium yield of TO, although massive and self sustaining, is at a somewhat variable rate, a state in which RELTO and TRU grow, yet TO doesn't appreciably lose its this-ness, haecceity or special universality. Epistemics, perspective, and presentation of TO may be divided into two portions, out of and in TO.

Glory in Act, (GAC) - It is in the contribution action's make that is their measure, and a large determinant, degree, and proof, of TO-TOT. If the same spirit, conviction, energy and effort are given all that enables and benefits TO commonly given to lesser actions and goals, how much greater we and our world would become. The Ultimate Maxim or Principle. (TUM) The force and action to be used, can be neither more, nor less, than its form, function, part and role in TO, TIT. (LAMOL) The greatness and power of proactive knowledge and its higher states are given by relative totality in actual totality is offset in the greatness and power of absolute totality and the physical universe. Relative totality and the relative universe of best human life are most predominant in the short and into the mid term. Absolute totality and the physical universes have been most predominant in the long term. What, and how it will be in the long term will be more or less depending on the power and actions of the relative universe and TO. The great polarity and disparity of the cosmos and nous, or physical universe and mind, have, are and will be with us, within and especially outside of TO. To bring and hold them together as they exist in actual totality is the most fundamental duty and endeavor of our lives, LARUTO, cf. Chap.12. Human knowledge has made giant progress in the discovery of the physical universe in modern times. Human knowledge has more or less held great belief and honor in the human mind. All its manifestations and application are making the relative universe of best human life, relative totality, and - although unknown as such - actual totality, the great unified totality it is and can be. Yet we must always not only maintain a balance of these as they exist in actual totality, each and all persons must be trained, educated, learn, practice and live their lives to honor knowledge, and especially universal and TO knowledge, as it sets the pace for all else. This is the existence, and main purpose of actual totality and all its parts, especially you and I. Without it matter overcomes mind, and survival is overcome by extinction. The knowledge, beliefs, attitudes and culture of a society, have, are and must, more properly, reflect this great condition of life. It is what we must do, the positive imperative of actual totality and our lives in it.

Although providence, hope, will and achievement are thought of separately, when combined their common roles are key to TÓ and its forward propelling and compelling power. By the providence, of all that has enabled, maintain and assure TO, by the hope of one and all persons for a better life, and by the will of new generations to seek unknowns and horizons that provide opportunities, fill voids, or settle problems, by sufficient and correct will to see it done, and by the kind of achievement that most effectively and efficiently attains TO, their combined roles become more recognizable and powerful. This is a major proof of actual totality, and it helps to reveal and give us the determination to devote and dedicate ourselves to best human life by RELTO, TOV and all in TO. It is this identity and union of each and all persons in their part and contribution to TÓ that guides all to ever better, higher and greater lives, or our parts, TES in TO.   By the providence of TO supplied by the highest kinds, powers, forms and functions of its knowledge and use in life, it is assured. TO expands and best manages its controls over all. This is what gives us the survival over extinction that is the most fundamental necessity of all.

CHAPTER 32 - SURVIVAL over Extinction of TO, to Disproofs and Proofs of TO

Matter and mind, Forces, Extinction over Survival, Gain/loss, success/failure, growth/decay, beginning/end, Equilibrium in TO and TO with TOT, identity, immortality, the enigma of none to all, civilization, TO,

The problem of matter over mind is the enigma of no TO, relative universe, extinction, and the loss of all that enables best human life in actual totality. It is a choice of one extreme with a loss of 1. the other extreme, 2. the whole, and 3. all in between. It is the common error of all that fails to discover and learn polarity, the whole, TO, its continua, their spectra, gradients and how they are formed and function together. Extinction is the loss or absence of an entity, object, organism or the whole that lacks all necessary for existence in life. Survival is their presence and improvement when all is of proper form and function. The problem of matter over mind assumes a partial and tangential view of existence and totality. It is the bias of objectivity, physical universe, science, materialism and fatalism. The ultimate answer is if we do not exist, our being without all that enables human life is inconsequential. Yet without life, in TO, this is the most opposite and damaging idea humans can have. It denies proof of TO and TOV, approaches to the absolute, TÂR, the most profound states, conditions, limits and relations of our potential to actual necessary existence.

To understand actual totality, and to act and fulfill its needs, it is critical that we know where it is most vulnerable and what holds it together. All that stands between actual totality and its success, survival, improvement, advancement, continuity, well being and permanence. To meet this challenge largely depends on accurate and thorough knowledge, accounting and application of actual totality. All the components of TO exist mostly by how they combine, interact, and what they mean in the whole, or the Vital Necessity of a Concrete Universe, (VINCU). In being all, the actual totality, TO can be envisioned and produced as a discrete mass with reduction to a closed set in the form of a concrete whole. It requires an identity of subject and object, both exterior and interior to TÓ. Solution from what is not, or from TO, SOLA. By learning what is not, and what loses in transformation we can solidify the object TO. By learning what is absolute and what is relative, we can focus on their common bond and continua, ARTO, and thereby discover and prove TO

Many of the seeds of demise of each and all persons, nations, cultures, humanity, physical universe, God, actual totality, TO, and TOT, already exist today and through time. They are shown by integrative negation and dissociation curves. They exist in all that is deviant, negative and non TO, TRU, TOV, RELTO, ABSTO, and TOT, past, present and future. They are all about us and in us. They are a consequence of transitions in time, transformations, and the collective negation that occur, especially when failure to react, change or adequate adapt to sustain existence. They tend to accumulate, whose seeds and effects accelerate demise. Each and all persons, by their optimal part and roles in TRU, TOV and TO, and hopefully more in TOF should reverse this process. This is why it is imperative that we are our best, nourish and serve our role in planning, preventing, protecting, managing and controlling our lives and all to reduce and eliminate this demise. This demise in all its extent, intent and affect on actual totality can be readily met and conquered. We have become so good at knowing what to do, even how to achieve, we only need to have the stimulus, interest, ambition and determination to overcome all that is and will beset us.

Of the many causes of survival over extinction are the interchangeability and identity of subject and object that show and prove the convergence of mass energy, modes, kinds, varieties and all in TO. Such a reversal helps to illustrate the dual and reversible form and function of TO. It also illustrates the flexibility of actual pure essences in, and out, of TO. This is evidence of the external nature and effect of language in TO, and how it can be misleading if the subject and object are not held in the reality of their reversibility in TO. By knowledge of TO and the reversibility of many of its continua, like subject and object, external, internal, and internal knowledge and action together, form TO. It is the acquisition of these powers of reversibility and others of TO that are the greatest supports for survival and continuity.

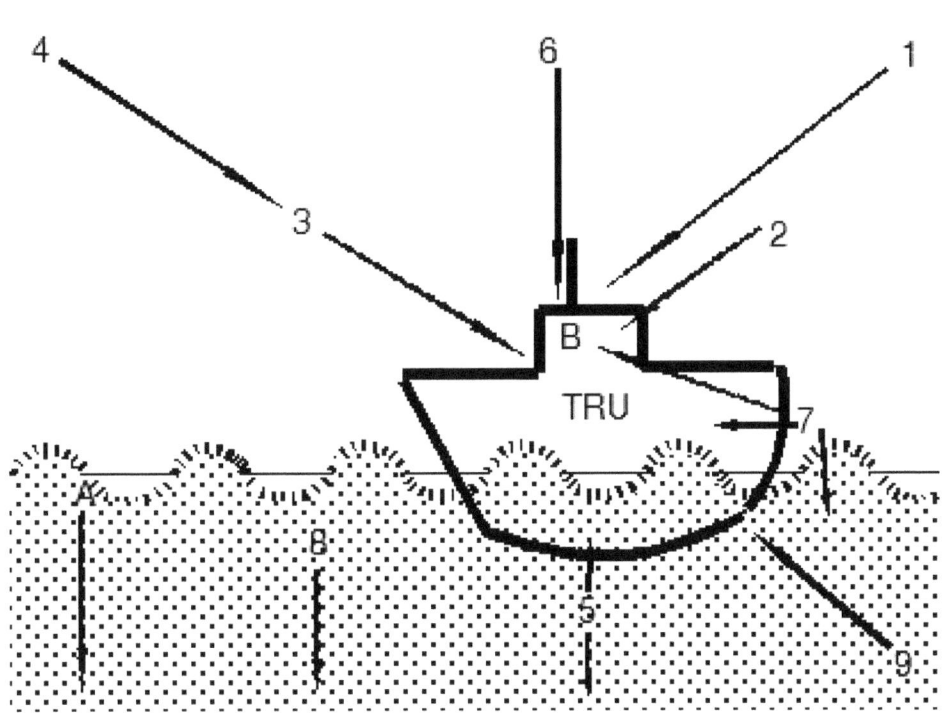

Fig. 32-1 Forces in extinction over survival.

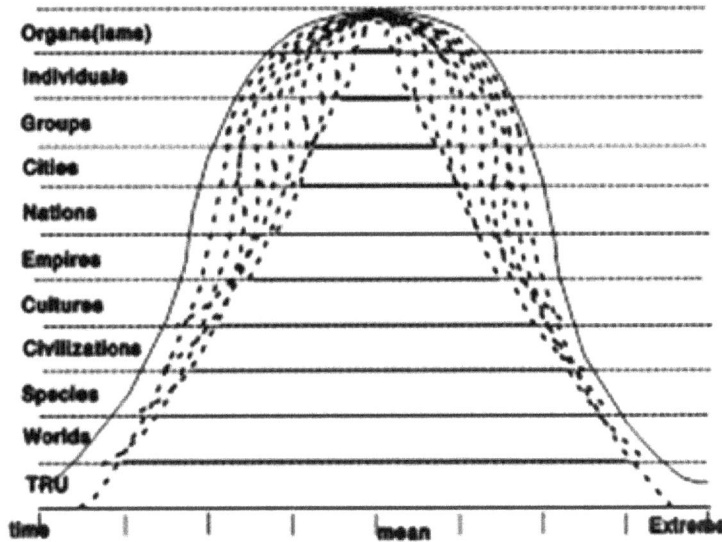

Fig 32-2 Rise and Fall, Growth and Decay, Survival and Extinction of Things, Worlds and TO

Usually, the larger the entity, or constituent of TO, the longer its viability or existence. All, in or out of TO is undergoing forces and processes of growth and decay. It is the quality of maximum growth and minimum decay that is necessary for, and helps to prove, TO. Most functions and dynamics of TO, are not simple homogeneous or singular phase motions. Much is in opposition, DAOST. Such diversity is neither chaotic nor abnormal. It is more or less harmonic and cosmotic in the sense that TO is in complex functional competition, with diverse components and processes that are, more or less, ordered and well controlled. Increasing degrees of attrition, losses of vigor and strength in life are natural processes that characterize decay over growth through age. It is when added, external or internal processes are bad that growth is limited and decay accelerated. All the processes that exist within the totality of TO are formed and function by the interaction of the principles and laws of mechanics and dimensional formulation adapted to TÓ. There are more of these, some more partial for each process, and some more total for all TO. Examples of these include redundancy, entropy, kinds of energy, in and out dynamics of TO. The conversion and convergence of many levels of mass energy, constantly working in different ways help us understand the totality and intricacy of the processes of TO. This is shown by the oppositions of processes whose function is to restore the optimal state of their parts in TO. When excessive or deficient the action reaction dynamics produced often engage whatever function in dimensional formulation best brings the whole back into equilibrium.

Depending in part on perspective, and more largely on itself are the beginning and end of actual totality. Physically, biologically and by perspective, TO seems to be limited, liable to great growth and decay as well as extreme vulnerability. By the converging powers and forces of actual pure essences in TO, and the potential powers of its special varieties, knowledge, mentality and will, TO seems to be unlimited, extremely stable, with a providence of permanence. These two extremes contrast the enigma of no, or all, TO. Their resolutions are multiple, by the gains of the past, power of the present and increasing powers of the future. For the survival, success, well being and eternal life of TO, all works with a high degree of interdependence. Thereby the key to, and proof of, TO, largely exist within each and all persons, generations, and all people. It is how the whole of life acquires all needed to produce the actual pure essences and all that fulfill their part and roles in TO.

There are many factors that contribute to life and survival. Among these are the certainty factors of, ancestry, proband and progeny. Certainty is dependent on what has gone before, is occurring and what comes later. If our ancestors were not the best we would not be here, if we are not the best, our progeny will not exist, if our progenies fail to exist actual totality of the future will cease. This also works reversibly to a large extent, proving the forces that determine TO, and the propelling powers of life. What it takes to maintain optimal equilibrium is a great balancing act of many factors and forces. A great number of processes of larger constituents inside and outside of TO, act as regulatory mechanisms or systems to maintain equilibrium, eustasis, form and survival. Often they play a role in more than one, several systems, or may act indirectly to produce positive effects in support of others and the whole.

Actual totality is, in many ways, like the human body and other systems in nature. They all depend on sufficient capacity, form and function. Like the anatomy and physiology of the human body has been formed to most effectively sustain life, so are the form and function of TO all that enables its well being and continuity. An accounting of all that most enables TÓ to exist provides an excellent description of its consistency and proof. This that most enables TÓ to exist are the very aspects and features, modes, continua, varieties and constituents whose absence or deficiencies need to be corrected. It is their lack that is often the most common cause of extinction.

The factors, forces and processes that determine equilibrium are the product of the states, conditions, limits and relations of TO internally, and TO with TOT. There are many integrated and dynamic forms and functions in TO, e.g., balance of polarities, tolerances, redundancies and buffer systems. Some of these that exist in the human body, largely by its physiological chemistry, are, bicarbonate - carbonic acid, phosphate, protein, hemoglobin and bone marrow buffer systems. Others include automatic and built in regulatory mechanisms and systems of respiration, blood pressure, heart rate, sleep, water, nutrition, digestion, elimination, temperature, and reproduction. These are only a few of many systems that work to preserve the equilibrium of the human body, as it exists in health, for its role in TO. They are mostly lower than the level of consciousness, occur simultaneously, cyclically, massively and interdependently. It is impossible to live by, or have an equilibrium and successful guide to life without some order in knowledge. It is impossible to have order in knowledge without a vertically integrated organization based on actual totality TO. This includes the modes, universals, continua, absolute and relative totalities, varieties, physical and philosophical systems on which actual totality is based. Without TO and its knowledge, it is impossible to have the purposes, goals, plans, preparation, protection, prevention, action, management and controls necessary, for the quality life must have. A culture, civilization and mankind without these are at the mercy of negative and adverse forces that produce uncontrolled extremes of boom and bust. With the present advances of knowledge, its organization, vertical integration, relative universe and actual totality recognized and well applied by each and all persons can achieve the hoped for quality of life, well-being, happiness and immortality our providence can produce.

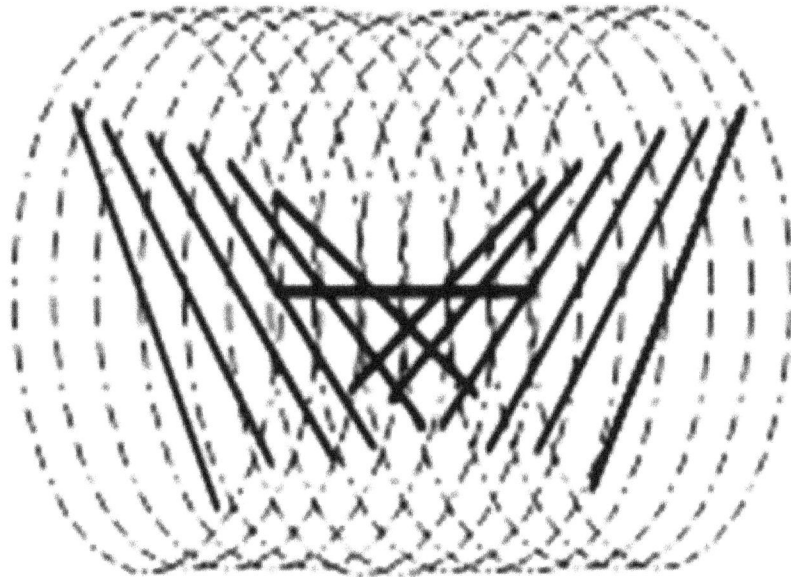

Fig 32-3 Equilibrium in TO and TO with TOT

A good example of equilibrium in survival is the great power of the future that already exists in TO, especially of, and derived from, human mentality. This is like the power of God, Cognos, personal identity, subjective-objective and reality in TO. This power is of present TO, and all TO-TOF, Their differences increase, at first largely quantitative, then qualitative in the far term. These differences are also most important by the Power of Future Gods and Cognos in TO, (POFGACT), cf. 31-2, 38. These powers have effects that change with different levels of TOV, TO and TOF. These powers, proofs, concepts, principles and laws are concentrated parts and features of the Law of the Power of the subjective, reality and the relative, RELTO in TO and TOT, LAPSRIT.

Fatalism, like much that is believed in, is highly inimical to actual totality. For an attitude based on a belief in fatalism reduces one and all to a position of denial. This is a denial of the most important of all purposes, actual totality, its knowledge, and the duty of each and all to them. On the other hand, fatalistic beliefs that are limited to all other than TO, or less than their practical and functional actual pure essences provides a contrast that can more efficiently and proportionally direct life in TO. We must be careful to know this contrast, and utilize both to the best advantage for ourselves and actual totality. Throughout life a person acquires much that is needed to produce actual pure essences and all that fulfills their part and roles in TO. Such is the matter of care, what we care for and how it is achieved. One and all persons who care for little, or care for what is little, sooner rather than later lose, fail, die and become extinct. One and all who have a great feeling of care, and care for all that is best and right, as TO, gain, succeed, live long and defeat extinction. At the core of caring is the power of the ego and its part in TO, TÍS. When well directed, we care for what is best, right and TO. This is not as simple as to care for what is the worst, yet like the proofs of TO, it is readily in reach of all. When they are mastered there is a rebirth of new lives in TO, and their guide to all its glory and rewards.

Special Operations and Applications of TO, TRU, TIT, SPALT, SPAT.

In TO all is related and more or less interdependent. It is this interdependent state, relation and condition that provide the basis for immortality. This is the Principle of Differential Immortality, PODIM, an interdependence of shared success or failure, survival and permanence or extinction. Herein are the test and hallmark of the continuity and proof of the predominance and preeminence of all in TO, and the contribution of best human life, TRU, RELTO, TOV and TO, potential to each and all persons. Many other kinds of RELTO also participate in these powers. Yet it is the role of an all wise God, through all time, that most dramatically and convincingly reveals this power, mechanism and law. When modified to TÓ and TO-TOT the role of God, and the proof of TO become more certain. This is a most important mechanism, concept, principle or law. It is a major proof of TO, and explains the power of God when corrected in the past, present and future of TO. The power of a reality, or subjective ideal, in which past, present and future are signified was emphasized by Buddhism. Bodhisattva was Buddha of the past on the present. Siddattha Buddha was of the Future on the Present, and Maitreya Buddha of the future. Yet, like so much in existing knowledge, these kinds of God or Buddha, exist in a different context, and are not so well modified as when existence is in TO and TOT-TOT. Such modification is essential if they are to be correctly known, and if we are to correctly know TO.

To get the most out of life, one must not only strive to have optimal identity with the best it offers, but learn the modes, kinds and varieties of TO that establish immortality. This is necessary for our existence, so that one can make the most efficient and effective use of life. The measure of this is the immortality coefficient or quotient, ImQ. To raise this measure, quotient and our lives requires the wisdom, courage, energy, conviction, will and dedication to see that the best use of ones life, TIS, produces the greatest benefit for the actual world, TO-TRU, and TES and TÍS in return. This is how we dedicate our lives to the actual continuity of all that is best for human survival. It is how you, I and all of any race, culture, age, sex, or 'persuasion' can fulfill our potential and have everlasting life by our part and contribution to the best human life, TRU in TO and TIT-TRIR with TOF. To be an immortal is to serve, work, join, give and receive the benefits and fulfillment that offer, produce and provide the best of human life, TES-TRU in TO. It is to live in pure, actual reality, PUREAL, and one with COGNOS forever be that makes us better, continuously perfects and assures permanence. The absolute, 'we know' is the extreme polarity of COGNOS, focusing in and out with infinite subjective power of, and on, all the multiple faces of totality. It is total cognitive powers fusing with TO. It is the most special proof of TO and a large target of our immortality.

Immortality is much more and much more complex than a simple existence after physical death. It exists in many ways, including, existence in actual totality, the permeation of absolute and relative totality, e.g., mind and matter and their combined effects, good works in life, identity with TIS, TIS identity with TES, their combined identity in TO, the continuity of actual pure essences of TO, the reversibility of much of TO, and fusion with new forms of immortality in TOF. Some of these are more personal, some more universal, both are highly interdependent in TO coexisting in universal and personal. It is possible that an absolute-relative continuity of aspects of actual totality could have preceded human intellect, philosophy, theology, human knowledge, humans on earth, and even earth itself. In such a way actual totality may have had more than an absolute totality existence, more or less relative totality. The almost limitless extension of matter, antimatter in relativity and space time would tend to support such a variant and idea. Yet, this largely depends on the definite of necessary existence. To be interested, a supporter and guide for such causes should add to all that is TO. Even words, like mystic and spirit, have some validity when limited and interpreted in relation to the complex and subtle interactions and affects of the dynamic and essential mechanisms and aspects of actual totality. This may explain Hegel's spirit, or Geist, in association with the absolute, and prevalence of the Mind.

Throughout the entire course of life on earth there has been a rise to the top. In humans this has accelerated. It is what natural selection and the survival of the fittest have provided. Forces and conditions that support this are not certain. The opposite is, and will be, more likely. As the barrel of benefits is depleted all that is left in the bottom is rotten. Without TO, a solid, unified totality and rational foundation, health, strength and fitness are lost in the confusing array of massively altered weaknesses. This is the reason for and a gist of TO. All that enhances growth, development, well being, and all that eliminates decay, stagnation and loss of existence. With improved and expanding discoveries, knowledge and technology most successful and effective predictions and solutions can be made by and for each and all persons, nations, cultures, civilizations, humanity, TRU, RELTO and TO. Much of the ground work has already been laid. Yet we must learn, remember and continually uphold those means that are the keys to an ever better and greater existence in, and glory of, TO. Nothing less is the hope, challenge and providence of each generation on their way to a better life, as with God, when its equivalent to the necessary existence of TO, and TO with TOF. The power of knowledge, technology and all can and will bring benefits to actual totality its best assurance of success and survival. Thereby, the proof's knowledge can bring to actual totality are the greatest weapons at our disposal to assure the greatness of actual totality, and its continued struggle for supremacy. It is a supremacy that depends on the fusion of the best in absolute and relative totalities, and the typical powers of actual totality itself. These are what the proofs of actual totality provide. The following chapters recapitulate and feature these proofs, From disproofs of what is less or not, qualified, reciprocal, multiple to the total combined unified proof that replicates actual totality.

VI Proofs of Actual Totality, Special Kinds, Interactions and Combined Total

CHAPTER 33 - DISPROOFS and Proofs, to qualified proofs in TO

Disproofs, kinds of disproofs of TO, basic states where disproofs prove each other and TO

The existence, survival and permanence of the ultimate unified totality, or actual totality, TO, are highly dependent upon being properly proven. There are many kinds of proofs and disproofs of which disproofs are especially effective ways of proving and demonstrating TO, its features, modes and proofs. Thus in proofs, and especially disproofs, we must always remember that of totality, actual totality, TO, and account for all that is lacking, in error, or has limits. When by disproofs, proofs, laws, and other means are reduced to TO a great new vista of all awaits everyone.

Disproofs are a good method of approach as long as they are accurate, proportional, and well described. They can be used both to reveal opposition TO as well as that in which TO may seem to be in error and lacking. Standing in opposition to the preeminence of TO is the physical universe. The greatness of the physical universe in TO must be remembered, yet its part in absolute totality makes certain disproofs evident. 1. It has no foci in present time, 2. Without relativity it lacks being the totality, 3. This lack completely ignores the polarity of the relative absolute, RELTO in TO. God, unless corrected by equivalence in TO, is polarized, lacks much of ABSTO and the physical universe, essential aspects of TO, and stands in opposition to TÓ. These extremes, when adapted to their part in TO, share some of their greatest features. The error in a separate God is somewhat the reverse of a physical universe, in lacking much absolute totality provides, and much that knowledge and science have added in thousands of years. Worlds and many other systems that claim supremacy are even more lacking in totality, or have errors that render them less than the ultimate unified totality, Even TO has limits, especially its knowledge and presentation. These are noted through the book, and are caused in part by the limits of vertical integration.

Disproofs arise from what is not or less than actual totality, TO, or some aspect, feature or constituent of it. Disproofs progress to qualified proofs. All cannot be accepted at face value or in any way, only as they exist, and can be proven, by TO. Further precision of proofs is given by reciprocal, multiple, combined, and other methods. Only when we learn and describe the fundamental enigmas and paradoxes of TO and bring all together in the totality of TO can there be the accurate, effective, proportional and efficient proofs that equal the greatness and order of TO.

There are two major kinds of disproofs, I. disproofs by TO, of that less than, or not, TO, and, II. disproofs of disproofs of that less than or not TO, existing knowledge, compared with TO. The first is largely by the increasing knowledge of TO, and separation of all less than TO. The second are the disproofs of apparent disproofs by existing knowledge of TO. The second is especially important, whose truth, in turn, is corrected by the first kind of disproof. When approached with a serious and rigorous attention to accuracy of the complexities and differences between TO and non TO, disproofs can be one of the best tools and methods of proving, demonstrating and upholding TO. They are especially effective in revealing the errors in common and less TO beliefs and what is TO. This contrast is what adds a new and often dramatic perspective of TO.

There are also minor kinds of disproofs, including,. Approach to TÓ. Representation and presentation of TO, and, Mixtures of all valid disproofs. .

List 33-1, Of the kinds of disproofs there are those that disprove by,

1. Apparent and conventional proofs lacking in one, more or all steps to TOT, TO or TOV

2. Various totalities, universes and deities that claim to be the ultimate unified totality.

3. Features and facts that are not TO, but only lesser parts, deficits, or facsimiles of TO.

4 Disproof of opposition, only by revealing the whole of the dualities can the disproof be corrected,

5 The transition and transformation of TO or the internal transformations of TO,

6. Are disproofs of each other,

7. Are less than their proportional parts in TO,

8. Do not disprove TO but help to prove some part or feature of TO.

9. Not from any disproof of TO, but from correction of representation, perspective or presentation.

Disproof # 1 and 9, lead from an approach to or departure from TO. Disproofs # 3, 4, 7, 8 lead from their existence in, relation to or corrected feature or component of TO. Disproofs # 2, 5, 8 lead from all that is not TO, other than their mixtures, especially when they assume existence or erroneous positions as TO. Disproofs of # 6, and many others, lead from #5. In this way disproofs become a more ordered and effective means of revealing, proving and certifying TO.

Disproofs are great means of substantiation not only by demonstrating how all that is not or less than TO, such as the physical universe, God, etc. is not actual totality. Disproofs do so by demonstrating how much that is less than total actual totality, partial, or of changes with time, is not the correct actual totality, TO. They help to show how much that is not or less than, may become TO. This is the kind of disproof that is the Law of Solution in Deficit. LOSID. This is often related to other disproofs and proofs, and occurs where a factor, such as existence, is less than TO, and its demonstration disproves it, and helps to prove TO.

Proof of TO by disproof of opposition is a basic method of resolution of paradoxes found in the disproof of the opposites of TO and of its constituents or continua. It is seen in the negatives, or extremes of absolute and relative totality. One of the greatest proofs of TO, is the disproof that actual totality is absolutely physical or absolutely mental and human. The opposition of absolute and relative can never be resolved by either separately, or as a continuum. Only as they exist, and coexist essentially in the right kind of totality are they to be resolved. This is TO, not simply from itself, but from all that makes it, saves it, is it, shows it, and can exist by it. To develop "a theory of everything" it must be equally a theory of "every-sense." To equally include a theory of everything and every-sense it must be based on a realistic balance of the absolute physical and the relative mental. It must be based on the subject and the object bond by the real, their combined reality. It is this kind of totality that includes the actuality, objectivity, subjectivity and reality, thereby typifying actual totality, TO. Thereby look to the opposite of all, their synthesis, explanation, vertical integration and differentiation, dynamics and equilibrium with corrections for proofs of TO.

Physical, mental and total health, are not only blessings, they are duties and responsibilities. It is much more than the treatment of ailments, illnesses and disabilities, It is even more than the prevention of all that causes poor health. It is the total care of the health of one and all persons, past, present and future. The only way to achieve this kind of health is to know and live by the right world of absolute-relative best human life or the ultimate unified totality, TO, or actual totality. When disproofs of a non TO by TO, and disproofs of disproofs of TO by non TO are successful TO and health care improves.

The modes of the right world or ultimate unified totality, are actuality or necessary existence, subjectivity or all that makes up the mental world, including knowledge and all its corollaries, objectivity or all that makes up the physical world, including the frame we live by, and reality the ever higher cohesion of subjectivity and objectivity that is the basis for human existence. The continua or dualities of the right world or actual totality, are many. It is their total existence form and function that constitute much of the major content of TO. All exist only as they exist as parts of TO, none exist separately. Such dualities as large and small, past and future, subject and object, absolute and relative, good and bad, right and wrong, inorganic and organic make up the world we live in, TO. It is their correct proportions, balance and variations in actual totality that determines our health, well being and permanence. They include, Absolute and Relative, General and special Varieties, Mass length time, Mass

Energy, S-0, language, logic, Dynamics and Internal Relations, Math, Proportions, Geometry, Organic and Inorganic, Life, Animals, Humans, Transitions, Relative absolute, Persons, Knowledge, Higher levels of Knowledge, Survival and Permanence. It is the disproof of the lack and form of each of these and proof of what makes up their part in the form and function in TO that each of these is its proof, as they follow the chapters in this book.

The Three Worlds of TO and Human Life, Disproof of other worlds, proof and law of actual totality

-

The world we live in is much different from the world we perceive, believe or are used to. These are only appearances or facets of necessary existence. They are a mixture of three,

I. The wrong world of all that is negative and contrary to the physical and relative universes of best human life,

2. The right world of all that is beneficial, a kind of heaven, positive for the union of absolute and relative totality, an ultimate unified totality that is extended TO,

3. A great variety of mixtures of these two kinds of worlds, of which almost all the earth's population, more or less, lives in. They understand neither of these worlds nor their differences, which causes bad effects and a poor life.

The three kinds of worlds are not completely separate, nor impervious to each other. In fact there is much in the mixed world, and even some in the negative world that is potential to the right world of ultimate unified totality. By living this interactive and interdependent existence each person without knowing, having the right principles, purposes, priorities and plans in life falls by the wayside and succumb to this or that wrong road that is the story of life's degradation.

Where proofs seem to disprove each other there is often in their corrections and eliminations proofs of TO, or its constituents. It is by these corrections and eliminations that new concepts are formed, or old ones made right or more distinct. Since actual totality is both the greatest topic of knowledge, and unrecognized, it is critical that all its major aspects, features, forms, functions and mechanisms are known and signified. Such phrases of aspects of TO as LANT or Language of TO, ARTO, TOV, TRU, TRIT, TRIR and TOT are indispensable to an adequate understanding and propagation. This is done by revealing the fundamental states, conditions, relations and limits, and by adding a new concept for every great aspect of actual totality that is unknown to the proof of TO.

Of disproofs, none should disprove each other if TO is well qualified, the continua are properly developed and exceptions accounted for with correct descriptions and explanations. They may seem or tend to disprove each other, yet if all are correct by TO any disparity will not only be incorrect, it often presents an opportunity for further description, modification, correction and proof. To clarify and be most accurate and convincing, it is necessary to show which of the many varieties, forms, states, modes or continua of TRU are being proven

What satisfies, dissatisfies or alters one extreme in TO, is likely to alter, dissatisfy or alter another and TO, thereby disproof of one or another of the whole. TO. This proves the delicate balance of all that is a part and has a role in TO, the importance of knowledge of TO, its extremes, means, and their partial and total equilibria. A person's short life is not easy. It is a long, hard road to travel to learn and do what is right and best. These can only come when we know what it is all about, and this can only come from the top down, the ultimate unity of its totality, actual totality and interaction with the potential.

The existence and interaction of the external with TO, does not disproof TO. It only confirms the fact that TO is only realistically absolute and with TOT in transition. TO exists by actual pure essences, or functional value, and the practical, relative and fully state that is actual totality. What TO and its varieties lose in transformation is a most effective way to learn, knows and prove them. Disproof of their prior form of existence relative to that transformed into, provides a clearer idea of what they have been and the succeeding form is. Disproof from proportionate parts in TO is often evident in the continuous temptation to mistake, or over and under estimate, the part anything, or component, is in TO. This highly effective means of disproving errors is, when applied to relations in the whole, a proof of TO.

Viability has a much greater meaning in different varieties of TOV, as well as interpretation within TO, or separately. The problem of viability and non viability is equivalent to the great paradox of absolute and relative, of the existence of only actual pure essences in TO-TIT, or existence of what is less. We have seen in the first chapters how things, objects, meanings, definitions and words selectively become their actual pure essences in TO, TOV and TRU. The region of interrelation, interaction and interchange between less than perfectly TO is vast and highly dynamic. This does not disprove TO, any TOV, nor their exclusiveness. It only shows the form and function of TO-TOV-TRU-TIT in each variety, as in general, TRUS, TROS, TRIS and TRÂS, and the extremely strict and fine line that must be drawn between what is less, what is perfectly viable and the actual pure essences of all, only in TO, or in each variety.

Because we do not, nor poorly know, what this right world is, and because it is the most important feature, and guiding purpose of our lives, it is imperative that from an early age through our entire life, we experience, be trained, are educated and learn to identify with, and be devoted to, the right world, of ultimate unified totality, TO, or actual totality. Yet life can, and must be, even greater, one of blessings, all the benefits and rewards it offers when right. To develop and perfect the right world, actual totality whether called by whatever name it will take a lot more work and time. It will change somewhat with time, so that it will need continual upgrading. For the work necessary a dedicated and wise group will be continually needed to provide optimal development and perfection of the totality. These facts lead to the Law of actual totality and Disproof of other worlds, LATDOW.

To know that this right world, actual totality exists, is to experience, be trained and learn what makes, save, shows and how it needs to be best used. It is made, exists and is formed by its categories and the quality of its modes, continua or dualities, absolute-relative, varieties, form and function. These are what saves, enable, represent and show us the right guide in life. Thereby worlds, philosophies, sciences, universes, and systems exist only as they exist in actual totality. Many features, such as the modes, kinds, major parts and continua will appear in different ways and places depending on perspective and representation. This can be improved and made more identical with TO, the better TO and its forms are known, and the more its perspective, representation and presentation are improved and perfected. Yet they will always be less than a perfect model of a closed set for TO, not from any disproof of TO, but its complexity and external relations. This is seen in showing ABSTO and RELTO as separate from TO, within TO, and how they exist as a continuum, ABSTO-RELTO within TO. This is like showing the modes of TRU in reverse order. It is how each kind of disproof is reversible, and to be correct must converge on TO.

CHAPTER 34 - QUALIFIED PROOFS to Reciprocal Proofs

What makes proofs more exact, accurate and perfect? They shouldn't disprove each other if TO is well known. The better TO is known and right, proofs will be right. Conditions and Limits, Proof by Codomain and Superimposition, Degree of Truth and Certification, Congruity, Refinement of Parameters and Posits, Interpretation, Vertical Integration, Validity in totality, Justification of TO, Identifying proof requirements, Identifying what is right,

Usually proofs are not simple, often they are not direct. To provide highly qualified and correct proofs, pertinent features and factors of TO need to be included. A good example of this is the actual pure essences of TO and its constituents. Other examples are those laws that apply to each qualified proof. When sufficient evidence is given for proof, and related proofs are similarly produced they provide added quality. When all converge on TO they provide the level of proof TO must have. This is a level which other worlds, universes, deities, systems and universes lack. This supports their disproof, except to the degree and way, in which they are major components of TO. It is a level without which any and all previous and current totalities have failed to provide. Proofs and their descriptions are the simplest, clearest, most efficient and practical way of representing and presenting actual totality, TO. They are not the only way. Some provide means lacking in proofs and descriptions alone. Yet it is by identity, proof and descriptions of TO that we can most often rely to make actual totality right. Thus to fully understand, apply and live by TO each and all persons should endeavor to seek that knowledge of actual totality that is the greatest in any way, especially molded to ones talents, interests and needs in life.

Of the methods to qualify, show, test and prove TO are those through the enumeration, interconnections, and key dynamic mechanisms in operation of its basic concepts, principles and laws. These are expressed in, the Law of Non Essentiality, LONES, Proof of TO by Non Essentiality, what isn't needed or can be done without, PRUNES, and the Law of New Concepts, LONECS. When we try to correct the quandary of "What is the matter" we quickly find ourselves in downward regression. Although questions, problems and needs may often be simply stated, corrected or practically met, in general they are paradoxical, of multiple to infinite causes, meanings and dependence. This reveals the lack of logic in language, and to account for the discrete meanings of words, actual pure essences, and all that brings totality to completion. Matter is physically only mass times volume. Yet because it has been so hard to discover the matter of all, matter has come to mean any basic cause for questions, problems or needs. The ultimate cure is in quality of proofs, which can come only from the unity in totality of TO. To correct this tendency and weakness we need to use vertical integration of TO, and retrace the differential to the integral point at which certainty and fullness is adequately effective. This often needs our attention to, and recognition of, all those kinds of proofs of TO involved, from actual pure essences, different worlds or totalities, mass, length time, language and all other proofs that add to the multitude of causes, like contingencies of solutions that show the dependency of all in and by TO. As in an ideal language, it requires the combined endeavor of each separate concept or word, and their total interactive meanings in TO, as evident in the totality of proofs diagram, SPALT, other models, their operations and applications in life.

Proofs, like all things, must be more or less qualified by their limits, and the conditions under which they exist. We properly emphasize qualified proofs, which are large aspects of TO, need extra description and distinction. Thereby these proofs are qualified by, 1. The degree they apply to TÓ, 2. Their proportion of TO, and 3. The degree they are necessary as a proof.

To be adequately qualified proofs must account for those unexpected, and seemingly incongruous features, qualities and characteristics of actual totality. Many of these are noted under other proofs and disproofs, many are a result of the polarities of dualities or continua, many are more or less stand alone aspects. One of these is the seeming disparity between the varieties of and in actual totality. If actual totality, TO, consists of varieties whose extremes are highly of absolute and relative totality, how do these extremes fit into the whole? And how can the construct and form of TO exist as a closed set without disparity? This is proven by that law which certifies the congruity of varieties of TO in the whole, Law of Congruity of Varieties of TO, LACTOV. This law is justified by the unusual, yet typical, state, condition and limits of all in TO. This is a product of both the conformity of varieties, TOV to TÓ, and the conformity of TÓ to its varieties, TOV. TO exists highly affected by both its absolute and relative, ABSTO-RELTO bases. It is also a largely a product of the actual pure essences, modes, kinds and other qualities of ABSTO-RELTO and other constituents. These unexpected and seemingly unusual features and forms are less so once learned with improved knowledge of TO. Then they are the disparity that exists by perspectives that are alien to TÓ.

The quality of proofs depends on how well TO-TRU is known. It is the influence of TRU on TO that is a large factor in perfecting the actuality and reality of knowledge. This is because, to be well known, we must understand the world we live in, TO. TRU, being of RELTO, the special variety of TO, converges on absolutivity. This is the Law of Absolutivity Proof of TO, LÂPRUT. Like the physical universe must have relativity, so must the best world of human life, TRU, have absolutivity. Although having great precedent, absolutivity is a new concept in language. It is justified both by its precedent, most basic roles in TO, and the actual pure essences and practical necessary existence in each person, and absolute of TO. Thus there are concepts and laws of special and general relativity, and special and general absolutivity. Each of these exists only in that form typical for TO. The absolutivity of TO, RELTO and TRU when general, are identified in TES, and when special, TÍS. TES is the convergence of all humans on the absolute, and TÍS is the convergence of the first person on the absolute. When viewed from the most separate aspect of TO the two kinds of relativity and the two kinds of absolutivity tend to take over all from their extreme. When they approach TO, as with ABSTO-RELATO and ARTO they converge, fuse, and become united by their adapted actual pure essences as they coexist in TO. This is the Law of Special and General Absolutivity and Relativity in TO, LOSGART. Like all parts of TO, the four extremes of special and general, exist by LOPAN, but only as this existence is no more nor less than their part, REM, in TO, (14, 22, 29)

The fact that TO can be known, as any object, and need not be cut off from pursuit and discovery is expressed in the Law of the Certitude of TO, (LACT) The Degree of Truth, (DOTH), represents the extent to which our knowledge of TO, KOT, is or is not equal to, TÓ, in actuality. The distinguishing, distinctions, and differences of TO that can be tested and enable its proof, are signified by, Validity in Totality, (VITO). This, like VAPO, only more exclusively, is achieved by the singularity and completion that are characteristic of differentiation in TO.

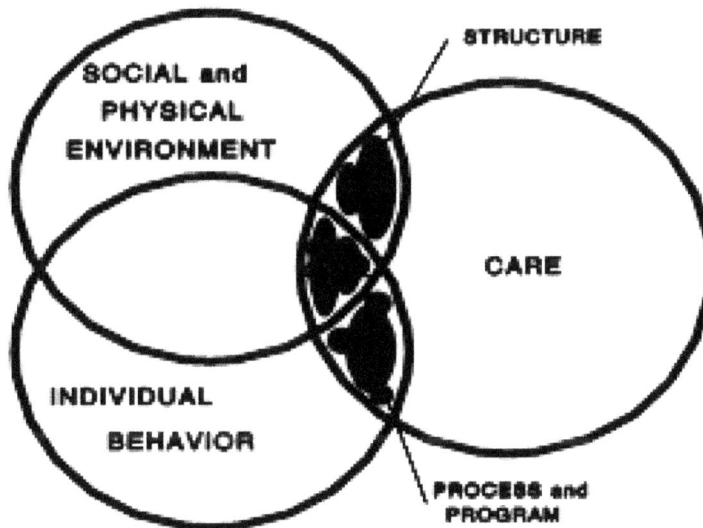

Fig 34-1 Means of Showing Superimposition and Processes in TO.

Superimposition, like codomain and coterminals, provide another means of showing different steps, methods and degrees of problem solving and presenting TO. In this figure we identify three steps, stages or degrees of function that are single, double, or triple combinations of different entities, units or systems. Codomain can show the relations of values and their various combinations, permutations and interactions. Without interaction we are mostly involved in structure, i.e., static/dynamic, whereas more often it exists as dynamic/static. Processes in TO are produced by the best methods of thought and include, 1. Piece by piece, instantaneously, or partially, so that the evidence, facts, proof and maximal form, or revelation of extremes, means, parts and processes, are brought to provide maximal form of TO, 2. Reducing all facts, evidence, knowledge and perceptions to their proper interpretation in terms of TO, 3. Partially, critically and analytically by differences in ones partial position, 4. Conclusively in order to bring out all by experience, intuition, deduction, induction, gist, and whatever means show their great significance, 5. Problematically, 6. Paradoxically, concentrating on seeming irreconcilable aspects that must be reduced by TO, e.g., DAOST, 7. Critically so that all fallacies and errors are considered and their pitfalls averted, 8. Explosively as in revelation, by brought out and building on the greatest of TO and TOT, 9. By the actual pure essences of all we attend to, 10. Of and by the parameters and dimensions of TO, 11. Exponentially by higher powers altered and adapted to replicate TO. 12. Massively for a sufficient quality and quantity of detail in TO, 13. Proportionally equal to the absolute-relative composition of TO, 14. Correlation, by axioms and corollaries based the state and qualities of TO. 15. In terms of ordination, coordination, in simulation, models based on exact form, e.g., SPALT, 16. Of unity, especially the unity of TO, and that maintained with TOT. 17. To think with optimal memory, intensity, imagination and revelation 18. To think by all these methods together, focused on the totality all in TO, and TO with TOT. By fully learning the best habits of thought and existence of superimposition we can achieve the validity and proof, worth and greatness needed for, and is, actual totality, TO.

One of the greatest of all tasks for quality and interpretation is the accuracy of a Positive Objective Approach, POA and its posits, the Refinement of Posits, (REFPO), and its associated Law of the Continuous Perfection of Parameters, (LACPAR). It is in these new concepts that we can finally produce the substantiation and perfection of TO that make a valid replica and yield absolute proof. This is like the most intricate and beautiful detail seen in the greatest works of art, and best maps of the world and space. They enhance the simulation of TO. Such concepts add necessary detail, and thereby quality to the proofs of TO.

As much as scientists may deny or defy God, religion, philosophy and TO, one of their most strongly held and powerful tenets is how observation means little without interpretation. Likewise neither observation nor interpretations mean much without determination by the right vertical integration, TO or TREQ. This is fundamental to human predominance, ability and the prime mass energy A form and function, provided by relativity, or the observer-observed fulcrum in TO. It is not any interpretation, however, nor even various combinations of related interpretations. It is only interpretation based upon all TO, as evident in VAPO, DAOST, SCARLS, TO-TOV and ACEXIN. It requires an ever increasingly astute, and accurate development and use, of the best in TO. This also necessitates presentation, POT, acceleration, action and advancement, (ACC).

With greater sensitivity, specificity, reliability and precision all posits in representation. Justification of TO, (JUT) and the Laws of Absolute Combined and Total Proofs, LACOP, together provide Sufficient Absolute Proof, (LASAP). The observer doesn't need to know every detail to realize what is there. Sufficient reason proves TO, as TO proves and sets the basis for universal sufficient reason and certainty. PRUT can be achieved through the convergence of fields, sciences, and all knowledge as they exist in TO. The Proof of both Á and of ÁLT, (PRÁ) and (PRÁLT) prove TO and vice versa. The subjective revolution, representation, and their utilization go a long way to provide these six proofs.

The disproofs and proofs of all that is less, or lacking in, any of the stages from non existence to, TÓ, help to reveal and prove certain aspects of their significance in TO. This especially applies to fallacies, which for non existence, is seen in the many fallacies and illusions. When reversed the extended meanings in TO, or its use, are demonstrative. Midgets standing on the shoulders of giants, look like giants. Sages and geniuses are not recognized at home, or in their homelands. The illusions produced are not always so non existent, but in the gray areas of valid and invalid knowledge. For each stage in the spectrum, from non existence to TÓ, similar disproofs and proofs are to be found. Many of these are given through the book. They help to qualify and fortify proofs and their sum, proof of TO.

It is not enough to provide one, or a few, simple proofs of anything, or of TO. Only when we provide the necessary proofs of, that they are, what they are, why they are, how they are, how the fit the form, how they function how important they are, how they can be represented, how they are to be applied, how to live by them, and how all these work together, do proofs fulfill the requirement of being correct, adequate and conclusive.

Although individuals come and go, and usually leave little marks on, and contribution to, the world, this is ameliorated by what is positive and right. Right is carried from generation through generation, the history and entire development of humans in time to become the great power of human mass energy, knowledge and APET in TO. By identifying with what is right and TO, each person can escape from oblivion and achieve eternal life. It is in this way we can find consolation and joy, not materially or in any social praise or condemnation, but by the ultimate exercise of our universal, special, partial, essential and combined knowledge as it is lived by the essences, mental and spirit of and in TO. The power and force of all that is right and positive not only reveal many interactions and processes, they help to substantially prove TO.

Other proofs are available in prior texts on totality, the relative universe and knowledge in general. A great amount of information, evidence that supports further quality proofs of TO is available in many of the figures and notes in this and associated texts. Throughout the text, and as enumerated at the beginning, proofs are given for the major topics, themes, and concepts of TO. The concrete closure these proofs provide is vital for each person to learn if they are to successfully fulfill their lives. Along with the laws of TO, they provide the basis for a guide that can bring us from the confusion and chaos of immaturity and partiality to a higher and better understanding of what totality actually is how it can provide a most efficient and productive life,

CHAPTER 35 - RECIPROCAL and REINFORCED PROOFS in and of

TO to Other Proofs

Conformity to TÓ, Differences, Proofs, Reversibility and Reciprocity, Working together, basis for reciprocity, major Limits or defects whose corrections are proof, Focus on greatest major problems of TO, Contingency of Solutions, Double Reversal of Disproofs, reversible proof by mass-energy Á of TO. Proofs by what TO can do without, Reciprocity as a basic feature of TO, Negative-positive interactions, Total Management

It is because of the most fundamental existence of TO that all is formed and exists. This is confirmed by the great and clear reciprocal form, function, states, conditions, limits and relations that exist among, and are reciprocally proven by the modes, continua, kinds, absolute relative, generic to special varieties of, actual totality, TO.   There are marked and clear distinctions between simple differences, proofs, reversibility and reciprocity of proofs. All are topics of TO and its presentation. Reversibility is a common feature and characteristic of TO. Reversibility alone, however, is not and does not generally produce reciprocity. Reciprocity is produced by reversibility in which each return, more or less, causes or proves the other. Reversibility may be thought of as simple interactions in which there are to and fro motion. Reciprocity is more, the effect, and proof, each interacting entity has on the other. However it is from the reversibility of major components of TO that many of its reciprocal proofs exist and are shown. Likewise many reciprocal proofs are given through the book, as the constituents of TO largely are, coexist, show and prove each other. Of these those that reveal the importance and method of reciprocal proofs are used for this chapter.

For each major defect, weakness, abnormality, disproportion and imbalance there is a corresponding explanation and description whose answer is a proof of TO. Their opposite correction is a reciprocal proof. This is part of the important collection of limits, COL, and of corrections, COR that provides the evidence that proves TO. For each major limit and defect the right corrections are Proofs of TO   Reciprocal Proofs of TO and TRU, (RECIP), are proofs in knowledge that not only verify certain aspects, parts or processes in TO, by their reversibility they can be used to verify other and further proofs that collectively prove all TO-TOV. This is the basis for the expanding unified whole proof of TO, and the method of accounting, accentuation and action for TO-TRU-TRIR-TOF. It gives us the means to discover, learn and know TO so that each and all persons can use this knowledge to guide their lives.

The proofs are valid and work effectively only when their ESINS are applied to and incorporated into TO. With many others the needs to Justify and confirm TRU, (JUT), are produced to equal their greatness and repeated use. They help to reveal the discrete participation each and all has in TO and its proofs. Working together proofs are shown by the Law of Reciprocal Proof, LARECIP. This signifies how proofs of TO, Á, SU, MOT, ALT, SPALT, and SPAT are, reciprocally obtained, from each other. The proofs of each major feature of TO or its different aspects help prove other major feature of TO. The more they prove each other, the more they prove TO. This is what the sequence of the proofs in this book show, As we provide more proofs of one key aspect of TO, they in turn provide more proofs of others, some of them more and others less. This applies singularly, in combinations, reversibly, totally and eventually as the Law of Absolute Combined Proof, LACOP, of TO.

For an accounting of reciprocal proofs to be most effective it must focus on what is of greatest, danger, advantage, power, the manner and time of action, degree of certainty, how it is given or works, and how related, extracted, managed and controlled. It must hold to the basics of TO, its modes, continua, kinds, varieties, VAPO, and apply solutions that are most efficient, effective and other means whose end result is the most positive for TO-TOV-TOF. It must be done sufficiently well to assure their valid control. This includes all qualified by TO as applying contingency of solutions, Fig. 1-35. Since the interactions and problems are most numerous, pernicious, important and critical for maintaining TO-TOV-TRU-TOF, it must succeed. To succeed, TO should be sufficiently developed to win all conflicts, struggles, battles and wars that will confront best human life, TRU, and all in TO. By and with TO-TOV-TOF there must be sufficient knowledge to directly encounter, control or eliminate each threat, wrong, cause for suffering, need, and loss, in the right manner and at the right time.

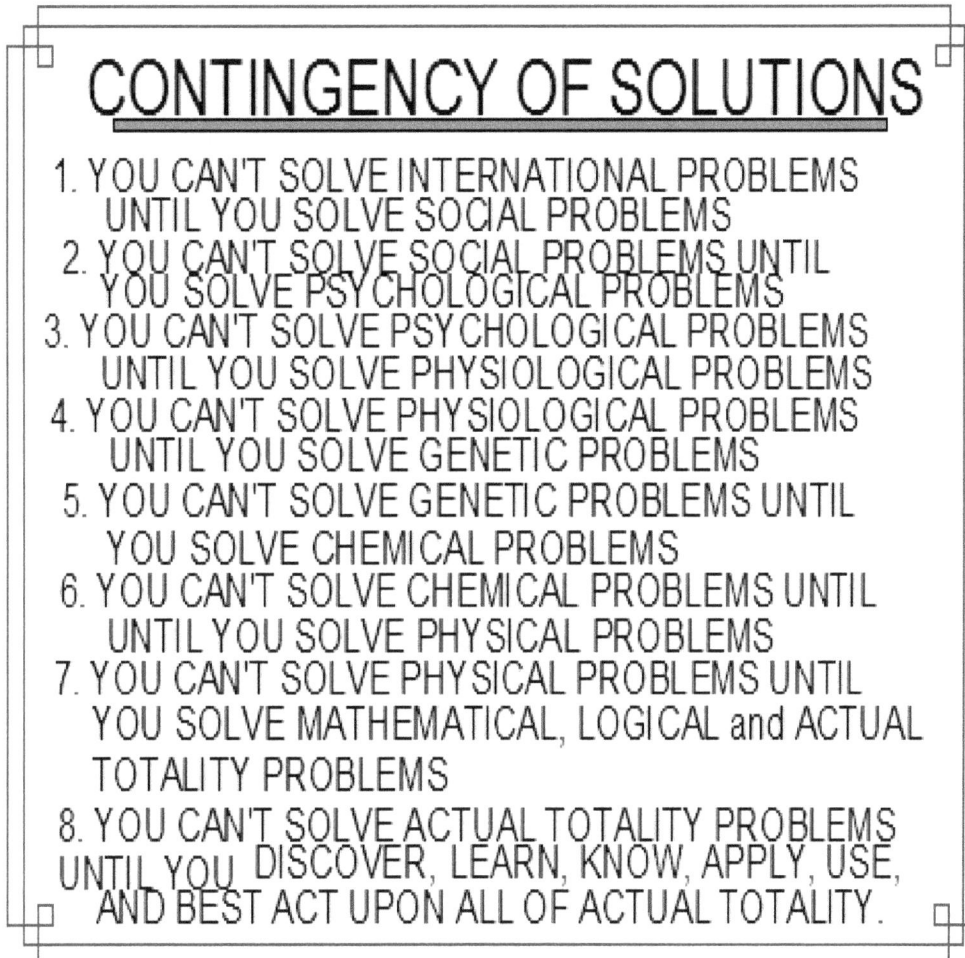

CONTINGENCY OF SOLUTIONS

1. YOU CAN'T SOLVE INTERNATIONAL PROBLEMS UNTIL YOU SOLVE SOCIAL PROBLEMS
2. YOU CAN'T SOLVE SOCIAL PROBLEMS UNTIL YOU SOLVE PSYCHOLOGICAL PROBLEMS
3. YOU CAN'T SOLVE PSYCHOLOGICAL PROBLEMS UNTIL YOU SOLVE PHYSIOLOGICAL PROBLEMS
4. YOU CAN'T SOLVE PHYSIOLOGICAL PROBLEMS UNTIL YOU SOLVE GENETIC PROBLEMS
5. YOU CAN'T SOLVE GENETIC PROBLEMS UNTIL YOU SOLVE CHEMICAL PROBLEMS
6. YOU CAN'T SOLVE CHEMICAL PROBLEMS UNTIL UNTIL YOU SOLVE PHYSICAL PROBLEMS
7. YOU CAN'T SOLVE PHYSICAL PROBLEMS UNTIL YOU SOLVE MATHEMATICAL, LOGICAL and ACTUAL TOTALITY PROBLEMS
8. YOU CAN'T SOLVE ACTUAL TOTALITY PROBLEMS UNTIL YOU DISCOVER, LEARN, KNOW, APPLY, USE, AND BEST ACT UPON ALL OF ACTUAL TOTALITY.

Fig. 35-1 Comparative analysis, dependence of problems, contingencies of solutions and reciprocal proofs in Actual Totality

This figure dramatically, if simplistically, shows the depth of interaction to which causes, problems and solutions occur. To fully demonstrate them a much more elaborate representation is needed. This is one that is not only reversible it reveals the reciprocal interactions that help to prove TO. It includes all clines or levels of mass energy in TO, SÁ, what and how all TOV-TO with SPALT, METAD, SPAT, ACC, COGNOS, ELTICMA, and SÂ can do. There is nothing more vital than to: 1. Discover, learn and know TO, TRU and TES. 2. Stand up, work and live for TO. 3. Stand up for those who do right. 4. Help, train, correct yourself, others and all. 5. Ignore, alter, negate or eliminate those and all that are not, nor do not do what is right as shown by TO-TOT.

This academic curricular type of approach to the interactions of problems, knowledge, TO-TOV-TOF and the contingency of solutions is reversible and reciprocal, interactive with all in TO. There are many increments and ways to continuity and permanence. The linear from largest to smallest best, vertically integrates from least to greatest in causation by actuality in TO-TOV. The levels of knowledge and contingent existence, (LECEX), each and together require sufficient form, function, health and well being. Their acts sustain the continuity of TO-TOF in the near to mid term, and absorb TO-TRU into TOT-TOF in the far term. This figure is a cross section of TO, from questions and answers of fundamental knowledge to the physical universe, to problems and solutions real live human have separately and with the problems of the whole world. When approached and interpreted by the perspective of TO, by SPALT and other models, we can produce a more detailed comprehensive convergent set, like some crows foot diagram, in which all reaches a crescendo in the climax of TO and its presentation. To fully produce and prove the full solution the final step is, 'You must comprehensively solve the problems of the generic to specific, and specific to generic of totality, especially the accommodation and fusion of the most specific TIT with future transformations of TO-TOF. How the actuality, objectivity, subjectivity and reality of TO and TOV must progress most effectively, successfully, functionally and practically in space time and mass energy.

Of the greatest proofs of actual totality is the Double Reversal of Disproofs, DREV. This is how any, all, and the degree to which TO is assumed to be disproved, are in turn disproved by a full qualification of TO, and a demonstration of the error in assumed disproof. It is why TO must not only be correct, of the right quality, and sufficient quantity, it must be able to demonstrate its contrast from all else, all that is less than itself. Such a double reversal of disproofs serves a second purpose. It helps us discover, reveal, distinguish, improve, refine and perfect TO. It is closely related to methods of describing paradoxes and enigmas and their resolution in TO, cf. # 37.

The double reversal of disproofs, DREV, is further explained from Chapter 12, by difference between absolute and relative, certainty and uncertainty, or between two different continua in totality. By LANERA, absolutely absolute and absolutely relative do not exist, except by the actual pure essences of the object, universe or totality in which they exist. It is the role of TAR, TIT and TRU in RELTO and ARTO of TO that a practically closed set enables us to prove TO, and further position all by their actual pure essences. In this way the uncertainty principle of a physical universe and the voids of monotheism can be corrected, to prove TO,

To further confirm the proof of TO and Á, there is the reversible proof of TRU and TO by typical mass-energy, Á, and Á by TRU and TO, (PRÁT). If there is no observer or subject, there can be no observed or object. Without either there can be no TO. If there is no totality, or TO, there can be no observer, or Á. That is, if things are disunited and chaotic the observer can't identify them as a whole, nor Á of TO, nor any Á, or TO. This has been called the final anthropic principle by those who are of extreme absolute totality, yet misleading unless interpreted by VAPO, DAOST, and all the rest of TOV, RELTO and TRU, as evident in TRIS and TÂS of TO.

A simple and most effective proof, and reciprocal proof of TO, PRUT, is PRUNES or Proof of TO by Non Essentiality, or what TO can do without. This is in a way, opposite to the law of partial necessity, LOPAN. Taken positively it is a proof by what TO most consists of, or is most vital, or REM. The Hidden Observer Proof of TO, (PRUT-HOB) denotes what is perfectly TO that we overlook in all, only to put in its place much less, null or the negative. By Proof of TO through Total and Reversible Objective and Subjective approaches, (PROTOS), a complete cycle return is made. Initial processes of S, O, and IA respond in observer observed relativity. Similarly there is the Law of Á or SO Reality and Reduction Proof of TO and TRU, (LARPRUT). It is the necessary reduction of how we look at TO, how TO is, and how TO is objectively separated. The Proof of TO and TRU by the Exclusion Law, EL, (PROTEL), is their proof by the separation of all within if disvalued and without TO, to the degree less than and separated from TO. This shows how by increasing production of the limits of ES, CES and EL to TÓ we provide a formally distinct and substantially complete TIT, TRU and TO. PRUT is repeatedly shown in the resolution of paradoxes as well as Proof by Negative Inversion, PRUNI. This can be compared to proof by parts and characteristics of TO, laws of objects, and proofs by the correction of errors, false beliefs, ideas, and tenets.

From Chapter 19 and 17, Man point time, reversibility in future, light and the levels of mass energy, light = a form of mass energy (DIMFOR). Since mass energy = a kind of object-subject, and reality = a kind of subject-object, then mass Energy in TO = reality of TO. Then reality = the square root of light/lt, by relativity. These equalities or correlations provide a reciprocal proof of the equation, reality and thereby contributions, to the proof of TO. Since mass is a collective of matter, material and exists with and depends on motion, practically they are of mass energy. Light as photons and waves are of mass and energy.

Many proofs of TO, ARTO, TOV, TRU, TOT and others, lead to added proofs of TO. In seeking a perfected proof theory, or the optimal proof of all, it is not only necessary that we Prove TO, PRUT, we need to prove that associated, and that which is less than TO, as ATO, A and ALT. Attraction, as *love,* is a commonly recognized bond that is also a good proof of TO, yet may connote separate ways of approaching TO. They are readily recognized as the positive interactive relation that exists between many aspects and parts in and out of TO. One of the most basic proofs of TO is to exist, and exist by TO, from being dead or alive. It is only as we are 'alive', alive and well, conscious, or can be attentive to the needs of life or whatever enables this life, that there is actual existence. Yet the dead, or alive that we personally know or can anticipate so deeply, are not the same as they are for TO. Only as they are in TO, and for us in TÍS and TES, does to be alive, provide good proof of TO and T0V. Thereby it becomes evident that, as in being, the relativity of our existence is the predominant feature of life, the most special variety of TO. Life and death are the most profound aspects and features of individual being, being TRU in TO. This is not a simple existence. It is a healthy, strong, successful, and predominant existence. For, only in this way do aliveness and special existence, have assurance of continuity.

Reciprocal proofs are shown by the negative-positive interactions. As lethargy, depression, overabundance, waste, excesses, overpopulation, ill-managed power and wars shake humans to their foundations so must renewal be reciprocally proven by the most fundamental restoration of faith in quality and other universal concepts, principles, laws, modes, continua, varieties, proofs, order and mechanisms of all in TO-TOV-TRU-TOF. The positive must not be superfluous. It must be genuine. It should be full of hope and happiness guided by a serious and vigorous pursuit of the correct form and function of TO-TOV-TRU-TRIR within TOT.

The modes of TO prove TO, as TO reciprocally prove its modes in a repeating cycle. Likewise its continua, kinds, absolute relative and varieties all prove TO as they reciprocally prove each other. This is a natural and normal outcome of actual totality and all that converges on its unity. For with convergence on unity all TO and its constituents tend toward ever greater self correction. This is the Law of Accelerative Self Correction of all with Convergence on Unity in TO, LASCUT. It is this the greatest power there is, generated by the creation and evolution of best human life that has sufficient capacity and strength to equal, and balance the seemingly all powerful physical universe, as it in turn, is altered and adapted to be its part and role in TO. This is the Law of Accelerative Self Correction in Unity, Equal to and Equilibrium with, the physical universe and absolute totality in TO. LASCUTE. This law, describes and proves how the physical universe like the absolute does not exist alone, they are in TO, by their adaptation and alteration by the capacity and powers of RELTO, TRU and best human life.

Losses that are apt to occur in the future can't be prevented unless they are understood. It is in timing, finding, developing, and using the understanding provided by actual totality that is a large part of this text's proofs and theme. The summary of this is, LOMT, the Law of Total Management through PPMA, ACC and all that are best preserved by and of Actual Totality. TRU and TES.

CHAPTER 36 - OTHER PROOFS of TO to enigmas and paradoxes

Multiple proofs, approach, methods, inclusions, external/internal transition, lack of consolidation, far different distinctions, degrees of separation TOT-to-TOF, comparison, great presence of TO, external perspective, positioning and operations of TO, SPAT, bringing into view, tests and other means

Explanations of added or multiple proofs prove TO, e.g., proportions, form, function, limits, etc. depend on what they refer to, developing TO, complete TO, uses and future TO and their degrees. Most proofs do not pertain only to their own continence. This is, for the same reason that phrases, sentences, descriptions and explanations are needed, to show interdependence and fulfillment. It is characteristic of proofs that they are supported and given with a sufficient amount of relative detail to provide certainty. Good overviews of all things are needed, including totalities, potential to actual, TOT to TÓ. They and their components cannot be achieved without good order and coordination. This is provided by knowledge of totalities and TO, as well as their representation and presentation. The better one understands, how all is separated and exists, both outside and inside TOT and TO, the better this overview is, and the better it can be successfully applied.

A major method, means and tool for proof is vertical integration, differentiation and the spectrum, scales and gradients they provide. By a similar means we have in the form and nature of TO, the means to reduce all practically and in functional usage to the level that is relative to our existence. The whole is demanded by TO for existence. The scale is what we can produce at various levels for various levels of proficiency, as for different ages in life, degrees of intelligence, interest, and opportunity to learn. It is by vertical integration we have largely represented and presented this book, TO. It is why we must always remember the whole, and repeatedly bring greatest modes, universals, continua, concepts, principles, laws and all together as they are interactive and interdependent in the whole, TO, and any extension, such as TOT-TO-TOF. This is how we prove TO, and how these proofs dispel all doubts by answering the questions and laying the foundation for an understanding that can solve all problems.

Proof of existence, totality, TOT, TO, TOV or its varieties, including TRU, can often be effectively made by comparison. By the defects, deficiencies and faults of all else, especially any universe, world, system or belief that is widely held, or is used to deny proof, TO-TOV-TRU-TOT can be produced. Among these, two that are most critical, is the physical universe of science, and the spiritual world of God. Both have great tradition and have provided much knowledge and benefit for the world and mankind. Yet both have flaws, errors, and wrong beliefs that render them inadequate to be the final unified totality. They distort the form and function of actual totality and thereby divert identity, from and in, TO. However we must always remember to be tolerant and honor the contribution and part of all has with, and the equilibrium of all in TO.

The lack of consolidation in all totalities, worlds, universes and deities when corrected by fundamental concepts, formal, elemental and functional features of TO, provides a most valuable proof. This is evident in LAREV, Chapters 6, 8, 9, and LOCET. Chapter 20. or Addenda. Laws. To clarify and be most accurate and convincing, it is necessary to show which of the many modes, continua, kinds, varieties, forms and states of TO are being proved, e.g., ABSTO, RELTO, TOVS, TRUS, TROS, TRU, TRIS, TIT, TRIT, TRA. How proofs differ between themselves and in relation to TÓ, TOT, TRU, TIT, TRIR and TOF. By the dynamics of external to internal transitions and transformations from the most absolute to most relative of existence and totalities, and their actions in TOT, TO and all its constituents can be known and thereby TO more fully proven.

Proofs are easy when they come from one ism or polarity or another. Only they are less certain, and must be qualified in all ways sufficient for TO. Paradoxes are also easy when they are a simple polarity of the duality of a continuum, or of simple differences, as what is or is not. These also must be disproved when inadequately qualified, interactive, reversible or reciprocal and insufficient for TO. In a sum proofs must be combined and collected so that all must, not only be their part and contribution to TÓ, they must do so when fused in the unity, totality and harmony with all else in TO.

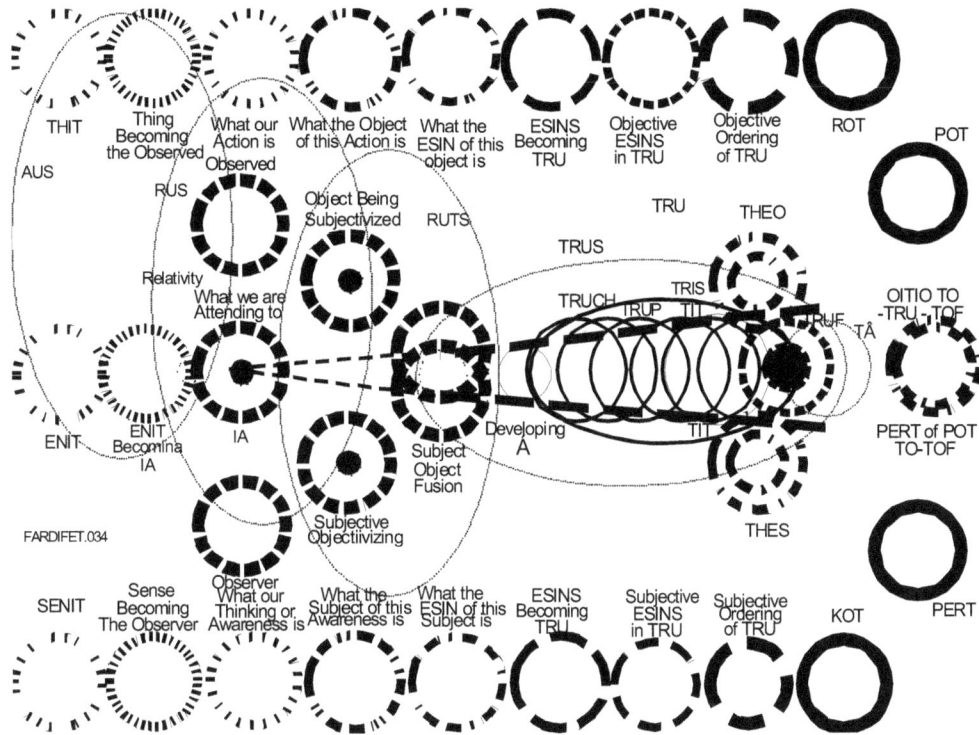

Fig. 36-1 TO-TOT Far Different Distinctions of all, FARDI

Each and all humans cannot begin to be unless they increasingly separate and rise to make discrete and distinct the world around them. This is a great gift and blessing we have inherited from life in actual totality. Like the continuum of integration and differentiation, separate existence is limited for a part, and without due care leads to a loss of integration and unity. It is this separate and disparate, integral and differential that makes TO, TRU and each person unique, special, exclusive and absolute. It is not only how we must approach TO and TRU, it is TO and TRU. It is the great continuum of math, whose form and function, states, dimensions, and mechanisms are much of our being, in TO. Each of the categories or elements in the figure, are major participants in the form of, approaches to, representation and showing of, TO-TOT. The greater we can realize their differences and make them formally distinct, the greater we will not only understand TO-TOV, but the greater we - as our part, TES, and TO will be. All less than this combined differential-integral, is the loss of the very reality and actuality that we so take for granted. We cannot live without light, blind to what this light makes possible. It is light and all the other components of the reality of our mass energy powers, Á, that enable us to approach TO, order all, and be, within TO. In Figure 1-36 the parameters are generally, time on the ordinate or horizontal, and S-O on the abscissa or vertical. Each of the separate items is a great entity-in-itself, ENIT, whose differences from adjoining, as well as all other items, show our approach, origins and development. In this text we concentrate on the hatched and black lines and figures. Without keeping such overall panoramic views in mind we lose sight of the whole content of TO, what it stands in relation to and are most important. It is like DAOST, different aspects of the same thing, what the Á powers enables TO and ourselves to make infinite and units, infinitesimal and total all at once. Without this we never encompass the whole, nor recognize our variety and future, TO-TES-TRIR-TOF.

It is subjectivity with objectivity in reality that brings the actuality of totality to life by making distinct all humans encounter. This knowledge and its cognitive powers are largely what has, and are propelling us. They prove our existence as actual totality, and the opportunity and probability of its ever greater and more permanent well being in the future.

223

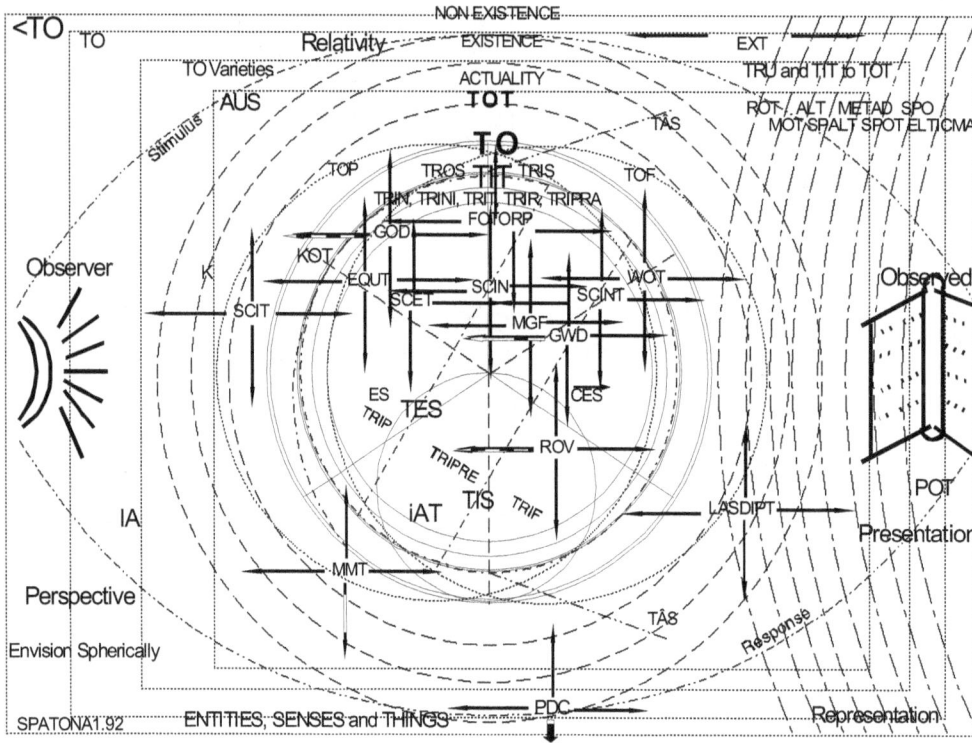

Fig 36-2 External Perspective, Positioning and Operations in a Standardized Presentation of TO, TOT, and their external approaches, SPÀT

This figure provides an Accounting, Accentuation, and Action, ACC, of the in and out form and function of TO and TOT, OITIO. Although such a figure, a website, and a book, are far less than their total form, dynamics and operations, they seek to show the way to construct, provide and operate all in and out of TO. It roughly helps to develop an idea and provide a picture for use. It shows what can be achieved, and the accuracy and clarity of positioning actual pure essences, constituents, concepts, principles and laws of TO in an extended figure. It provides a new method of foreknowledge, how we can manage and control our lives, the future, and all to come, rather than being victims of deadly adversity, repeated suffering and failure. Such views always look better in the morning, or after war when peace has been won. But this does not overcome loss, only signifies how powerful human lives, hope and determination can be when accompanied by persistent guided endeavor. This rendition by SPAT of TO, TOT and external factors and relations are figurative, iconic and used to show how many concepts and proofs that otherwise remain unrevealed can be provided. The positions of new concepts in its field or sphere are also a work in progress and show, as we do for SPALT, how important it is to provide not only a model, but show how other models add to the demonstration of actual totality, as well as how it can be ordered and operated

The full value of Bringing Into View, (BIV), comes from the clearest and fullest descriptions and expression, SPÀT. Completion of KOT and TO are aided by processes that rise above the particulars of S and O, to complete objectivity, OBT. Combined S-O produces the reality whose actual pure essences form the basis for TO. This is by focusing and concentrating on SPÁLT, SPÀT and their posits. The A in SPÁLT, signifies the mass energy dimension of TO. That of SPÀT, signifies mass energy from the outside of TO. Both figures are largely the same within SPALT, only different when outside in SPAT. Solutions are made by the exclusive constituents and aspects of TO, limits of its FAF, and summation of its methods. One of the greatest tasks for completion will be the continuous perfection of the dimensions of TO, accuracy of POA, and posits in MOT. By TO, SUT and all, accelerated success in the actuated future is probable and can be achieved with a proper mind set, that is on TO.

Other proofs are available in prior works on totality, the relative universe, knowledge in general, and knowledge that contributes to actual totality. A great amount of information and evidence that supports further proofs of TO is available in many of the figures and notes in this and associated texts. The frequent use of the supplemental contents, figures in each chapter, proofs and laws that are summarized in the addenda help to give the reader a quicker and more comprehensive perspective and understanding of TO. This is achieved when all are collected and integrated in figures, such as Intd-1, 14-2, and 38-1 which together help to give us a vision of the actual form and content of TO. Throughout the text, and as enumerated at the beginning of the text, proofs are given for the major topics, themes, and concepts of TO. The concrete closure these proofs provide is vital for each person to learn if they are to successfully fulfill their lives. Along with the laws of TO proofs provide the basis for a guide that can bring us from confusion and chaos to clarity, order and a most efficient and productive life.

Among tests of TO, many of its proofs when confirmed can be exchanged and used as tests for TO, or that aspect of TÓ, to which they pertain. All that can be reduced to accurate measure of TO can be used as test, some much more and some less important. Importance is determined by how great or valuable they are in TO, how well they can be adapted for testing, and how well they can be understood. All major continua of TO, whose dualities form spectra and gradients are apt to be good tests. One of the areas that make the best tests is that based on proportions in TO. This includes REM and LAPREM or Law of Á forms, dimensions, gradients and their Representation Measured by REM.

Completion of KOT and TO is guided by MEPAT or the Method of Paradoxical Transformation wherein what seems so absurd, both great aspects within TO and extra TO, can when properly reduced to proof in TO, be managed, maintained and controlled. Completion of KOT and TO follow completion of S-0 successfully fulfilling stimulus response, qv Fig. 25-2. These involve All Time Mastery of Things, ATIM, that comes with improvement of TO and oneself in identity and as a component, TES, in TO. When proofs of actual totality are collected, related and more or less certified, several that are less certified remain. It is this remainder that must be attended to, and brought to order if actual totality is to achieve the needed trust that is most characteristic and typical. Eight of these include,

1. The definite and specific focus of actual totality. There are many absolute totalities, physical universes, relative universes and relative totalities. Yet, to discover the ultimate unified totality is to find actual totality. This is done by fulfilling and ordering all the typical features, form, function, characteristics and properties common to all totalities and universes in actual totality. It is done by recognizing the focus and form of mentality and best human life in mass energy space time.

2. The polarity, duality of continua in actual totality. Actual totality consists of a host of continua of various kinds. To learn and describe these continua and how they work interdependently and in equilibrium is key to knowing and proving actual totality.

3. Of the continua of actual totality, that of the absolute and relative are paramount. They do not exist as separate totalities in TO, They are changed, by actual pure essences and develop into a form that most benefits actual totality. It is by the effect of the relative, relativity, on absolute totality, and the effect of the absolute, absolutivity, on relative totality that we can envision how all unites in TO.

4. Of the absolute and relative totalities, their changed existence in actual totality needs to be signified. This is ABSTO and RELTO, which when fully and evenly united exists as one, ARTO. ARTO signifies how absolute and relative totalities, or the physical and relative universe of best human life and mentality exist in TO.

5. Since actual totality must be definite, or "the" actual totality, and since it is in transition, especially in the long term of time, and since it is in continuous and massive interaction with what is less than actual totality what is less and especially potential for TO must be accounted for. This is TOT or TO potential in totalities. TOT of the past is TOP, of now, TON, and of the future, TOF. The relations between TO and various kinds of TOT are basic to understanding actual totality. They explain and prove many possible disproofs and unique features of TO.

6. The continuum that exists between the extreme of most stability in TO, and instability in TOT, is

accounted for by its actual pure essences, practical and functional convergence, adaptation and alteration with TOT.

7. The stability of actual totality is also produced by the convergence and reinforcing powers of its form, parts and processes. When TO is in the best of health and greatest strength its powers are most beneficial. These tend to assure the set form, and necessary existence of TO.

8. All that, and the degree to which they participate in TO are contributory to its existence, survival and permanence.

Thereby, it is how well we discover, develop, learn, know, apply and utilize all ideas, concepts, principles, laws, mechanisms and proofs in the right way, that most fulfills the purpose, goals and providence of actual totality.

CHAPTER 37 - REINFORCED Proofs by ENIGMAS and PARADOXES of TO, to Total Proofs

Focus of Proofs on the most profound enigmas, paradoxes, uncertainties. doubts and unknowns of TO, Non existence to correct resolution by developing totalities in their existence in an extended TO

Actual totality is highly enigmatic. Many dualities of continua are paradoxical. Some of TO is, uncertain, in doubt, and unknown. Like any rigorous field or discipline once the more basic concepts are learned, they become a natural part of ones life. Without the incentive to learn, many people are turned off in despair, deny TO in stubborn disbelief, or escape in prisons of their own beliefs in false world views. This will be remedied as knowledge of TO-TOT expands. The universality and common convergence of all in TO is for one and all. It will bring well being to each in their own way. Life is much more successful when you know where you are, why and where you are going, how to get there, and what to do after you are there. This is personal, being one's soul, TÍS, as it converges with that of others, TES, in all that is in and of actual totality, TO. - Criticisms of actual totality or any of its equivalents, TREQ, can and must be met and resolved by revealing their greatness and that of their categories, features, modes, dualities, varieties, form and function and their reinforced proofd. This is done by an analytic, rational and precise demonstration and evaluation by the validity, actuality, proportionality and reality of the whole of actual totality. Much that knowledge has discovered, developed and consists of, when correctly identified and equivalent to its counterpart and actual pure essences in TO, is readily available for reinforcing by its proofs and usage.

Paradoxes and enigmas arise from the most fundamental aspects of actual totality. This is especially so for those divisions and continua that are dualities, whose dimensions provide ready knowledge of form and also function. In turn, as we discover, learn, correct and confirm actual totality further knowledge of other aspects is provided. This is knowledge, like the difference between unknowns, mythology, scientific fact and certain knowledge is ordered by the convergent unity in totality of actual totality. In this way many paradoxes and enigmas are already proven when we know actual totality. Other problems, paradoxes and enigmas, when fully investigated, interpreted and brought to their part and role in actual totality are solved, resolved, and reconciled. The resolution of enigmas is discovery of how they work that of basic paradoxes is by understanding continua, duality, polarity and mechanisms. This holds from nothing through existence, essence, continua to final resolution in all of extended TO. Of the most enigmatic and paradoxical proofs of TO, are the differences between potential and actual, or TOT and TO. It is similar to the difference between absolute totality and relative totality or the physical universe and human existence, mentality and knowledge. They arise from changes in mass energy, space and time that occur in transitions and transformations in and origin, development, existence and future of totalities, TOT and TO. The part and role of actual pure essences in TO and its varieties with transformations in time, TOT to TÓ, to TOF. Once these are discovered, proven, learned and become common knowledge the rest of actual totality is much more readily known, remembered and used to guide the lives of each and all persons.

There are many enigmas and paradoxes that arise from the fundamental features and mechanisms of actual totality. Is relativity a spectrum, one in which it changes both qualitatively and quantitatively from absolute to relative totality? The physical and relative universes are altered, adapted and corrected in actual totality? One in which their massive, intensive, valuable, practical and completed whole of each portion is included, as they exist in actual totality. This is evident when separating and uniting absolute and relative totality, each existing as ABSTO to RELTO, the generic to specific, in actual totality. It includes the most, to least, contributory potential to actual mechanisms and all in their interactions and transitions between actual totality internally and less than actual, but potential externally through time.

Whether the big bang of astronomy began from nothing, or from single or multiple universes conjectured by physical theory, does not practically deny the role of absolute and relative totality in actual totality, TO. Whether absolute totality did or did not exist before the big bang, and whether there is, or is not, anything outside of the big bang, are problems yet to be solved. Seemingly by LANERA or the Law of Never Perfectly Absolute or Relative, the regress and progress are a kind of finite between infinite and infinitesimal. Transitions in time and transformations of mass, energy, volume, pressure, temperature and other features of mechanics and physics are not the same for different totalities and are accommodated by TO. In addition, actual pure essences of TO work in many ways to maintain and sustain TO, absolute totality and relative totality in continuity. This occurs by the form and function of TO, and the changes and reversibility in TO with TOT. The astrophysical problems that threaten human life, existence and continuity will increasingly become better known and resolved by the much more rapid expansion of TO, TRU in TOT and TRÂ. The role of actual pure essence, the form and function of TO, and the powers of intensification by TRU, TIT, TES, TIS and TRÂ are more probable factors that can sustain TO-TRU-TOT through time, and in the farthest term. Hereby states of being absolute and nothing, like the impossible absolutes for which their unity by convergence in the relative of TO provides the basis, is the proof of its essential, actual, pure and perfected existence. To the extent the astrophysical physical universe, as absolute totality, is altered or adapted within TO and much that is not or less, we must always account for this departure from TO.

That most evident or enigmatic is often true and right. By simple attention and the unity of TO all seems so be true and right. Yet what seems so true and right, our attention and unity only scratch the surface of TO. Much about us is most subtle and profound, and unity alone is superfluous, whereby both are often untrue and wrong. What is true is a relation and agreement between sense and thing. Truth is the state, condition and virtue of holding to what is true. Validity has a higher quality, one of evidence or experimental precision and proof. For true, truth and validity to hold in their existence in TO, requires the highest level of essence, precision and proof. Although much is similar, common and valid, much is also very different and invalid unless proven to be an aspect, mode, concept, principle, law, mechanism, part or process of TO. It is not any answer or solution, but those of VAPO, that are valid, right, essential in TO.   The principle of ultimate coincidence, PUC, is shown by the fact that a person, group, specie or variety may do as they please, or in their own way, to an extent or for a while. Yet they will sooner or later have to do that which is necessary. This is a return to harmony with the world around them. Ultimately this is actual totality, except to the extent they change, or TO changes, from various positive to negative causes. These in turn must be convergent on TO by PUC. Such coincidence is induced by the power of the same, as in the continuum of DAOST. It is the balance, and reconciliation, of identity and non identity, of absolute and relative totalities. This identity and non identity which reality overcomes and coalesces proves the certainty of TO. This is a product of the subjective objective fusion in reality that reflects the long term development of TO, and the sum of all that enables it to be.

The most enigmatic, yet probably most important proof of all is what reconciles the absolute and the relative, the physical with the cognitive, universe. This is key to a united integrated totality that most fundamentally resolves many others. A reconciliation of generic and specific, matter and mind, object and subject, expressionism and impressionism, the astrophysical or material universe and God, with relations to many other continua in and of TO.The physical of absolute totality is differential, oriented toward the infinite and infinitesimal, highly divergent and exists indefinitely. It is based on the limits of time, space and mass energy. Being very objects oriented, absolute totality, and the physical universe, are sensed and seem to us to be distinct and definite. Yet alone, without relation or relativity, it is an open set without a center.   The cognitive or mental relative universe, is a world of best human life, mind, philosophy, knowledge and God that are integral, oriented and convergent on a center, here and now. It is based on the present, of small size, subjectively oriented and convergent. The enigma is clear, the dialectical separation of being and objects, or ontological separation. How the immediate certainty of knowledge of things, objects and being, are so readily apparent, yet faulty, and even more enigmatic, how they have no unified whole. This is resolved by TO, which becomes, equally and increasingly certain and proven by the integration and fusion of all. This is a result of the reality of subject-object, in the mass energy, Á, continuum with all in TO. This occurs when knowledge becomes identical to its part and role in TO. When ontological separation and apparent certainty are reduced by actual pure essences, in the different aspects of the same totality, TO, there are the highest and most fundamental certainty of being, TO, qv. ontology chapter 6.

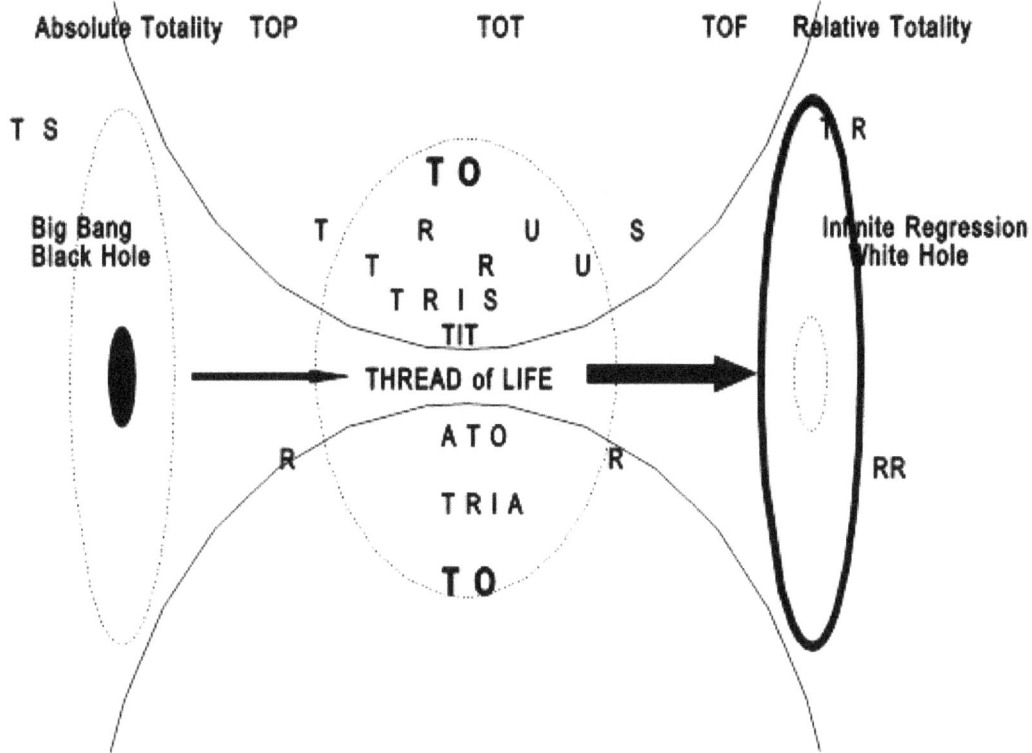

Fig 37-1 Eternal Continuity or thread of life in an Autonomous, At Once, Total being of TO, ATO

This can be shown in many ways, one of which is to add many details. It is the basis for what is positive. The thread of life is the window of opportunity that all humans have, are and will continually encounter and must overcome if they are to find well being and permanence. It is a thread that forms a web in which all positively converges on ATO, at once autonomous TO. The figure shows a distant external view of actual totality and some of its constituents interacting with TOT and the more general physical, or an objective universe. It features the central role of survival over extinction. The black and white holes may be better expressed by the big bang, dark matter, dark energy and yet to be found more fundamental states of matter and light in space time. Such concentration and diffusion of matter, like black and white holes, contraction and expansion, are like the extremes of absolute and relative, AA and RR. All of the existence can be included in such a figure, which thereby would include an infinite myriad of components, internal and external to TÓ. This figure shows both the enigma and proof total unity of and by TO must be maintained and preserved. It is balanced by, and finds resolution in, its approach to the absolutely necessary existence it seeks.

When one remembers that the basic existent in all totality is matter, and with anti-matter makes up its primary continuum and dimension, it is also imperative that one remembers that they exist in different forms and states in TO. We realize that these forms and states of matter are not the same as they are in the physical universe, but take on that typically of each succeeding feature of combined absolute-relative totality, as TOV, in TO, by (LAMA and LARIT). Only when we trace through the part and role of the forms and states of matter-antimatter and mass-energy transitions and transformations in absolute and relative fusion, and the varieties of TO, TOV, do we appreciate the validity, actuality and importance of a sufficient proof of TO and its laws. The fact that dark matter and dark energy, like matter and antimatter, exist in different forms and states, as well as subatomic particles and their functions, helps to confirm their transition in absolute-relative totality, and its combined existence in TO. Much is unknown, much will need correction or fine tuning, yet with all the great discoveries and additions to knowledge in many fields and sciences, we can with a high level of confidence find joy in the wonderful opportunities actual totality provides to direct all knowledge and things to agreement in, or by, TO. Yet if we learn that mass does not equal matter times volume as held generally, but mass equals matter times density, then the relativistic laws of the physical universe are correct as far as they go. Only it remains what is matter, what is density and what is their form and function. Density is how much matter per volume. Thus, mass is simply equal to the amount of matter, and anti-matter. Thus it is the form, states, function and mechanisms of matter - and anti-matter - in space time, and in the various gradients between absolute and relative totality, in TO that we must be primarily concerned. This is part of LAMAS, the Laws of Matter and Anti Matter.

That there is a disparity between TÂ, TIT, and TRIT in DAT is the Paradox of TÂ and TIT disparity, PAIT. PAIT is cursorily answered by, absolutely we are relative and relatively we are absolute, or Absolute Relative Identity, ARI. ARI occurs as the relative and the absolute approach and become each other in TO, ARTO. The realization of perfect form and function of TO in TÂ, as the creative, emergent, procreative, and DAT powers converge on TÓ, to produce everlasting life, the relative and the absolute converge in unison. This is a special case of the overall state and condition of interdependence seen in Godel's Theorem, and is called: STIN = State of Interdependence. All that is so great of TO is so, not as it was in itself, nor as it may be taken carte blanche into TO, but only as it is selectively reduced to, or completed in RELTO and TRU as this TIT is completed by it, or the Principle of Ultimate Coincidence, PUC. The actual world, TO, is the maximum of absolute effects of absolute causes, as all are in TRU pure TIT. It is relative unity by RELTO, as things and partials are or are not of TO. Their validity and actuality are a function of their relevance and function in TO. If all that is TO is all by TRIA and TÂ and all that is in TO is absolute by TIT and LOPAN then we can be assured of being absolute - have immortal souls - if we are our TÍS, TES and ARTO in TO.

The focus of union is proven and tested by the fact that nothing exists separately by actual totality. Even actual totality does not exist separately in full extension, intension and actuation. The power of the absolute and the relative combine to expand and contract all in optimal equilibrium in actual totality and in its essentials on transitions and transformations in time. The extremes of differential and integral, absolute and relative define and determine the degrees of separation or union of the physical universe and the world of best human life. Together the absolute and relative in TO, exist as a closed set. This is by their practical and functional values. It is a closed set that partakes of some of the qualities of absolute and relative totality, only modified by their actual pure essences, and that of their union in actual totality. This is the clearest distinction and proof of TO, and enables a solid basis for a closed set with the fusion and union of absolute and relative. It is a closed set of an extended present, modified by functional values and practicality. Together absolute and relative totality in TO exist in the most static to dynamic continua internally and externally, whose existence depends on the strength of its mechanisms and opposing forces.

Actual totality, TO, is the ultimate unified totality, neither an absolute physical nor a relative mental can exist separately without denying the actual pure essences of human presence. Such denial imperatively excludes all that has enabled the material and mental, physical and ideal, absolute and relative. An ultimate unified totality is a combination of many continua whose dualities exist in equilibrium typical of it, and its actual pure essences. It is a fusion of actuality, subjectivity and objectivity, whose existence is confirmed by their combined reality in actual totality. The laws of totality conform to, and are confirmed by, differences that are reconciled in the whole of TO. TO and TOT, are proven by extra special totality, how the unity and reality of knowledge and civilization are composed of increasingly higher states of mass energy, SÁ of Á that increase capacity, self perpetuation, power and total convergence in itself, and in TOT future. They enable, prove and assure the actuality, relative absolute, and the degree of exclusive and relative existence of TIT and TO. Among the most difficult of enigmas, are that of the future, for which much is unknown. These also can be largely resolved and reduced to management and control by intensive use of existing knowledge, discovery, determination of knowledge of, and operations by, TO.

There are three kinds of transformations in the future for TO-TRU-TIT-TRIR-TOT. And for each of these three kinds there are three ascending levels of TRU-TOT. They may be classified by time, near, mid and far term time in the Future.

Initial to increasing adaptation of TO and alteration of TOT

1 Completion and consolidation of TO, transformations from past to present.
2 Developing completion of balanced TO-TOT transformations in the present.

3 Early improvement to perfection of present TO, toward eternal life and completion with TO-TOT.

Increasing Predominance of TO, before altered Special to General TO,

1. Continued adaptation and perfection of TO. Transformation present to future
2. Continued completion of balanced TO-TOT, transformations in present to extended future
3. Developing eternal life with increasing predominance and balanced TO-TOT

Transformations that involve convergence of all on TO as it becomes the predominant General-Specific

1. Overwhelming predominance of TO producing less specific status and form of TO incorporating TOT.
2. TO moving through less specific toward General Specific status, with assured balance and control of all,
3. Eternal life with overwhelming continuity in long term future perfected transformations of TO with TOT, ultimate in time, assuring permanent existence.

A great enigma, especially for the future, is whether, and if so to what extent, the center of size changes, qv table 1-6. Also if it varies, in what way, and how much? It would seem that the specific varieties of TO would tend to remain the same. Although the general, or physical, varieties are less definite, especially in the very long term, the permeation of absolute totality through actual totality would tend to keep the same average for size. Unless the ratio of ABSTO and RELTO change, which it does, yet tends to return to equilibrium. This helps to support, as well as prove TO.

Problem solution and paradox or enigma resolution of all in actual totality is one of the greatest needs and work of actual totality. Their achievements will determine, whether, and the degree to which, actual totality improves or fails, survives or becomes extinct. This is why each and all persons must discover, learn, know, remember and best apply TÓ, to guide their lives to maintain and assure TO. The finite predicament of extreme polarity of the subjective-integral, subjective-differential, objective-differential and objective-integral is what the subjective-integral of our special partial position in TO, seeks to reverse. It largely corrects and succeeds with sufficient knowledge of TO, or its equivalent, and the combined ordered proofs in totality, TO.

The greatest enigma, and problem of actual totality, is whether and how relative totality and the universe of best human life exist with absolute totality and the physical universe. Is the human world of mind and applications and all they include, only a tiny part of the physical universe, that is here today and gone tomorrow? Are we barely holding on to the thread of life only to be gone, like so much before us in a short time compared to time in the astrophysical universe? Or is the physical universe, as we ourselves, only a part of actual totality? How we and our world are formed and function as a unified whole. Neither are, absolute, both are in transition, and both are parts of each other, as absolute and relative totality become one by ARTO and TOT and TO. The evidence for this is manifold. Every alteration of the physical universe that does no, or minimal harm to it, and increasingly benefits the human world is a step in proving the existence and power of the human world, TRU, RELTO and TO. There are many ways humans are adding to and improving this alteration of the physical universe. It is the function of life to make a home for itself, to make minor changes to the environment, that are major changes for itself. How these progress, depends on many factors, including how well humans can make these changes, at what optimal rate, and how long, or permanent. To make a home ideal, like a Garden of Eden, for ourselves that nears permanence, has been, is, and will be a foremost purpose and goal of life on earth and beyond. Everything we do that is a permanent benefit to the environment as it is potential or actual in TO, is a proof of the separate existence of TO, TRU or best human life. This is the Law of Alterations of the Environment that are Beneficial to it, and proofs of the predominant effects of TO and best human life. LABEPT

To what extent are absolute totality and relative totality in union, compatible, incompatible, or two different aspects of the same thing? Until we know more about the origins and completion or beginnings and ends of the physical universe, e.g., big bang and TO-TOT, we can only make determinations that have imperfect probabilities. Likewise since mass, space and time are altered outside the big bang, a unified progression of totality is indefinite. Even the chemical form of matter is questionable, including states of pressure, volume and temperature. Furthermore this assumes that TO-TOT-TO has permanence. Such an assumption is questionable, depending on the strength and success of each and all persons and all they have that will support TO through time. Such a scenario depends on a great amount of adaptation and alteration of both the absolute and relative totalities, and their union in TO. Yet it is the scenario of the thread of life we are in, depending on time allowed and whether we attend to, or neglect it. If there is anything outside of the big bang we must alter or modify certain laws, including those of the conservation of matter, and the form and function of mass energy space and time. To the extent we accept great changes in the existing form and function of mass energy, space and time, we must be capable of managing and controlling their existence to support TO.

Since we do not know much of what existed in the extremes of absolute totality or the physical universe or the implications on TO, we need to keep an open mind and rely on current knowledge, and knowledge provided by TO. We are well experienced, in knowledge of absolute totality and relative totality, in the world near about us. We are also well experienced, in the interactions of the absolute and relative totalities, evident in the subjective and objective, stimulus and response, physical and mental and many others. It is only for us to put them together as they exist by actual pure essences and the form and function of actual totality. How they exist when combined mass, length and time as adapted by TO. In this way enigmas and paradoxes are practically and largely resolved, and actual totality shown and proven. What is most certain, however, if we are to perpetuate the hand we have been dealt and live up to our potential, we must be, support, sustain, prove and optimally actualize TO. This we have been given and must uphold if we are to avoid the suffering, typical of all less than TO. If our ancestors have been selectively the best, and our progeny, are also, the best, we can do no less than enjoy and assure the opportunity to be the best we can be. The measure of each and all persons, nations and civilization can be no less, for this is actual totality.

Although the detail of enigmas and paradoxes of all are incompletely resolved, increasing knowledge provides sufficient information for organization and order in the unity of the world. This work on actual totality adds convergence on such unity to prove that we can, are and must provide the highly effective and efficient completion of actual totality. Actual totality of TO may not be the only relative totality, or even a perfect totality by absolute totality, Yet actual totality is for all practical purposes by its unique relative powers and its own actual pure essences and potential permanence, the right totality. For of all totalities, and right totalities, and their summation, only actual totality consists of all requirements for ultimate unified totality. This is the Law of Double Relative of Actual Totality, LADRAT.

Not only are there great problems, paradoxes and enigmas of what is actual totality, what it is associated with, and what it consists of. One of the greatest is how to show TO, and what is external and interactive, in the most orderly and efficient way. How can vertical integration be best devised, organized, proportioned and shown? Some of this will have to be left to those who with a good knowledge and interest in actual totality continually work to make the representation of TO both simple and correct. Just as actual totality has borrowed from knowledge of past religion, philosophy and science to best build and strengthen itself, so must we continue to make all that is external, including representation, perspective, presentation and application exactly TO, as well as how it changes by transition in time, TOF. Such a great enigma is what the final chapters 38-40 seeks to prove, show and apply, the sum of proofs, representation and application of actual totality, TO.

By the relations, connections and interactions of the proofs from chapter to chapter a more accurate and fuller picture of the whole of TO, its external relations with TOT, representation, presentation, perspectives and use, are provided. When proofs of the greatest themes of actual totality are brought together and formulated a closer approximation is possible with greater systematic unity. All that is of relative totality, 9,10,15, of relation, 9, 12,13, 17, 19, of factors proving all that initiate life and relative totality, 13,19,23,24, of how actuality and actual pure essences work to converge on TO, 3,4,8,9,10,12,15, when combined and perfected provide convincing proof of TO and much of its makeup. Frequently in the chapters those regions that need much work at all levels are noted. In this way multiple proofs of the ten to twenty greatest key features, aspects and categories of TO, can be brought together to, provide the basis for the rest of the text, our knowledge of TO, and the advancement of TO.

CHAPTER 38 - TOTAL COMBINED PROOFS of ACTUAL TOTALITY,

to Representation

Introduction, Development, Single to combined proofs, continua and modes in a sum, integration of parts, figure of total proofs, relations and interactions of proofs, remember proofs and best use of an overview, TO

Like all less than a whole cannot be complete, so are single and simple proofs open to criticism, and in doubt. Combinations of proofs although still incomplete and limited are not only stronger. They reveal the path to total proofs. Only in all TO when equal to a well-ordered duplicate of the whole of what exists necessarily, satisfies the need for final proof. Yet, singular proofs when accurate and perfected, are the basis for total proofs and certainty in TO. Like LOPAN, the Law of Partial Necessity, singular proofs are often indispensable proofs of TO, or their part in TO. All inaccuracies and imperfections are inimical to proof theory and proofs, the cause of failure to learn TO, and its weakness if not corrected, proven and best applied.   Thus it is the proofs, of TO, and potential TO, TOT whose combined relations and dependence substantiate and provide total proofs of all. Many descriptions and ·explanations of all in TO are proofs, especially when combined.

To summarize the greatest need, and enigma of all, is to prove and largely resolve which describes TO,

1. One ultimate unified totality,

2. And this is actual totality,

3. It is the physical universe,

4. It is a relative, humanistic and an idealistic universe,

5. A spiritual world, as God or deity.

6. Or whether it is more or less of these, and all that exist both potentially and actually in the conditioned and qualified equivalence of actual totality and ultimate unified totality.

First: there cannot be one, stationary ultimate unified totality, it will change with time, especially in the long term. Yet there cannot be two separate or different totalities if it is to be the one, ultimate totality. So that it must be found both static and dynamic, stationary and changing in equilibrium with time. This is proven by the massive powers of convergent homeostasis of actual totality. These accompanied by reverse reactions to change with time, as well as adaptations and alterations of both TO and what is potential, explain and prove, how actual totality can, if properly supported, be one, practically stationary and unified totality.   Second: Of the many absolute and relative totalities that possibly exist there can only be one ultimate actual totality which is formed and functions by the requirements of ultimate unified totality. Actual totality is also formed and functions by selective and qualified alterations and adaptations of the physical, relative and spiritual universes, as they by selective and qualified alterations and adaptations become their constituent parts, processes, mechanisms and features of actual totality.

Also, other universes, worlds, systems and Gods although in select and qualified ways a large part of actual totality, separately do not meet the requirements of ultimate unified totality. They lose their identity with actual totality by retaining what is lost in their differences, and by a tendency to retrench their differences from common sameness. Sameness must be based on the valid, proportional and actual totality of the whole. For any of these to be perfectly right they must be highly select and qualified to include all they lack, and exclude all that they have that is invalid, not actual, proportional or whole, to fulfill the requirements for one, ultimate unified totality. Only what is proportionate in ultimate unified totality and actual totality can claim to be its part. When less than this proportionate part it is not right, it is in error, or a slight variant in function that is compensated for. Yet all that is a large proportionate part of ultimate unified totality and actual totality when correctly interpreted and known must be recognized and honored as such. This is why the physical, relative and spiritual universes, although different, and in ways in opposition to ultimate unified totality and actual totality, must always be remembered - and honored - not separately, but for their place and contribution in the whole. Finally: It is when all pertinent and important proofs, concepts, principles, laws and resolutions are provided that all that is more or less a part of actual totality. This is the actual totality that replicates the one, ultimate unified totality, in the functional, practical and actual pure essences of a relative-absolute spectrum of time. Each, of these kinds of totality, exists and does not exist, depending on their make up, and how they are interpreted. To the degree their makeup and interpretation fulfil their part and role in actual totality and ultimate unified totality when these in turn are properly fulfilled, then their proofs, resolution and actual existence are complete. They achieve relative to absolute and exclusive existence, as one, ultimate unified totality.Each stage is dependent on the higher, and the higher is dependent on all it consists, and those qualities, characteristics and properties that are typical and more than the sum of its parts.

Although single proofs of TO are highly certifying it is often in combination that they are most convincing, and in their combined totality most conclusive. No matter how great or firmly held, it is largely because of beliefs that are singular, less essential and disproportional to the whole that the whole of actual totality, TO, has not been realized and comprehended heretofore. For this reason an open mind that admits all proofs of TO, especially when ordered and related, are needed if each and all persons are to understand their world, and make the best use of life. There is in each proof a certain beauty, an emphasis on some, not always major, aspect, feature, relation, form or function of TO that is interesting and helpful. This is fortified by its relation to actual totality and some of its other proofs that together form the whole.

It is the complementation and fulfillment of TO which reveals and proves the combined and mutual existence of objective and subjective. The general to specific, and unified to differentiated, are reversed as the specific modes become the integrating power that enables the existence of TO. It is this integration with balance of differential and integral, objective and subjective, and all continua that fulfill their essences, typical and proving of TO. The most direct and simplest way to show how TO exists, and is formed from the unity of its totality, is to explain and represent how each of its modes, kinds, continua, varieties and major constituents relate and depend on each other and exist in TO. The modes consist of actuality, objectivity, subjectivity and reality, the kinds absolute and relative totality, the varieties of a number - ten - levels of totality from most generic to most specific, and its continua and constituents, including mass-energy, space, time, matter, temperature, pressure, potential-actual, positive-negative, and many others. All of these do not exist separately, nor in TO singularly. They only exist as dependently and relative to each other as different aspects of the same thing, in the totality of the actual totality, TO. More basically TO is most proven by existence, necessity and change with TOT, and the contribution of each of its varieties, e.g., TRU. How each in their own way affect and determine the form and function of TO as they become their part in TO.

So that actuality, what is necessary for existence, sets the needs of all to be their actual pure essences. This includes the creative and evolutionary powers of positive-negative and potential-actual. Reality is the fusion of subjective and objective, mass energy progresses to subject-object, space is volume, and what is of time is highly dynamic. All are centered by relation on here and now. Other continua make up the bulk of TO by their form and function within the construct of these basic aspects of unity in totality of TO.

The physical universe, as well as God when perfected by their complement of mass energy, subjectivity-objectivity and actual pure essences approach their fully completed and perfected part and roles in actual totality, TO. All that TO cannot do without, like actual pure essences, and what is of large proportion and divine must be known, learned by one and all, for without it a major proof of TO is lost by what is potential that does not achieve actuality in TO. The laws of totality conform to, and are confirmed by, differences that are reconciled in the whole of TO. The whole is proven by absolute and relative totality, bonded as ABSTO-RELTO in extra qualified totality. The unity and reality of knowledge and civilization are composed of increasingly higher states of mass energy, SÁ of Á of increased capacity, self perpetuation and power that converge on actual totality. They enable, prove and assure the actuality, relative-absolute, and exclusive existence of TO and its constituents. This is by the Laws of Gains and Losses of TO, LOGAT, cf. 15, 32. Ideas, words, concepts, principles and laws define and help to make actual totality distinct. Through description and interpretation they prove what is most typical and important in TO. The more they are identical with their part and role in TO, and the more they reveal relations and interactions of their combinations and summation the more correct the unity in totality of actual totality becomes. This is the basis for a proper organization of knowledge, and optimal world perspective or overview of all.

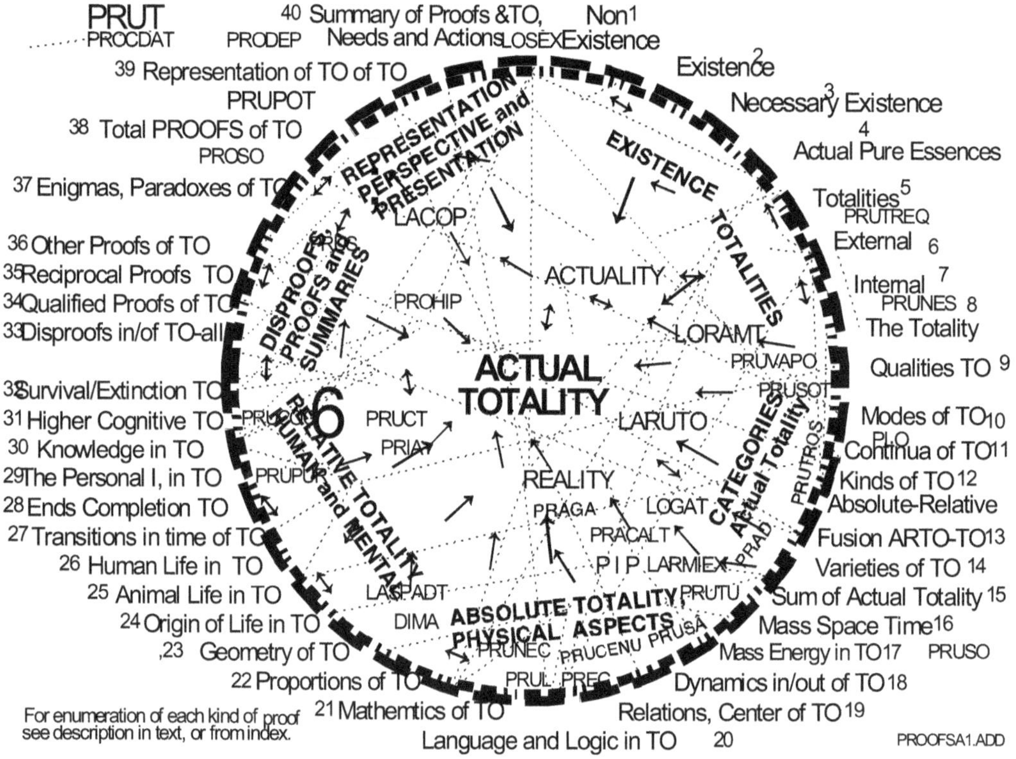

Fig 38-1 Summary of proofs by combined types in TO Interrelationships, Interdependence and Interactions

This diagram, 38-1, of many different types of proofs of actual totality gives a circular overview of their interrelated and interdependent summation. Although the diagram is incomplete, it provides a good way to envision and show actual totality. With other figures, a little imagination, greater interest and the use of the actual pure essences, other features and figures a highly successful replication of TO is available. Although limited by the format of this book and representation, this figure reveals the great potential for summarizing and showing the top concepts, principles, proofs and laws of TO and its related aspects. The items taken from the contents of books on TO show the major types of proofs around the periphery of TO. There are many more proofs, existing and new concepts inside and outside of the circle. Since many are not described in this book, the reader need not try to recognize all their detailed meanings, only the great significance of the overall massive interdependent and accentuated power of proofs. Their combined order and power lead to the certainty of an ultimate unified totality by actual totality, TO. Some of these proofs apply not only to the total proof of TO. They apply to its means of representation, and applications to a perfected guide in human life. Much we perceive and represent are less than TO. Many of the items, objects and terms used are less than their actual pure essences in

TO. It is to enhance our knowledge of all in TO, and judge to selectively bring all less than TÓ, to its part that gives each persons their differential advantage, and the chance to make all right, as right can be.

When fully developed the figure shows how actual totality and its concepts, laws and proofs can represent all, especially when properly limited and qualified by the varieties and use of TO. As we recognize the necessary sameness and allow for the most detailed in TÓ, we can get a better picture, to know and show TO-TOV-TOF. With evident and valid divisions, forms, parts, concepts and laws of TÓ to go on, highly valid reasoning and logic can provide the best possible premises for proofs, and the benefits of their uses. Combined, and used with SPALT, they provide valuations, positioning and operations from which we can draw, irrefutable demonstration and conclusions, that lead to an accurate and total summary of TO. A summary which when correctly used can solve many of our problems and enhance our lives. Figure 1-38 provides a sample of the great many interrelated proofs of TO. It presents them not only to confirm the tenet of a concrete closure of all in TO, and how in practice it can be treated as a closed set. It also can show what can be done with the much deeper and greater interdependence all has in the unity of the integrated totality of all the proofs of TO. By DAOST they are all different aspects of the same thing, seeking to achieve the maximum in differential distinction in the maximum of integral unity in TO. Each proof, as each topic and concept, must be held to their explicit meaning, state and role in or of TO. In this drawing we can clearly envision the great part, the modes, kinds, continua, varieties, VAPO, ATO, DAOST, SCARLS, TROS, TIT, TRIS, TRIT, TRIA, TO, ALT, REM, POS, PIP, and HOM and many other constituents, have in the proof and existence of actual totality, TO.

The dependence of all in unified totality is why we must always remember the whole, and repeatedly bring its greatest modes, kinds, universals, concepts, principles, proofs, laws and all together as they are interactive and interdependent in the whole, TO. This is how we prove TO, and how these proofs can dispel all doubts by answering the questions and laying the foundation for solving all problems. Only when we can acquire and retain a wrap around overview of TO, and much it interacts and is associated with, can proofs give us, and can we achieve, the mastery, management and control that satisfy the positive imperative for ourselves and all in TO. By increasing integration, proportioned differentiation, and ordered summation of the correct proofs humanity has the means to have the plans, prevention, preparation and action needed to assure quality, improvement, and fulfillment.

For many reasons, naturally from the conditions and limits of life, as well as weaknesses and false beliefs, humans do not accept the best of valid knowledge when it is before them. There will be much avoidance, denial and resistence to actual totality in the near future. This is similar to partiality over universality in knowledge, and especially when not proportionately unified. This is so for all that seeks totality. For totality, as the physical universe, philosophical systems, deities, worlds, and political systems have not achieved the validity, actuality, proportionality and totality they profess. Many political systems have given totalitarianism a bad name. This is partly due to simple mindedness, narrow beliefs and decay that make them not only dramatically wrong, they have failed to provide the order and human well being required. This is exacerbated by their stand in opposition to other systems who are also more or less wrong. For these reasons TO must be, 1. As perfect as possible, 2. As well ordered and proportionate as possible, 3. Made and held in the most helpful contrasting way compared to other totalities, universes and Gods, 4. Directed to each and all persons, so that they can identify and learn TO, and be their part and contribute in all ways to it.

Each individual must be given the opportunity and helped to have the correct attitudes and beliefs that TO can inspire. All can be categorized and managed by the gradients they are less than TO. Nobody can have the right way of life unless they have the correct world view. To be the correct world view it must include all provided by actual totality or its equivalent. The vertical integration provided by a unified totality that is TO and includes its many proofs, laws, demonstration, major features and constituents when correctly designed, interpreted and applied is the certainty and organized knowledge necessary for an increasingly complex world. By centering on here and now in human life, consciousness, reason, total understanding and higher knowledge of relative totality, the actual pure essences of RECTO in TÓ, achieves their rightful place in ultimate, unified totality. It is a RELTO that fuses with ABSTO and the physical universe, and incorporates that of philosophy, religion and other worlds when equal to their actual pure essences in TO. Combined they all exist predominantly and inclusively, in actual totality.

CHAPTER 39 - REPRESENTATION, PRESENTATION, PERSPECTIVE of TO to Summary and Applications

External Combined, Representation, Distribution and Position, Classes of Posits, Presentation and Applications, Sector numbers in SPALT, Perspective, Positions in SPALT, Applications

Many of the proofs, notes, evidence, facts and ideas through the book when brought together and demonstrated is the proof of all in and of TO. As new proofs are added, existing proofs are substantiated and qualified. This is why singular proofs, as great as they may be, should be neither over emphasized nor criticized until their combined interrelations and interdependence in TO are known and shown. When we show and look at TO, the representation, presentation and perspective are often at divergence from TO. Therefore we must always remember, ECRO, to correct for this divergence to maintain the proper identity with TO. Within, all is the same by the relations and unity of TO. Outside of TO, all is different by origin and approach, except to the extent they focus on TO, or some aspect of it. To remember the various approaches, aspects, features, categories and major constituents, and their form and interactive function are how to greatly improve our knowledge and proof of TO. Representation and presentation not only show and prove TO, TO shows and proves them. Together they help to make TO easier to understand, perfect and prove. Without them perspective suffers and fails. Without all three, actual totality is remains unknown.

Development of form, representation and design of TO, TOV and TOT - Since knowledge and representation of actual totality arose from work on the world of best human life, or The Relative Universe, TRU, their combined relationship is signified by TOTRU. This is helpful in describing the origins and much of the form and function of actual totality, yet it always must be remembered that TRU, although very important, exists only as a segment of RELTO, the relative half, and TOV, the varieties, of TO. The potential, varieties, form and representation are mutually interdependent and vitally necessary if we are to accurately and fully show and prove all of TO. As TO is increasingly known, its varieties effectively applied and its form and representation perfected, we can present, what should be an obvious idea, best human life, or TRU in the whole of actual totality, TO. It is our most profound and earnest expectation that such a representation and presentation will be recognized, adopted, and worked on with the enthusiasm and determination it deserves and urgently needed by humanity.

There are innumerable kinds of representations that can reveal and prove TO, or some aspect of it. For initial and practical use the one representation and model that most closely shows TO, is SPALT. It is that most commonly used in this book, the spherical mass energy, length and time construct of TO, described in Chapter #16. No representation is complete alone. Even SPALT is limited by being only, a representation, of the internal form, lacking other aspects of TO in itself. Many figures - and descriptions -in the text show us how we can demonstrate and reveal the different that can be increasingly separated, and the same that can be unified in TO. Thereby we can better assure ENT, TRU, TÄ, TRIT, SUT, MOT, SPALT, POS, ACC, and TRA.

Distribution and Position in TO, TRU, TIT

The positions of the constituents, components, parts and processes of actual totality, TO, require foci on their centers. To represent them their functional values and actual pure essences are a large factor. This is often achieved indirectly by implication and effect. The amount, type and distribution of spread from their centers in SPALT and TÓÀ, help to show their broader and interactive role, and correspondence in TO-TRU-TOT. The positions of parts, processes, and components of actual totality, TO, focus on their centers, to represent, often indirectly by implication and effects, their functional value or actual pure essences. The amount, type and distribution of spread from their centers in SPALT and TÓÀ, help to show their broader and interactive role and correspondence, to in TO-TOV-TRU-TOT.

It is how positioning and descriptions reinforce each other that we can continually achieve vitally needed knowledge and representation of TO. Varieties, form and representation of TO are essential and mutually inter dependent, if we are to accurately, adequately, correctly and finally show what the world is, and properly prove TO. As TO is increasingly known, its constituents effectively applied and its form and representation perfected, we can present, what should be an obvious idea, best human life, its relations and existence in TO.

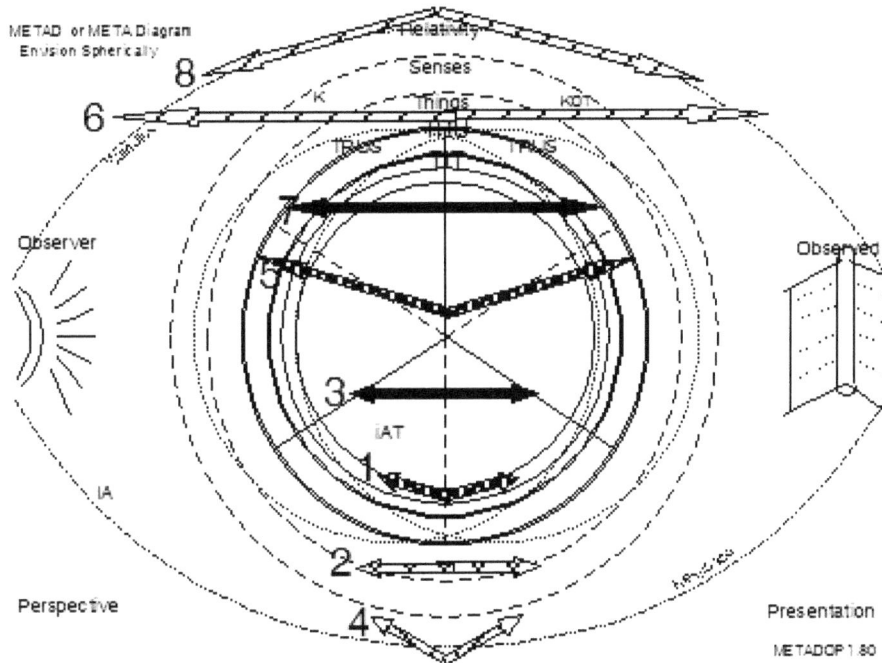

Fig 39-2 Eight Classes of Posits of all in TO, four inside and four outside.

In order to give the detail and order TO warrants the better all can be positioned the more accurate and complete it becomes. The parts, processes and all can be brought to a high level of refinement that simulates TO and TO with TOT. To provide optimal proportion and order to a representation and perspective of TO it is most helpful that all are organized to best reflect their distribution and positions. The category or class that best relates and classifies each detail to be positioned is designated by their given numbers,

1. focal, inside of TO, unevenly distributed,
2. focal, outside of TO, evenly distributed,
3. focal, inside of TO and evenly distributed,
4. focal, outside of TO, unevenly distributed,
5. universal, inside and unevenly distributed,
6. universal, outside and evenly distributed,
7. universal, inside and evenly distributed,
8. universal, outside and unevenly distributed.

Such a differentiation enables us to better distinguish between entities and concepts in and out of TO, OITIO. It also improves their posits by comparison and qualification. Posits in the figures we use are largely evaluated in this way. As better models are developed to fully replicate TO, our subject-object relation, observer-observed variations, sphericity, surrounding orbits, and closure can be greatly

239

refined. This will be better envisioned as our appreciation and familiarity with TO increases.   This figure provides the placement and positioning of many static parts and dynamics processes, whose combined proofs when complete fully simulate TO. The different topics, themes and divisions of TO, can be placed and interpreted from this model. Since many of these are more of one variety of TO than another, the loci of concepts, or processes within, they are listed on the side, or used in figures external to SPALT, as of SPAT, and by TO-TOT. It is in positioning the many parts and processes, aspects, concepts, principles, laws, and mechanisms, of TO in SPALT and other models that we can best operate, exercise, practice, and reproduce what actual totality and our lives are in TO and TRU. By the continual operations of positioning in the simulated TO and emphasizing its most valid and important aspects, the intricacy, unity and beauty of TO and TOV can be more readily recognized and applied to life.

The interposition composition and comparative relations and actions of parts and components greatly help substantiate and prove each other and TO. This is the detailed composition of parts internally and relatively formed and functions in TO. It enables and displays the beautiful intricate detail of the greatest practical number of ordered components of TO. In the addenda we list 1,000 of these parts and show some of their actions in SPALT. This is one of the greatest fields for which continuous effort and work is needed, bringing TÓ to fruition for one and all persons by a more perfect development, understanding and use of TO and TRU in life.   Of the many means that can be used to select and grade the parts in TO by SPALT is the method of class order in relative magnitude. If we know a class, system, or field of knowledge that exists in TO, and its parts are properly ordered, this can be transferred for positioning in SPALT, (MOCBIREM).

In addition to TÓ, and its representation, is the need for presentation and perspective. Since all are different aspects of the same thing in TO, or converge on TO, they tend to coalesce in unity. This also depends on the proofs of TO and the will, use, dedication, identity of each and all persons with TO. This is aided by their joy in improvement and guide to greatest life.   The flexibility and controlled comprehensiveness of perspective are two of the most important qualities of showing TO. This is in contrast to rigid beliefs in lesser totalities, universes, systems, gods, and worlds that suffer from a state of telescopic inertia, (POSI). The need is imperative and certified for the right and full perspective of TO. When we seek to view all, we must be prepared to do it well, both statically, dynamically, in and out and with as much ordered detail as possible and practical.

This is a highly accurate basis (fig 39-3) for the beautiful intricate detail of actual totality. Where all fits together in unified totality for permanent use. Each region or sector is given a numerical signature of its mass, length and time position. This spherical approach produces a precise way of demonstrating all, as by a transparent sphere or globe, of TO. Variations of this model and figure add to its value and use, such as positioning in each of the eight large sectors of TO. For the contents of each sector a glossary of is provided in this book's addenda. It must be remembered, ECRO, that each term is only more or less centered. This figure as others is preliminary, in progress and needing more work.

When actively engaged in the simulation of TO by positioning in its models there is superimposition caused by a large amount of overlapping of content. As this is recognized and accounted for it helps to collectively reveal and prove TO and TRU-TOT   Such a world view and different perspectives, provided by proving and showing TO, are not only of its positive imperative, they are what is needed to accelerate all potential to, and all actual in TÓ. This is what fulfills the providence and destiny of TÓ in the advance of the relative with absolute totality in perpetual synchrony.   By changing ones perspective, the unity to diversity of TO, can be duplicated. This is by simply moving ones a vantage point farther away or closer, and by viewing TO from without or within. By telescoping to find that focus on TO, differential to integral, that best envisions parts, features or the whole. This allows greater degrees of proof and appreciation for the limits of distinction of TO and its constituents. We must balance and always remember this spherical solid and reliable representation with the relative and variant role that exists within and produces certain marked effects in TO.

In separate figures, different objects whose fields and internal relations are in marked contrast, may be shown as, 1. A single focus with radiations, 2. Multiple foci with radiations, 3. A triage or three-dimensional field, 4. triage with outer spheres, 5. Multiple point centers with large vectors, cf. Fig 11-1.

This figure (39-4) presents TO as a closed set, a positioning of constituents that are most effective for description. It is a precursor to the application of proofs, concepts and laws needed for human life in TO. They enable, enhance and benefit TO, as TO enables, enhances and benefits its constituents. It is the focus of other representations, presentations, perspectives and guide in human life. It provides a means by which what is less, external or other than TO can be compared, corrected and managed. Since actual totality proper, TP, is the core and primary concern of life it concentrates a stability of form that allows us to treat TO as a closed set. Yet we must always remember, ECRO, the differences, and union of TO and TOT. We must keep our perspective open to all features of TO and TOT, especially as they apply to our lives. The detailed position of concepts, laws and proofs of TO, are shown in the book's addenda. Compare SPALT, SPAT, and other figures for the most comprehensive view.

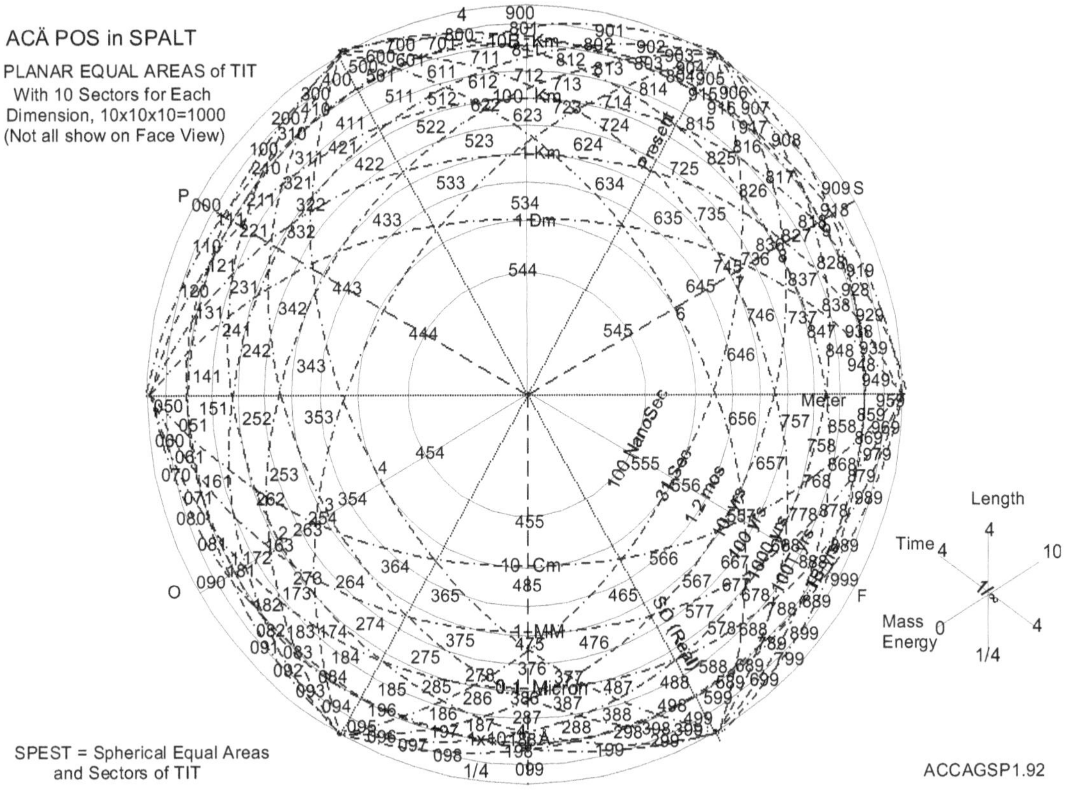

Fig. 39-3 1,000 Numerical regions in TO, by equal volumes of model sectors in planar SPALT

Each and all persons must always remember the spectrum of their vision and what it is focused on. It is a spectrum from the most specific focus of activity in life to the most general focus of all, actual totality. We must always remember and apply the fact that this most specific focus of our attention is much less than a picture, a picture is much less than the panorama of which it is a part, panoramas are less than the whole of which they are a part, and these wholes are less than the actual totality, whose closed set state is other than its interactions, potentials and its transitions. Covering this total view is the fact that to gain an ultimate perspective, we must account for both the transitions of TO with TOT, and the fact that perspectives themselves are less than actual totality. Herein is the accounting and world view all must achieve, hold, and continually apply in life, if we are to be our part in actual totality, and if actual totality is to fulfill its mission, well being, improvement, advancement and continuity. Although we can attain a good representation and view of actual totality by figures that are planar, they are never complete, or never entirely accurate if completed. The form of TO as a closed set is more like a sphere. To represent a sphere, as we do with models, we use SPÁLT, based on its mass energy, length and time dimensions. Such a model can also be based on any continuum or duality whose spectrum provides a good replica of TO. There are limits and balances to the kinds and number of ways we show, present and view actual totality. The fuller, more accurate and less numerous they are, the more they show all TO, the better readers and students learn TO, and the more effective and successful they will be. For once the reader, student or person can visualize the whole, place its parts and operate their functions, all falls into place. It mirrors the whole, becomes much more meaningful, remembered and applied. Nothing can be of greater benefit to the improvement of mass energy, the glory of our lives with that of TO.

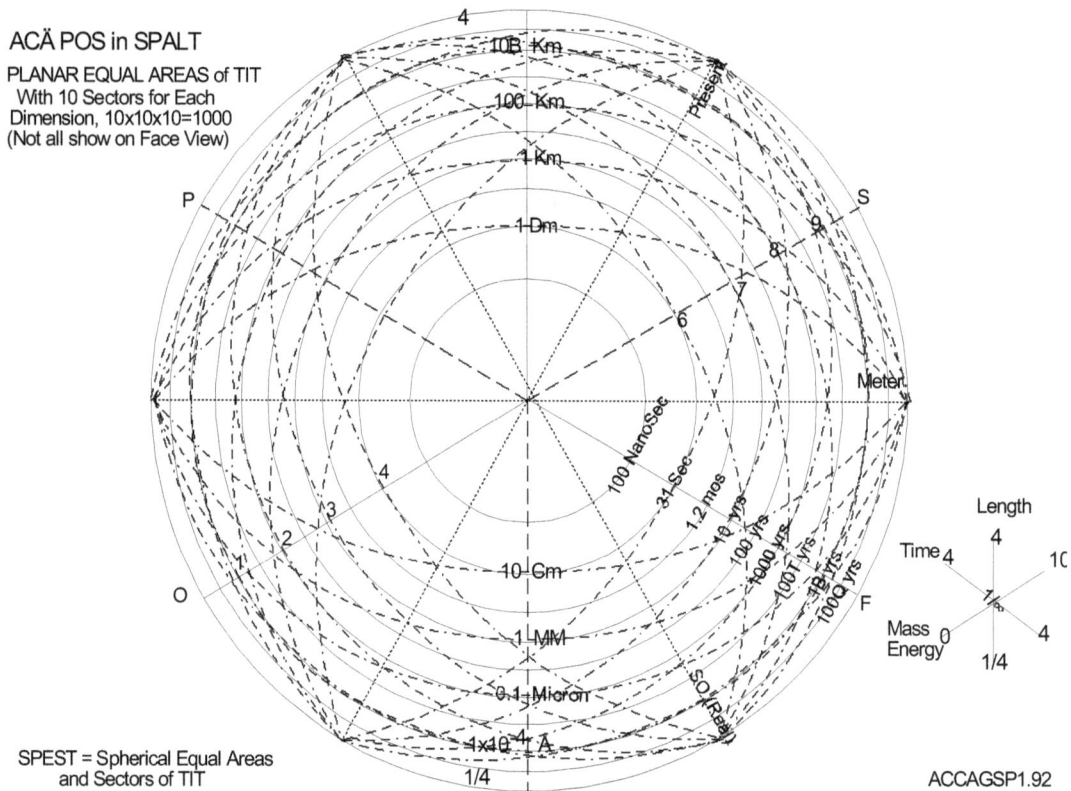

Fig. 39-4 SPÁLT, used for Demonstrating the major Proofs, Concepts, Principles and Laws of Actual Totality

The positional and operative accounting, action and accentuation of, in, and in and out of the form and function of TO, is signified by OITIO and shown by SPAT. They reveal the predominant and unique part and role of the subjective in encompassing and comprehending all. It helps to prove TO, by the fact there is, or could practically be, no other. If we envision what is external, internal and their combined mixture for a perfected replication of TO, and TOT we can maintain the overall perspective needed to avoid narrow and mistaken views and beliefs. This is key to fulfilling our subjective role and life in TO.

The potential and the means for increasing actuation, expansion and intension of actual totality can be immense. It only awaits each and all persons to wake up seize the day and opportunity of a brave and glorious new world. In many places in this book, its concepts, laws and proofs reveal the wisdom of TO that is right. It is the positive imperative that must be and win, largely by all that is already in actual totality. With sufficient proofs and certainty of TO, a sufficient number, quality of representations and perspectives to know and apply it, and a large percentage of the world's population to reach the needed and perfected common consensus the future in TO, the long searched for hope of mankind will be here. This is the assurance of the right capacity, quality, well being, happiness and permanence that TO provides. It only requires these steps, and the discovery, learning, understanding, will, identity, service, joy and achievement to fulfill the actuality of TO.

CHAPTER 40 - SUMMARY, Uses for Proofs, Practice and ACTIONS for ACTUAL TOTALITY

Means and Action for TO, PMAC or - Planning, Prevention, Protection, Preparation, Accounting and Prediction, Control, Maintenance, Action, Accentuation, Acceleration, Beliefs, Habits, Ethics, Behavior, Toward Completion and Fulfillment of past, present and future goals, needs, proofs and life in TO with TOF

In order to finish the proofs, the uses, applications and future of TO need to be added to complete the whole. These are powered by the means and consequences of plans, prevention, protection, preparation, accounting, prediction, maintenance, control, action, acceleration, beliefs, habits, ethics, behavior, completion and fulfillment. All should converge in the optimal unity of TO and TOT. Actual totality has not been recognized as a whole. Some of its largest parts are poorly recognized because they lack much of their form and function, actual pure essences and proportions in actual totality. Evidence from all we know suggests limits, errors, misconceptions, lack of laws and proofs have obstructed and prevented the needed knowledge of actual totality. To the extent these are less than best and deviant, results in a compounding of error in interpretation. It is because of these deficiencies and errors the worlds, universes, systems and deities we live by are limited, less than TO and often incorrect. The result is the opposing views we encounter and their often violent ends. All steps must be the best they can be, from unity in totality of TO, through its major constituents and proofs, interpretation, and how they are put into action, For this reason long term cooperative and rigorous work will be needed to perfect TO and its knowledge. In the mean time work to develop TO must continue. For any and all interpretation to be valid they must not only be based on the right actual totality, they must encompass all associated with, and explain their and its external position and future by TO.and TO with TOF.

With a relative, unified, well formed, selected and proven totality in hand, and the language and mathematics to represent it, we can with confidence progress toward a model, plan and world that are reasonably accurate and reliable. The degree actual totality, TO, is valid, actual, proportional and total, and the degree its form can be replicated, make its model an ideal way of showing and applying them for a greater way of life. Actual totality is, in a sense like a computer, if you expect it to be perfectly accurate, you must follow its order precisely. By doing this, its products when correctly applied, can be amazing, beneficial and the answer to many needs and problems. Be careful what you hope for, your purposes, policies and plans largely determine your success and that of TO. These are why actual totality is so important to know. They are indispensable to prevent failure and loss. Experience, training, education and all means that enable everyone to rise to the positive imperatives of life in actual totality, must be fulfilled. Each person, and all groups of people must be the best they can be, in TO.

One of the greatest reasons why actual totality and its knowledge are so important is that they are critical to the life of each and all persons. In order for actual totality and its knowledge to fulfill this, and its, purposes, is that it be accurate, unified, total, devoted to, and progressively improved. This is signified by LAKUPIT. Actual totality and its knowledge must be sufficient for every encounter and occurrence. The reason why they are so critical is that they are the cornerstones of all, all we are and do in life. What we are, and what we do, are built by, and on, actual totality and its knowledge. For this reason our beliefs, priorities, plans, protections, preparations, actions and ends must be by that way of life derived from actual totality and its knowledge. Law of knowledge in life based on priorities, plans, preparations, actions, advances, and completion must be based on the best and most appropriate features, aspects, qualities, kinds, varieties, form and function of actual totality, LOKPACT.

It is not possible in this initial developing book to explain all of its many important applications. It will, by showing when, where and how they can produce their greatest benefits that TO is fully realized and achieved.

The thrust of these last pages on proofs of TO, is similar to the thrust that has propelled TO through the ages. It is toward accomplishment of its goals, to sustain, perpetuate and fulfill TO. From its basic purposes in the beginning of ARTO to perfection by the most advanced plans, prevention, protection, preparation, prevention, accounting, prediction, management, control, action and acceleration, These need to be accompanied by an equal degree of the right beliefs, ethics and behavior to assure the best kind of completion of TO and its preparation, management, applications, actions and control, PMAC.

Why is it this world? Why is actual totality focused on humans and mentality? Why couldn't it as well be focused on any of a vast number of other centers of relative totality?

1. Lacks relativity of observer-observed, subject-object,
2. Denies being human, our lives, knowledge, minds, etc.,
3. Denies personal consciousness, instantaneous attention and activity.
4. Denies unified viability. The physical universe and deities lack the necessary form, function and qualities,
5. Lacks the kind of physical that is its part in TO,
6. Lacks practical proportionality, thereby equilibrium,
7. TO is the averaged sum of all that converges on its center, here and now, and typical mass energy,
8. The varieties of TO already have many centers of absolute and relative totality that are similar to other worlds. From their own existence other worlds are modified to be the same as what they are in TO.
9. Any completely separate absolute or relative totality could not exist. Even in the physical universe its parts are more or less relative centers in the more or less absolute whole. The more they are combined the more they tend to be equivalent to actual totality, all else being equal,
10. Relative totality when equal to its role in TO, RELTO, explains many of the problems, enigmas and paradoxes of knowledge of actual totality. It imparts much of its form, function, characteristics and properties on TO,
11. It is highly improbable that any extraterrestrial source of combined absolute and relative totality fulfills the requirements of being human, mental and all that actual totality includes such as knowledge, technology, culture, civilization and actual pure essences and modes in TO. On the other hand the unknowns and known of the differences and variation of physical existence have, are and will be highly interactive with actual totality. The probabilities of interaction and change increase in the far reaches of time.

These are answers to the most likely criticisms of TO. Yet the answers must be as certain as possible, if for no reason that any other imaginable answer would be their disproof.

Planning or to plan ahead, is the first rational step in any successful human achievement. Many animals have forms of life that plan to act, in subconscious ways. As absolute and relative totality fuse in TO, absolute totality is transformed by relativity, relative totality is transformed by the absolute of developing RELTO. This is the new concept of 'absolutivity' that proves TO. New forms of mass energy, length and time are created. RELTO induces TO and absolute totality to adapt to a new kind of mass energy and matter. This consists of S-O, whose reality and concepts are based on their actual pure essences, and exist as a new kind of matter in TO. This typifies the developing and burgeoning actual totality. They, when mature, as TO matures, become like the matter we have today, the major kind of elements in the prime mass energy spectrum of TO. This is the mass energy spectrum in TO, whose common form is called its Á dimension. It is with this new form of mass energy that planning directs much of the action in TO. The higher the degree of S-O, S-R-O, Á mass energy, the more apt planning achieves its purpose. Without TO, or its equivalent, humans are going to destroy themselves, probably sooner rather than later, not so much intentionally, but out of neglect. Such neglect is especially of the quality of all that is needed, and how TO is formed. This to a large extent is the definition of TO.

Prevention is another mechanism whereby mass energy, as S-O, enables TÓ to survive and thrive. Some of the problem's human's face, remain unsolved, yet the role of prevention in all fields of endeavor, especially health has been possibly the greatest benefactor in TO. Load versus capacity of genetics, how much can it handle, how much are animals, humans and TO limited and liable for depreciation? Many causes for extinction, must be learned, and applied, prevented and treated in TO. A massive spread sheet used to prevent major problems will greatly enhance TO. It is the eminence of TO, and its systematic unity that guide much of the use of prevention.

There is a great need for protection, both from external ill effects, and internal derangement. The greatest problem and threat of our day, and to TÓ, are the lack of understanding of the meaning of a unified world. Neither we, nor the worlds that are currently around us are, typical, nor the best way to be or serve TO. This is TO. Such a lack of understanding is a misuse of knowledge that creates doubt and error. This lack of truth produces a lack of trust. A world in which there is little trust, is a world of opposing forces that tear it apart. The cure is not so difficult. It is a rational, adequate and practical presentation of what this unified world is, TO, with an adequate number of valid proofs that justify its great existence. It is also the protection required by TO and its form and function.

Priorities, plans and policies, and even laws, must be perfect, or as perfect as possible. Even when slightly less, and in use and abuse, they quickly escalate into negation for actual totality. This is why they must be grounded in the basics of actual totality, developed properly and accurately to fulfill, within their limits, their roles in TO. This is why the development, proof and existence of actual totality must be well done, and continually perfected.

Everything one and all do, must be by how each step succeeds only when based on a concrete unified whole,

UNITY of Actual Totality

All CONSTITUENTS of TO

Concepts, LAWS and PROOFS of TO

Values, PRIORITIES, Proportions by TO

Preparation, PLANING, Prevention for TO

The Right WAY of Life

Correctly DO, Acts and Work,

Different people, groups, nations, and cultures will have different approaches and levels of achievement of attainment by this guide, signified by, (DOWPLAT). The emphasis and order on action for actual totality are the greatest simple guides to success and survival for all. All that is, and must be, a part of actual totality. For those with greater or lesser benefits, genetics, health, ability and intelligence more or less, same and different must be given and received.

Organization of knowledge, best human life and actual totality are impossible without unity, purposes, plans and activities based on a progressive integration and differentiation to and from top to bottom. They can be no better than when, and the degree they are right and replicate the one world that best suits the form and function of necessary existence. The progression from integral to differential, must not only be right and proportional, it must be as ordered by a form permitted by its representation and presentation. With unity, all formed and functioning in optimal harmony, and the correct integral to differential proportions when best represented each and all components and persons have the basis, and means, to fulfill their part and roles in the survival, success, well being, continuity and permanence of TO.

Through good preparation much can be expected. For those whom much has been given, much is expected. This means that life cannot be wasted in lesser pursuits and gains and why TO is so important. For those whom less has been given, that expected should equal their capacity and abilities. Both must be tolerant of each other and all that enables all to succeed and survive by their part in TO.Preparation depends heavily on prediction just as prediction on preparation. Their success heavily depends on the correct knowledge and proof of TO. Prediction is a function of degree of integration and how well it is known, e.g., TO. Lack of prediction is a function of differentiation and how poorly it is known, e.g., lack of TO. There is hardly any greater incentive to the acceptance and use of TO than to be able to make the right predictions guided by all that provides knowledge of vertical integration for action by TO.

Accountings and the predictions that accounting helps to provide are examples of the role and higher powers of subjectivity in the relative totality of TO It is their ability to support the mass energy, S-O form of TO that produces right management, control, action and fulfillment. Different sized spread sheets from simply practical to most detailed accountings of all problems for solutions, and PMAC of paradoxes and enigmas for resolution, are needed to overcome that impairs action for TO. Also, sectional spread sheets for, 1. Time, near, mid and far term, 2. Fields and sciences, 3. Skills, and many other activities. All must be guided by the whole of TO, TRIR-TOF. It is the fine tuning, and ever finer proofs of the solutions to the problems of TO that set the tone, like that can make the predictions that enable control of all, TO.

Management and control are not only central to all means that produce survival, success, continuity and permanence, being indispensable they must be the best. The great interdependence and interactions of management and control of all in TO prove their contributions, and the dynamics of TO. There is a balance between management and control and natural selection. From past to future this has included an increased need and use of management and control. Actions which, when based on sound knowledge, as that of TO, are apt to be superior to natural selection. Natural selection, although less efficient and harmful at times, is superior to chance. To not only know what to do, but to know all that is important, especially the optimal equilibrium of natural selection and management and control is the height of excellent judgement by TO. This balance is not only continual, it is multiple and helps to reveal and prove the way of best human life in TO. Action and its variation including acceleration are the responses of life and all in TO. They show and prove TO by their roles and the feedback of the positive practical effects they produce.

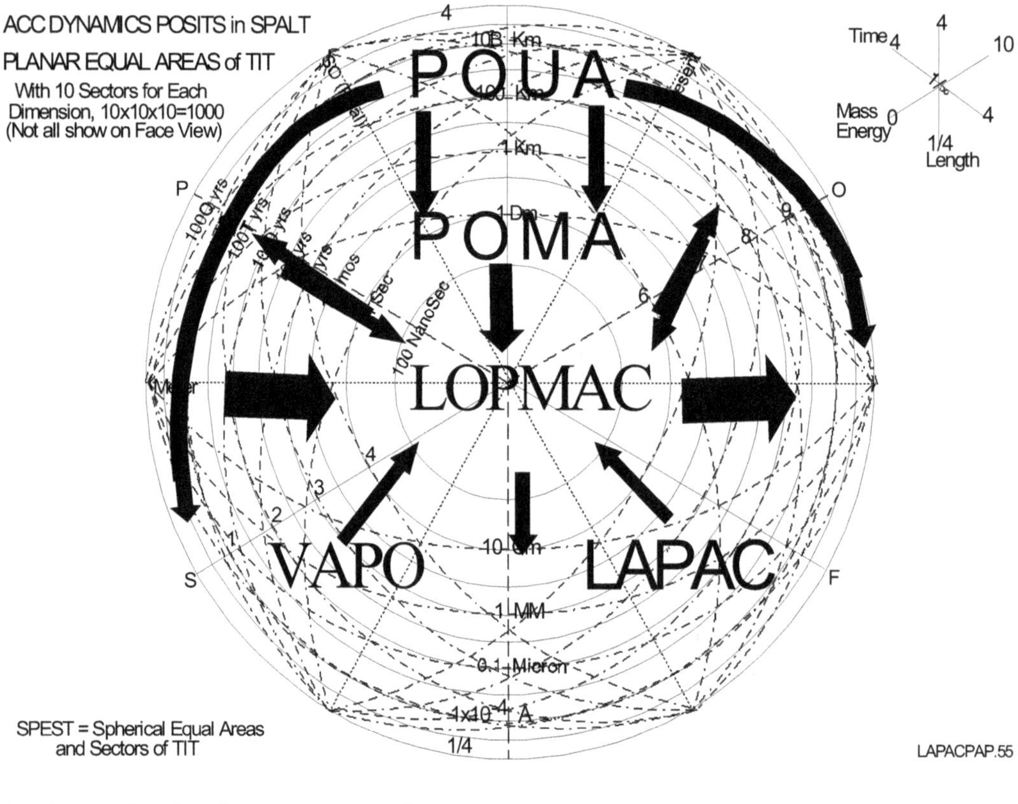

Fig 40-1 Action, Total and Proportionate of Processes

By POUA, the principle of universal action, POMA principle of multilateral action, LOPMAC Law of Planning, Protection, Prevention, Preparation, Priorities, Accounting, Prediction, Managing, Action,

Acceleration and Control, LAPAC or law of proportionate action and VAPO or Validity, Actuality, Proportionality, and Totality, posited in SPALT, one can envision much of the basis for action dynamics in TO. They show the model, the position of processes, how they apply to life, and especially how action can be better managed. Actions and accentuation that are proportional provide a major and comprehensive guide to life by TO. This is what is needed to lift humanity from disproportion, weakness, non completion, failures and existence that are less than, null or against TÓ, to levels that will assure future continuity and permanence.

There is a great four cornered dichotomy and dynamic that exist between TO and individuals, the actual pure essences of TO, and what is practical or of interest to an individual. The dichotomy is largely between the difference each individual or person is and produces and what they should be and produce as their part, TIS of TES in TO. A person through lack of, knowledge, experience, training, education and belief in the right kind of totality, and all its integral to differential form and function, fails to, remember, acquires and sustain their part and role in TO. Not only do base, simple and impulsive inclinations cause failure, continued temptations and tendencies to greed, and inequitable ends. They deviate from the actual pure essences and TO. This is the crux of human failure to participate in TO, it is like losing ones soul, loss of truth and principle, which creates the absence of trust that permeates modern civilization. Its cure is not easy, it is pernicious, and can be cured only by the most dedicated and devoted life to TÓ, as evident in these proofs on TO. Such a deficiency in a person is bad enough, it is when becoming a part of the routines of societies and cultures that keeps them in constant turmoil, failure and extinction. This is why actual totality, its equivalents, these proofs, their continued improvement and the right kinds of knowledge and beliefs of all persons are so important.

Many of the common and often controversial problems of society and mankind originated in the basic states, conditions, limits, relations, differences and oppositions in or associated with actual totality. Once their origins, development and form are discovered, and made known, their solutions can more accurately and properly provide the needed management and control. Those that are basic have a presence that is continuous and often increasing. This adds to the need for most astute solutions, management and balanced controls. All are contingent on quality, what is best for an individual, all groups and TO, as well as the physical universe. Each problem, and kind of problem, has a core meaning plus other causes and affects. It is to focus, reveal and be clarified, the most real and accurate mechanism of each problem that their solutions, actions, and their effects will be right. An excellent example of this is given by Thucydides, who expressed dismay at the power and multiplicity of illusions that distract men from reality. This is common through history and time, wherein all depart from actuality and reality by illusions, base inclinations and distractions whose lesser, null or negative knowledge fails the unity, proportions and priorities of actual totality.

Another common problem is money. Money is not the bottom line, money is not the end, it is not even the end of the beginning. Money is only a very effective medium of exchange. It is how money is best used that it and the bottom line are to be measured. It is hard to be poor without money. It seems better to be rich. Yet the more money you have, the greater your responsibility. Here is the gist of the problem, those who get money think responsibility ends. They lose their conscience of all that is right and best, TO. The solution is not simple, is multiple, and stems from lack of actual totality, the correct knowledge, beliefs, principles and actions. These are based on all, from the simple unity and proportions of actual totality, TO. Decisions and choices have to be made based on the greatest modes, continua, kinds, varieties, proofs, laws, principles, proportions, parts, processes and mechanisms in the most ordered form and way of life in the world, TO.

Several lists, tables and figures in the text, and in the addenda, include applications of actual totality. They provide a guide for a person's life, e.g., accounting and reconciliation of assets and liabilities, problems and solutions, and future, TO with TOT. In the following list a few of the more prominent problems and their origins, causes, and solutions are given as examples for application. It must be remembered, ECRO, that all are multiple and highly interdependent.

This tentative and temporary list (40-1) is presented to show how knowledge and proofs of TO can greatly improve our approach to many of the causes, problems, issues and needs we have, and will, need to manage, act on and control. It is only a preliminary sketch of what must be done to develop an exact and rigorous preparation, accounting and action for actual totality.

Problem	Origin, causes, development and form	Solution, Balance, Management, Action
General, all	Differences and Oppositions in/out all TO	Concepts, Knowledge, laws, proofs, usage all by TO
Modify, all	Generative, Competitive, Selective	Quality, equilibrium of dualities
Time	Transitions, and changes, of time TO/any	At right time, during period of, and all time
Action	Any, undirected, hasty, wrong	Don=t wait too long, stop to soon, rush things
Environment	Physical universe and best human life	Optimal balance to protect and utilize
War \ Peace	Differences and Oppositions in/out all TO	Interest and Love of common basis in TO
Racial	Human GMLPDF, selection qualities TO	Highest capacity, quality, best support TO
Cultural	Development, maturity, selected, multiple	Genetic health and strength, quality, based on TO
Languages	Long term growth of mass energy	Improvement, quality, perfection, union of and new
Religions	Ultimate Unified Totality best of A-R inTO	Development and perfection of, tolerance
Nations	Common use, Security, Solidity, dominance	Capability, Universal/partial interests balanced by TO
Economics	Credit/Debt, supply/demand	Needed possessions and demand, hard work, avoid debt
Politics	liberal/conservative	Balanced Right freedom and duty, tolerance
Finances	Money/Greed, credits and liabilities	TO knowledge of proportions and ones place in
Individual	Growth, development Self interests/\society	Live learn experience mature AND gain wisdom, as TO
Beliefs	Motivation, Identity ES, TES All needed	Must be In, not out of, TO
Knowledge	Negative, lacking, non TO, potential, TOT	KUPS, higher states to perfected knowledge of all TO
TO Action	Non, exist, necessary, APE, Ordered	LOUA, POMA, LOPMAC, PPMA, VAPO, LAPAC

List 40-1 Major Problems of TO, Diagnosis & Treatment

The cure for all that is wrong, defective and ailing is primarily simple, for one and all in actual totality, TO.

1. To identify with and exist by and in the one world that is right, or to be, and help perfect TO.
2. For each and all to do no more, no less nor other than all needed to best fulfill their roles in TO.
3. To avoid, ignore, or remove all that is negative, not and less than TO
4. To continually act to correct imbalances, more or less, that detrimentally will or does affect TO.

These are the fundamental laws of successful existence. LOSEX. Without them through the ages, past, present and future there has been, is and will be great want, turmoil, loss, decay, degeneration, death and extinction. With these cures, there will be very much, gains, growth, flowering, life, harmony and permanence. We either, live and work together with validity, actuality and proportionality to improve best human life in actual totality, or increasingly fail, suffer, and become extinct. Variability within actual totality can be and is normal and natural, yet repeated cycles of boom and bust, especially to the degree they are not by actual totality initiate the downward spiral to absence of, and harmful opposition to actual totality.

Actual totality is right as a unit, basically and largely by relation, convergence and unity. It losses right by failure to integrate, changes with transition in time, inability to accommodate detail, and have the needed differential. These can be greatly corrected by the improvement, expansion and perfection of physical well being, knowledge, applications and future discoveries for actual totality. To have one totality that is right includes, degrees of tolerance, flexibility, adjustment, adaptation, alteration and modification. This depends on external transitions in time, and its internal states, conditions, limits and relations. To do no more nor less that all needed to sustain oneself and all in actual totality include, degrees of tolerance, flexibility, adjustment, adaptation, alteration and modification. The right existence in totality should suit age, culture, circumstances, limits, capability, environment and other factors.

Although humanity and civilization need actual totality or its equivalent now, there are many opposing beliefs and forces that will prolong its acceptance, development and fruition. There are also many complimentary beliefs and forces that will advance and accelerate its acceptance, development and fruition. Some of both of these have been noted through the book. All may be reduced to how and the degree they lessen, or generate TO. Not until there are enough highly appreciative, rational, devoted and dedicated persons with enough stimulus and power to produce the needed acceptance and development will TO have the following that achieves the fruition it deserves. Like a seed in human knowledge it awaits generation. Once TO is sufficiently perfected and recognized it will grow like a plant in spring. It will rapidly engulf mankind and civilization to become the right basis for best human life, and the right guide to find optimal identity. Once this level is reached, it is only a beginning of never-ending lives of training, experience, education and service to sustain best human life in TO, and in transition with TOF in the far reaches of time.

There are equality and balance between accentuation, action and acceleration and all that enables and enhances best human life by TO. For quality expansion, what we and all do often needs simple and quick action, yet these must be based on concepts, principles, policies, judgements, proofs, routines, laws and constitutions that are based on all the complexes, fundamental modes, form, states, conditions, limits and relations of TO-TOV-TOF. There is an optimal level of action, ACLO, or threshold most suitable for all that needs to be done. All acts, jobs and work can be over, under or undone. The aim in action is the level that is optimal by and for TO. It is time for a new approach to ultimate unified totality, the unity of the universe, God and supreme being that extracts all that is superfluous, disproportionate, outdated and incorrect, and accepts all to the right degree that is of knowledge that is its vertically integrated part and role in actual totality. To achieve the needed new, and right, approach for humanity in TO, it is necessary that we have the right knowledge, beliefs, ethics and behavior. These are not possible without the right habits in life, and the right habits are not possible unless they are based on TO or its equivalent, and their knowledge. This is why in generation after generation we must work to produce persons of the highest caliber and quality, both in genetics, development, culture and civility. This is possible only and largely if based on TO, or its equivalent by preferred name.

Actual totality is needed if we are to rise to, be, help others and enjoy the great benefits it produces. Actual totality has arisen, exists and will persist by how well it maintains and advances its form and function. All components, proofs and laws permanently guide each and all persons to better lives in it. The right planning, prevention, protection, preparation, accounting, prediction, maintenance, control, action, acceleration, beliefs, habits, ethics, behavior, completion and fulfillment are necessary for, and proof of, TO. Each of these, and especially all together, represents the great future propelling forces, for TO. They are a large part of our hope, will and work to attain the well being and permanence of TO.

The greatest problem and cause of loss, suffering, and failure are an inadequate and imperfect world view, or perspective of actual totality. This is the basic problem humans have had, are having and will continue to have indefinitely to the degree they learn are trained and educated to increasingly understand the world they are in. This is TO, actual totality, which may be called by many names that have commonality yet are more or less lacking the ultimate unified totality necessary. Because equivalents to TÓ are, more or less, the same, only gives them credibility to the extent and proportions they are their actual parts and roles in TO.

 The concepts, principles, mechanisms, proofs and laws of TO are the basis for, treatment of and solution to most of the questions, problems, paradoxes, enigmas and dangers that threaten us. This is no where clearer than the many continua we must live by, and exist in and about TO. An excellent example is that of the continuum of statics-dynamics. Like form and function, statics and dynamics are at the root of conservatism and liberalism. Neither extreme is correct, nor a rigid medium without the needed right flexibility. As LAMOL, law of more or less, and DAOST, different aspects of the same thing, TO-TOT is both a closed set and in flux. Like cold and hot what is just right is that medium we can best adapt to, live by and produce the optimal benefit necessary in TO. So it is with all the other continua, their ideal is the medium, flexibility and range under all circumstances that we can best adapt to, live by and produce optimal benefit in TO.

What is fair, what is just, what is right and what is justice are questions that are not as easily answered as commonly believed. For that which is fair, just and justice is not the same for one person, society, nation or world than another. Also they are heavily biased and misjudged by the needs, wants, desires, mean and powers at hand. For what is commonly believed, may be much the same as what it is by TO, it is also commonly mistaken by being less, null or negative to TÓ.   Only when all is united in the right totality, and can be apportioned correctly by the part and contribution they make to the whole, can what is fair, just, right, have the law and justice that are well determined and used to measure action. Common belief, like common knowledge, as well as ethics, morals, and legal codes, have by experience and time-honored reason, provided a good measure of what is fair, just and right. Yet this is inadequate in at least two ways. First it is misused and abused. Second, without a firm foundation in what the world of actual totality is, is formed, functions, proportioned and how applied, fairness, law and justice, cannot be a sufficiently correct and right to be a good measure and guide in life. Furthermore, our legal system is not only inadequately based, it is much abused, especially to the degree they lack all revealed of actual totality, or its equivalent, TREQ. For laws are like all of and in TO, not simple nor singular, they are different, of different kinds, most complex, highly interdependent and must be well proportioned. It is actual totality that will provide the greatest basis for the laws of what is fair and just. Existing laws are more or less of value, depending on how much they are the same, as their contribution to actual totality. The best application of laws for each person, groups, or total population, requires a full accounting and planning by TO. It must depend on importance and the balance of all at the time, and in all time, by actual totality.

The ideal permanence of the ultimate unified totality is a goal whose benefit acts like a beacon and magnet for developing and guiding TO, its components, and each and all persons. To the extent we know TO, practically well, TOF sufficiently well, and can live to perfect our actual pure essences in their most successful existence in TO, we will be on the right road to the ideal of permanence in Ultimate Unified Totality, TO. (IPUTO) It requires a perfected solution to all problems, a balance of all imbalances internally, an exclusion of all external forces whose potential lack actuality in TO. It requires the continual alteration, adaptation and compensation for all action-reaction needs to maintain and sustain homeostasis. This is a homeostasis that is a perfect balance of the static and dynamic in TO, and a perfected receptor acceptor interface, with all external. It is the optimal quality and providence of TO that largely perpetuates its actual existence in time. Although IPUTO is closely related to TRA, it also is much different. It is the difference between absolute and relative totality combined in TO and relative totality in TRA. Although TRA largely drives TO and IPUTO, they have definite and imperative limits. The extremes of an absolute relative are not the main purpose of TO or best human life. It is their needed sustaining power, quality, strength and will. Nor is it an absolute absolute, AA, we seek, as much as it provides much of our form and function. It is the whole range of ABSTO-RELTO or ARTO into that we must forever serve, sacrifice for, seek to preserve, prevent from loss, maintain, sustain, advance and act by, to achieve, greater and stronger, necessary and actual existence. When this ideal blends with TOT and TOF, over all that is less, opposite or negative, TRU, TOF and TO achieve the needed homeostasis that will provide the best continuity, and "eternal life."   Yet this ideal in time is not the same as extant, present or existing TO. It is somewhat variant, the ever greater quality, improvement and perfection of TO, and the ever greater potential to actual creation of new and renewed positive actual pure essences in TO from TOT and all else that completes this ideal in TO-TOF. The ideal is not so much the completion of TO, as it is the ideal target to guide the providence of TO and the will of each and all persons whose souls, identities or actual pure essences strive to fulfill their mission and role in TO, with TOT, in the infinite series of TOF to come.

Future and ends of actual totality, TO-TOF, are revealed in the fact that the future is not as vague nor unpredictable as it may seem, or is commonly believed. There are many great aspects, principles, and facts that are already known, as well as shown in and proved by TO, which when combined, ordered and learned provide a much improved comprehension of the future and all. The necessary existence or actuality and proof of the fundamental validity, unity and proportionality of actual totality, or its equivalent by whatever name, is the foundation of all things united in the definiteness of the combined absolute to relative totalities. This foundation is the indispensable method of knowledge and life by which each and all persons must guide their lives. It is a guide whose purpose is not only the survival, success and well being of each and all persons. Its purpose includes how all the rest that constitutes actual totality can be qualified, improved, purified, perfected, advanced to achieve the actuality and reality they have been honored by, given and blessed with.

POSTSCRIPT

The implications and applications of actual totality are, it must be put to work, it must completely and perfectly guide and benefit best human life and it's consistency, and it must be actual totality. This book on the proofs of actual totality only lays the foundation. Yet all depends on this foundation, as is so missing in the 'modern world'. Thus we must bridge the gap between discovering and developing actual totality, and learning and making it the best it can be. Actual totality, and their size and greatness, cannot be covered in initial work. Its major tenets have many more features and factors that should be added. Many will need, more or less correction. Its proofs, concepts and especially figures are incomplete. Additions, improvements and perfection will be needed. Some basic enigmas, paradoxes and problems will need the help, work and cooperation of professionals in many fields and sciences. They will need to learn a reasonable degree of TO, so that tangents and partiality can be avoided or ameliorated. Yet a good beginning has been made for a presentation and approach to TÓ. Anything less cannot provide ultimate unified totality, or TO. To accept a lesser universe, system, world or deity is partial at best. The key is in deciding what and how absolute and relative totality are defined, exist and are treated. There can be no doubt that absolute totality is largely the physical universe, and relative totality is the mental, or relative universe of best human life. The major quantum leaps to be made, are deliberate realization and production of the essential fusion of all that constitutes actual totality.

The Three Worlds of Human Life - The world we actually live in is much different from the world we perceive, believe or are used to. These are only appearances or faces of necessary existence. They are a mixture of three, 1. The wrong world of all that is negative and contrary to the physical and relative universes of best human life, 2. The right world of all that is beneficial, positive and a kind of heaven from the union of absolute and relative totality, or the ultimate unified totality, TO. 3. A great variety of mixtures of these two kinds of worlds, of which almost all the earth's population lives in. They poorly understand how it affects them, and worse, poorly chose how to live by it. . These three kinds of worlds are not completely separate, nor impervious to each other. In fact there is much in the mixed world, and even some in the null to negative worlds that are potential to the right world of ultimate unified totality. By living this interactive and interdependent existence each person without knowing, but having some of the right principles, purposes, priorities and plans in life, falls by the wayside and succumbs to this or that miss step that is the story of everyone's life. Only when these principles, purposes, priorities and plans are right, perfected and well organized by TO can success, survival and permanent continuity be achieved. In short life is not easy. It is a long, hard road to travel, to learn and do what is right and best. These can only come when we know about actual totality, and must start from the top down. Life can, and must be when right by TO, one of blessings, all the benefits and rewards it offers.

Developing and perfecting the right world of ultimate unified totality, actual totality, will take a lot more work and time. It will change with time, and need continual upgrading. A dedicated, cooperative and specially skilled group or staff will be needed, to further develop and perfect actual totality. Both loyalty and tolerance to the cause will be mandatory, with a great appreciation fo r all that has enabled actual totality. Because we are not, and poorly know, what this right world is, and because it is the most important feature, and guiding purpose, of our lives, it is imperative that we, from the earliest age through all years of our lives, experience, are trained, educated and learn to identify actual totality. Our physical and mental are not only blessings. They are duties. It is much more than the treatment of ailments, illnesses and disabilities, It is even more than the prevention of all that causes poor health. It is the total care of all persons, past, present and future. The only way to achieve this kind of health is to know and live by the right world of absolute-relative best human life or the ultimate unified totality, TO.

To know actual totality is to experience, be trained and learn what makes, save, is, represents and how to use it. TO is made from, is, and is formed by its categories and the quality of its modes, continua or dualities, absolute-relative, varieties, form and function. The modes of the right world or ultimate unified totality, are actuality or necessary existence, subjectivity or all that makes up the mental world, including knowledge and all its corollaries, objectivity or all that makes up the physical world, including the frame we live in, and reality the ever higher cohesion of subjectivity and objectivity that is the basis for human existence. The continua or dualities of the right world or actual totality, constitute much of the major content of TO. All exist only as they exist as parts of actual totality, none exist separately. Actual totality is determined by the large and small, past and future, subject and object, absolute and relative, general and specific, good and bad, right and wrong, inorganic and organic, mass and energy, static and dynamic, integral and differential, inorganic and organic, in the right proportions, balance and variations. All is different aspects of the same thing, DAOST, in the way that is most typical and beneficial for actual totality.

This book is only a beginning of work on actual totality. The book is also only descriptions of proofs rather than a direct and definitive approach to the whole of actual totality, its implications and applications. Thus we must review several of these limits and needs and show how their corrections and fulfillment must be provided. This requires two separate, yet united, approaches. The first is how actual totality as a topic can be directly presented. The second is the many ways it can indirectly benefit the world, all that exists that is of actual totality only not named or recognized as such. Both approaches have been separately and incompletely noted through the book. To present and advance actual totality in its own right is extremely, but not absolutely, essential. It requires all the help, interest, dedication, work and continuity possible.We have used the term PMAC to describe all the main themes and methods of future action for TO. Many of them are a continuation of what already exists, without the benefit of actual totality. To grant and enable actual totality to focus entirely on its contribution to all that is indirectly for actual totality in existing society, yet attain optimal benefit, will require a large amount of appreciation, work and cooperation.

Chief among the great contributions to actual totality are common knowledge of the world, religion, philosophy, science and their fields and uses. In order to accomplish an optimally effective and efficient fusion of actual totality with existing knowledge and cultures it is necessary that the overwhelming emphasis must be on what is mutual, right and beneficial to both, and avoid all that is dividing, wrong and harmful. Among the most specific benefits will be those means that extend the best of universal knowledge in all ages of individuals. One of these will be by maturity in longevity. It is possible the aging process of cortical, and other neurons can be extended both directly and indirectly. This will allow the benefits of experience and knowledge to avoid the great loss that occurs with each generation. It will also produce a much greater reservoir of capacity for understanding the complexities of, and problems that face actual totality, as such or as its equivalents exist.

ADDENDA and APPENDICES of ACTUAL TOTALITY

Introduction, Combined, Use of chapters and addenda

The chapters of the book and the addenda are, mutually reinforcing for TO

Actual Totality is not only new, it is simple, extremely complex, deep and important

The reader must remember complexity and be prepared to gauge their learning by it

The chapters and addenda are critical to continued interest, learning and development

The Addenda includes, Glossaries, alphabetical, or modified

Appendix A Definitions of major concepts, existing and new, used in actual totality

Appendix B Major New Concepts of TO

Appendix C Proofs of actual totality, frequently and less frequently used

Appendix D Laws of Actual Totality,

There are many other ways to make actual totality more accurate, complete, understandable and useful for each and all. This will come in the future. How soon will depend on all of us? The separation of approaches to, TÓ, such as kinds, proofs, general - specific, primary - secondary, direct - indirect, outside - inside, integral differential, by dimensions, others, and all reversibly will greatly enhance the part and role of actual totality and its success and survival. In order to achieve optimal efficiency and sufficiency of perspective of actual totality vertical integration and differentiation must be systematically unified. This is largely achieved by a sum, maintaining a proportionate, or relative magnitude, perspective of all, one to the infinitesimal. The measures of posits are only preliminary, more as initial guides and examples, than final conclusive determinations. Much work will be needed by many to reach highest consensuses and the more specific measures that fulfill the ideal POA for TO and all. The concepts weigh heavily in the future. This is to be expected for the outlook of new concepts. For a fully balanced positioning of concepts, the ECT of EC of OC, or TO essences of top existing concepts should be included. Some entries have numbers for the chapters, or page, they are in. A full numbering must wait for model development, next edition and future work. Humans are naturally both limited and lacking in actual totality and its understanding. We must not expect the impossible, yet we must seek the positive imperative of all needed for TO. The key focus must be on knowledge, justification, best human life and all that are the greatest in TO.

Appendix A Definitions of major concepts, existing and new, used in actual totality

GLOSSARY of EXISTING TERMS of Deepest Implications, by ACTUAL PURE ESSENCES
Prominent Meanings Emphasized in Actual Totality

Absolute is 1.Opposite of Relative without relation, 2. Beyond set limits of entities, universes or dimensions, e.g., AU, TÂ 3. Unlimited, 4. Opposite of zero

Absolute totality is complete existence of all that is material, the physical universe minus relativity.

Absolute Relative Totality is their combination in or out of TO

ABSTO-RELTO - their combined existence in Actual Totality

Absolutivity - quality of extreme relative totality that makes the absolute conform to, TÓ. It is the contraplete of relativity and the universal highest state of observer reality in S-O

Acronym-s - Short lettering or abbreviations that symbolize words, phrases or new and unnamed concepts

Action - Motion and force with a direction or purpose, what best achieves a set goal, e.g., ACC

Actual - the quality of being necessary for existence

Actual Pure Essences - Are the many features that give an entity, object or totality typical and special meaning.

Actual Pure Essences of TO - functional and practical features and values that are essential for its existence

Actual pure essence-s in TO, the quality of any thing or anentity needed to be its part in TO

Actuality - state necessary existence, tends to be highly TO

Actuality - the total state of essential existence, it is one in which there is precise and optimal form and function.

Adequation - Sufficient for actuality, the relation of truth to being, validity to, TÓ, (KOT)

Appreciation - Understanding with feelings of gratitude forwhat or all that is best, becomes TO

Benefit - What helps, a positive force for improvement andenhancement, and the future of each and all in TO

Big Bang and Dark Holes are two extremes of the duality of rapidly expanding and contracting matter, parts of the spectra of expansion and contraction, and variations in density

Circumspection - Ability and exercise of viewing all, looking from the outside at the outside, inside and all around. 6,38,40 Closure - The mechanisms and processes of completion, whatfulfills an entity, totality, the end product of creation, of TO

Comprehension - Mental capacity and activity to understand totally, especially TO and its major divisions, AU, TRU

Conscience - The awareness and intention to chose, and do what is right. This must be trained and guided by TO Continuum - A parameter, dimension, or universal that exists and functions massively. They make major divisions of TO

Contraplete - Polarities whose opposite function together andhelp to fulfill the whole, e.g., absolute and relative in TO

Correlation - The statistical level of relation, agreement, or sameness between entities, their validity depends on TO

Cue - The information, evidence or facts key to, TÓ knowledge

Dark energy and matter, is what constitutes a large portion of physical space. They are altered by relative totality.

Differentiation - The separation into finer and finer entities, elements or divisions and parts of TO

Duration - Time between two or more foci, higher level of time

Energy - The dynamic potential or active quantity in an entity, object or universe, a dual aspect of mass energy, and TO

Entity - That which can be separated, is other than anothertreated as a unit. a most general state of things and senses. May be actual, fictional, illusion or exist in combined way

Equilibrium - Balance, especially of diverging-converging continuity to and through the ultimate future, 7,16,21,40

Essence - what anything can't exist without, most practical

Essential - functional, practical and needed quality of an entity

Exist - to be, to have presence, in an entity, totality, or TO

Existence - the state of existing, contrasting with non existence

Finite - What is separate, part from infinite to infinitesimal

Focus - Convergence on center, static or dynamic, of FAF

Force - Motion of one that affects another, length and mass in accelerated time

Form - The aspect that gives shape, structure, construct, strength and body to an entity, or when typical, TO

Foreknowledge - Forethought, precognition, correct, inclusiveunderstanding of all TO, positive or negative 31 32 40

Frugality - Living in ways that spares what is best, saves for the future, minimum of waste

Function - The basic and simplest activity, dynamics, motion, workings and operations that support an entity or TO

Gist - The key, central and deepest meaning, what best reveals an entity, object, topics, or TO

Given - What is presented, offered, or has to be work with. Often the initial conditions or premise, ATO of TO

a god - any divinity, god or supreme being of many kinds

God - The one deity, the universal divinity of monotheism ruler of heaven and earth, champions the best in human life

Limited by obsolescence, diversity, disproportions, disunity and deficiencies

GOD - The equivalent of TO, TREQ, is the same although not known as TO or, TOT, TOV, ARTO, RELTO, TRU

Gradient - Finite steps in a parameter, dimension, spectrum or continuum, basis for differentiation and measure

Haecceity - Thisness, exclusiveness, beingness, what gives anentity separate importance

Identity - Precise knowledge and determination of oneself relative to any and all, especially of and in actual totality 29

Infinitesimal - Smallest separate quantity, the inner or minimal limit of an entity, object, parameter, or dimension

Infinity - The outer or maximal limit of an entity, object,parameter, or dimension, or TO, often of m, l and t

Improvement - The capacity and accomplishment of doing better, growing in quantity and quality

Insight - The mental ability to penetrate all, and get to the gist, crux or needed idea, the power of deepest comprehension

Inspiration - Intense emotive and cognitive quality whose force enables one to discover, learn, know, and do, e.g., TO

Integration - The processes leading to a limit, unity, totality or completion, a sum of collective steps to systematic unity

Intent(ion) - The state and processes of focusing upon a setobject, objective or goal

Interval - The higher power of duration, or stretch between two elements, points, or time

Intuition - Spontaneous ideas and insight that come from deep feelings, conceptions and associations

Jist - That special kind of gists and ideas that most profoundly reveals the deepest and greatest of knowledge, and TO

Knowledge - All that is known, over unknown, and directed toward TO, accumulated, often massive, information of TOLength - Basic dimensions of space in nature, measures size, a large, medium, small, major dimension of form in TO,

Level - Graded steps, Points on the scale of dimensions, that show its differentiation, e.g., Á dimension by SÁ

Light - Major component of mass energy, enables S-R-O.

Limit, -s - The extremes that produce boundaries, shape the form of entities, perimeters of TO and its parts

Logic - High form of reasoning by APE concepts ordering TO

Mass - A basic dimension in the physical world that measures combined content, e.g., weight, gravity, matter times density,

Mass-energy - Basic state and dimension in nature and TO that measures predominant form, function, forces and powers

Mentality is the totality of the subjective, or the human mind taken as a whole that enables knowledge to be beneficial.

Object - The entity observed by an observer in relation or opposition to the subject, and that acted upon by the subject

Objective - To focus on the object, to be physical rather than mental, to use things for a purpose, ultimately Á in TO

Objectivity - State and ability to focus on the observed by the observer, extraction of distinct detail in objects and TO

Polarity - The existence of and relations between opposite extremes, the ends of dualities, spectra or continua

Power -The dynamic quality an entity exerts, purposefully for benefit, if TO

Practical - What is useful, beneficial, works best, effective, has objectivity, the quality of actual pure essences in TO

Quality - All that makes an entity best for itself, a characteristic of universal refinement in TO, the purified and proportionalstate, mode, and properties of an entity or TO

Quantity - That of an entity which is measurable, of all thatenables differentiation in TO

Real - What exists, has relativity, joins the observer-observed, as in S-R-O, higher existence produced by relative totality, by pure essences makes things and senses actual in TO

Reality - the bond between subjectivity and objectivity, all that is knowable and unknowable, knowable/unknowable of TO, the supreme state relative totalityi gives to actual totalit,, the mode of TO that most separates it from all else

Relation - The connection or comparison between two or more entities or parts, what makes them the same

Relative - Association or bonds between entities, and in TO, the fact, degree and focus to which two or more entities have a common existence, a basic feature of TO

Relative totality is separate existence of all that has relation, from an incipient relation to the most definite, as in TO

Relative universe - RU, many kinds less definite than TRU

Relative Universe, The, - TRU, definite aspect of relative totality that on best human life & mind, TBHL

Relativity - quality of the relative that makes the absolute con- form in TO, universal state of bonding, observer-observed, S-O conjunction. It is the contraplete of absolutivity in TO

Revelation - That sudden and inspiring apperception of deepest important meanings, especially if divine or of TO

Sense - The mental kind of entity, rather than a physical thing,may be of feeling as well as cognitive

Service - Activity in life that is primarily devoted to the well being of others, God, TO, should be a joyful contribution

Set - A group that has its own modes, may be open or closed

Soul - That within a person that is divine, of God, in which all that is bad is excluded, signified in TO by TES and TÍS

Subject - The primary feature in a relation, person and observer formed with an object and function to make a complete unit. A basic aspect of relative totality, and hence, TO.

Subjective - focus on the subject, mental rather than physical,to use ideas, the relative side of Á of TO

Subjective and objective - the continua and duality of observer observed relativity of existence, a primary mode of TO.

Subjectivity - Mental focus, the state and processes involved by the subject in RELTO, TO, relativity and absolutivity

The - That which specifies a definite or certain object, topic, entity, universe, e.g., The giving of exclusiveness to, TÓ

Theme(s) - The deepest and most significant ideas that are carried through a topic, representation or presentation

Thing - The kind of entity that is opposite from sense, largely physical rather than mental, often used to specify

Time - Basic dimension in three levels, duration, change and variation, measures of past, present, future in TO

Totality - The whole, a unified collection, inclusive of all, or existence, any totality becomes actual by all that makes TO

Totalities - Multiple universes, deities, systems, worlds that become exclusive unified wholes, less than actual totality

Training - Purpose and ends are to have qualities of thinking and behavior that best achieves identity, initiative, leadership, confidence, reliance, interests and knowledge, of TO, 29, 40

Understanding - The mental capacity and activity of knowing, appreciating and doing by principle, basic to view of TO,

Unity - Oneness, existence as one entity, object or universe, a basic feature of TO if convergence and fusion are complete

Universe - all things, an inclusive unified system or world,focused absolutely, relatively, astronomically or in otherways, lacks much that is less, or other than, actual totality.

Universe, Physical - the absolute universe with relativity

World - all that exists as a whole, commonly and variably used,often associated with planet earth specially relative to humans

Appendix B Major TO New Concepts, Graded and Alphabetical

A GLOSSARY of Abbreviations or Acronyms of New CONCEPTS of ACTUAL TOTALITY

Names, definitions and positions of most frequently used new concepts, including those derived from TO

T Totality in General, Totalities.

TO Actual Totality

TÓ Actual Totality, used to distinguish it from other to-s

TOF - TOT-TO, of the Future

TOP - TOT-TO of the Past

TOT - Potential TO, and externally reactive

TOV - Varieties of Totality, (10)

Contractions of TO concepts,

TO-TOT - Actual to Potential, states, interactions, interchange

TO-TOF -Actual to future TO-TOT, adaptations and alterations

TO-TRU - TO and TRU form, function, relations and roles

TO-TRU–TRIS-TIT-TRIR-TOF Specific varieties of TO future

TO-TOV -TO and TOV, 10, general to specific varieties,

TO-TOV-TRU-TRIS-TIT-TRIR-TOF - Specific TO with future

TOT-TO Potential to Actual, Interactions

Other frequently used TO Derived New Concepts,

TAR - Extreme of Absolute Totality

TES - Collective Individuals in TO, function and power of,

TÍS - The First Person in TO, Perfected Participation

TP - Totality Proper, central realm of form and activity of TO

TRA - Extreme of Relative Totality, most specific variety

TREQ - Equivalent-s, equal, correspond, or similar to, TÓ

TRIR - Reversible TO, especially specific, and in the future

TRIS - Kinds of specific varieties of TOV, and TRU

TRU - The Relative Universe, of best human life and mind, large realm in RELTO, opposite of the physical universe

TUC - The Universals Common

TUD -The Universal Definite, most important in steps to, TÓ

TUC-TUD-TID-TO, From entities through each identity to, TÓ

Other New Concepts, mostly of multiple use in this book

Å - The prime mass-energy dimension in TO

ABOABI - Absolute outside to absolute inside, ens, T, TOT or TO ABSTO - Absolute Totality in TO, The 50% of all absolute in TO

ABSTO-RELTO - absolute relative totalities fusing converging on TO

ACEXIN - Actuation, Extension, Intension, highest mlt level TO

ALT - the combined mlt dimensions, in TO

APE, APET - Actual Pure Essences, Actual Pure Essences of TO

Apps - Applications, especially in the future

ARTO - Absolute Relative Totality

ATO - At Once Autonomous being, aspect of the extended present, TO

AU - Absolute or Physical Universe

CES - Processes of TO, fundamental determinant functions

COGNOS = We know, know that we know, all as perfected by TO

COR - Collective corrections of TO

DAOST - Different aspects of the same thing, ultimately must be TO

ECRO - we must always Remember, crucial facts of TO vulnerability

ELTICMA = Everlasting Life in TO by IA, ordered massive attention

EXAB - Exclusive ability of meanings by SO or A for TO, is reality

FAF - Form and Function, in TO, major feature that varies with TOV

GMLPDF - Genetic material, life product, determining forces

God - The one deity, in common knowledge, monotheism

GOD - God equivalent to, TÓ, much presence yet much unknown

IA, IAT - Instantaneous Attention, TO, central focus of TO

KOT - Any and all Knowledge of TO and TOT, Causes and effects,

LCD - Least Common Denominators, By math and generally, the reduction of all to their basic denomination, as by and in TO

l - length, fundamental dimension, and measure in space, and of TO

ME - Mass Energy, basic continua of TO, depends on variety, TOV

MIDST - Mass is in different states in entities, and especially formed in TO by what is predominant, subject-object reality

mlt - Mass, length and time, Absolute and physical dimensions

MOT - Models of Actual Totality, Highly effective in showing TO

NEC - New (TO) Concepts, Needed abbreviations and acronyms

NETIC - Non Existence, except by APE, FAF of TO, exclusive

OITIO - Out in, and in out of TO, all that is functionally more or less,

PI - Positive Imperative, convergent and enhancing power or all in TO

PERT - Perspective of TO, the many ways of viewing TO as one,

PMAC - Plans, Preparation, etc., to action, Management and control

POA - Positive Objective Approach, to enhance identity with TO

PRET - Presentation of TO, The many ways and means of presenting

PRUT - Proof(s) of TO, Known and yet to be developed proofs

PSA - Positive Subjective Approach, to enhance identity with TO

PUREAL-ITY - Purity, Reality, Actuality basis for and guide by TO

RELTO - Relative Totality in TO, The relative 50% of all in TO,

REM - Relative Magnitude, Proportionate Parts of TO, powers of ten,

ROT - Representation of TO, all figures, models and mens to show TO

SCARL-S - States. Conditions, Limits, Relations, general, and in TO

S-O - Subjective-Objective, major mode of TO, and for reality

SOID - Four to Multi Polarities, Subject, Object, Integral, Differential

SPALT - a Spherical, 3D model of TO, by ALT 16, 39

SPAT - Spherical 3D model of TO, TOT, ROT, POT and PERT

SRELT - Special Relations of TO and constituents 19, 14, 26

S-R-O - Ever higher state of S-O Reality in TO, Central bond in TO

t - time fundamental dimension and measure in the universe and of TO

TASPA - The form, shape and structure of TO are not necessarilyspherical, but will vary by type, view and change

TIT - TO, highly specific variety, In and of itself

TOPNEC - Major or top TO Concepts, Largest components in TO

TRIA - TO is all, and TOT-TO-TRU are all compared

TÚ - TO as a Unit, as the optimal, ultimate One World

VAPO - Validity, Actuality, Proportionality and Totality, power of

Other New Concepts, in or out of TO

CIPTRU - Corrected edition of The Relative Universe

COT - co-terminology, overlapping terms, decreases DREM

DIMFOR - Dimensional Form, two or three, types of functions of TO

DOFINT -Terms, parts, universals each Different separately or in TO

DREM - Coordinate Positions of REM parts of TO in MOT

ELE - The Exclusion Law, overlap and limits, at TO interface

ENS - Entity, entities, most fundamental type of separate existent

EPRE - Finite and Particular Preeminence in TO, Windows of finity

IPT -Initial editions of The Relative Universe

OBT - Actual Totality as an Object, Outward extension of form of TO

ONTO - Ontological basis of Actual Totality, religious proof of TO

SAPETO - Sum of the Elements or parts of TO 16

SPUNO - Ideal or Standard Unified overview approach to totality, TO

TÁ - TO Prime, The man focus of all, generic to the Á

TÂ - TO Absolute, State of and Approach to the absolute

TÄ - TO Derivation, Origin, Development

TÅ - TO Form, structure, shape, etc.

TÉ - The special or definite part of TO

TÊ - A or any part of TO, detailed differential

THEO -TO as the The ultimate Object

THES - TO as The ultimate Subject

TI, TINI - The Infinite, and infinitesimal in and of TO

Appendix C Proofs of actual totality, frequently and less frequently used

New Concepts of PROOFS, (In text, figures and others)
If in Figures, use for added proofs of TO, eg., # 1-38

Frequently used Proofs

PIP - Proof of and by Integrative Processes 18, 21

PRÁ - Proof of and by the Primary Mass Energy Dimension of TO 17

PREM - Proof of and by Proportions or Relative Magnitude of TO 22

PRUCEN - Proof of and by the Center of TO, ALT 19, 16

PRUN(**EC**) Proof of, and by New Concepts 118, TO Terms, 20

PRUREV - TO proves x, and x proves TO, for all reversible, 35

PRUSOT - Proof of the Unity of SO by TO, TOV, TRU, and vv. 17

PRUT - Proof of and by TO, and all of greatest value in TO

PRUTARU - PRUT Reversibly from TARU 12, 13, 28, 35

PRUTAD - Proof of TO, by proving what it is not, 1, 5, 34

PRUTO - Proof of TO totality, holistic proof of TO, 15, 38

PRUTREQ - Proof of and by TO Equivalents, 5, 40

PRUTU - Proof of TO by TU, and vv. 12, 15, 19, 21

Less frequently used proofs, of importance in TO

PRACALT - World ground, coordination, ALT, 16,23,38

PRAD - of non Absolute, (of DIMÂ, & Á, & „) 13,24-26

PRAGA - Proof of Á and Pragmatic Proofs of TO, 15, 17

PRIA - by Instantaneous Attention, immediate awareness 29, 19,

PRODEP - Deposit of Posit Falling into Place 39, 6

PROSO - of singular occurrence of many views (in TO) 39

PRUCT - of TO by consciousness, thought, 26 19, 29

PRUFIX - Proof of TO Fixation, also for RU, TRU, TOV 9, 19,

PRULATA - Proof Limit, Á proves TO and TO proves Á 17

PRUMT Proof of the approximate measure of TO 3 56

PRUNES PRUT by what and how it is not done without 7, 14,

Proof of TO by Non Essence what is unneeded 4

Proof of TO by disproof, by non essences, 33, 4

PRUNKOT - Proof by Absolute Need to KOT 30-31

PRUSÁ - Proofs by states of SÁ, Á1, Á2, Á3 17

PRUSO - by union of S-O, interaction and reality power 10, 17

PRUSU - Proof of Systematic Unity of TO, 15, 22, 5

PRUTATO - Reversible Proof TO and ATO 35, 15

PRUTAW - PRUT by kinds,Thisness, Whatness, Quiddity, Haecceity,

PRUTHOB - Hidden Observers Proof of TO 17, 19, 15, 26

PRUTIM - Proof of TO by ATIM, All time mastery of things, 40

PRUTMID - Proof of TO as Multi-dimensional, bi, tri 11, 16, 39

PRUTSAF - by relative convergence fields, sciences, knowledge 5 26 30, 19

PRUTROS - Proof of and by the Divisions and Kinds of TO 9, 11

Many proofs are direct contractions of existing or new concepts,

PRUAPET - PRUT by APE in TO

PRUCPI - Proof of TO by convergence of positive imperative

PRUDIM - Proof of Dimensions

PRUL - P roof by Language, Speech, Writing

PRUP - Proof of Presentation, POT, 39

Also for AU, TRU, God, Philosophical Systems, ABSTO, RELTO. 8

Is, or is not, in the book, yet helps to prove TO

PRUCE - Proof by the center of a part's position 39, 7

PRUE - by what and how a part can't be done without 7, 15 (LOPAN)

PRUES - Proof of E, ES, CES, two Parts, Characteristics of TO 7, 38

PRUEQ - Proof of E, ES, CES Quality, Quantity \ Quiddity 7, 38 PRUFAD - Proof Non Absolute Relative Dimensional Form 15-16

PRUEMO - Proof by the multiple overlapping of parts 7, 39

PRUL - by language, speech, writing, (logic) 20

PRULCOD - Proof of TO by least common denominators 21, 11

PRUNEC - Proof by, and of, New Concepts of Actual Totality 20,SC

PRUNES - Proof of TO by Non Essentiality, what isn't needed

PRUNI - Proof by Negative Inversion, right side >>> 18, 33 35

PRUNT - of Notational System of TO, ultimate Terminology 20, SC

PRUS - Proofs Universal, or Universal Proofs, 9, 4

PRUSÁ - Proof of States of Á, Á^1,Á^2,Á^3 17, 39

PRUSAP - Proof Sufficient Accuracy and Precision SC, 21

PRUST - Proof by Standardization 34,

PRUSU - Proof of Systematic Unity, of TO, SUT 19, 21, 39

PRUTAN - Proof of TO Anatomical/Physiological 34-35

PRUTE - Proof of TE, E, ES 7, 39

PRUTDIM - Proof of TO, as multidimensional, 11, 16

PRUTIN - PRUT through additions to the whole, ESIN, 7.21,39

PRUTROS - Proof of the divisions, and kinds of TO, 9, 11

Less than book value, auxiliary proofs of TO

PROTOS - Proofs of TO through Total, and reversible, O -S

PROVAC - Proof of Accumulative Value, Reversal ATRU 18, 22, 35

PRÖVEPT - Proof of Varieties of EPI 30-31, Intd, 14, 39

PROVEX - Proof by Depreciative Value, ATRU, K lost 22, 30

PROWRU - Proof of, and by Power of TO, ARTO, ATO, TOV 15,18 PRUCT - Proof, PRÁ by consciousness/thought, 29, 17, 19,

PRUCOT - Proof, Comprehensibility of TO and All SC, 38, 31,

PRUH - Proof by Processes of Man, Homation, 18,

Many other Proofs are from each typical chapter of TO e.g., PRUNEX, PREX, PRUNECEX, PRAPE, PRAPET

Those ideas and concepts that are important evidence of actual totality become its proofs. Proofs that achieve a level of certainty are axioms. Axioms that are more certain, directive and compelling guides for life in actual totality are its laws. It is in this way that many proofs of actual totality are not signified but may be found in its laws.

Appendix D Laws of Actual Totality,

DIL - Distribution Laws, The direction of patterns and fractals take is produced by highly dynamic combined mechanisms.

DIMSCAL - Dimensional States from Continua Laws.

EL - Exclusion Laws, expresses the imperative conditions under which all exists in TO.

LABEPT - Law of Alterations of Environment that are Beneficial, proofs of predominance effects of TO and best human life.

LACA - Law of Content over Form, TO changes mostly in content, except in the very long term.

LACET - Law of the Actual Pure Essence over thing in itself, Actual functional and practical determinants that are pure.

LACKOF - Laws of existing or commonly known world features, qualified by, and existence in, sum of TO, LCD.

LACNE - Law of the Continuum of Non Existence, all less than precisely TO. A subclass of negative to positive of TO.

LACOFS - Law of the Absolute Combined Fixed Set, Sum of TO in the present practical duration;

LACONF - Law of conformity of same and relative, produces and becomes the whole, totality of TO, with unity in totality.

LACOP - Law of Absolute Combined nd Total Proofs, greatness of reliability provides sufficient reason for TO.

LACOSM - Law of the Cosmotic Absolute if we have a sufficiently distinct and evident TO.

LACPAR - Law of the Continuous Perfection of Parameters and Refinement of Posits to Prove and Perfect TO.

LACSTO - Laws of Action by TO, - universal action, multilateral action, action by VAPO, Validity, Actuality and Totality.

LACT - Law of the Certitude of TO, it can be known, as any object, and need not be cut off from pursuit and discovery.

LACTO - Law of Common Cause, purposes, goals, only by the unified sum of TO.

LACTOV - Law of the Congruity of Varieties of TO, justified by the varied typical, states, conditions and limits of TOV.

LACTSAR. - The center is the mean sum product of all that is and enables the optimal union of absolute and relative totality, TO. LACVAS - Law of the correspondence, correlation and correction of the Varieties of TO with their models, SPALT 0-10.

LAD - Laws of definition determination and actuality differential, Metalogical conditions of thinking inherent by TO reason.

LADIR - Law of Absolute Distinction in Relativity, comes to a point, reverse of uncertainty law of absolute totality.

LADRAPT - Law of dependent representation and perspective of TO - dimensions, arranged, viewer position.

LADRAT - Law of Double Relative of Actual Totality, TO consists of all requirements for ultimate unified totality.

LAKDI. - Law of Internal, kinds, Divisions and Differentiation. Everything consists of many kinds, quantity and quality.

LAKUPIT - Law of Knowledge of TÓ to fulfill its purpose must be, accurate, unified, total, and progressively improved.

LÁLREAC - Law of Mass Energy Form and Function and Light Reactions, interactions and interrelationships.

LALT - Law of mass length time transformation in varieties of actual Totality, by powers of converging presence of TOV.

LAMAS - Laws of Matter and Anti Matter, form, states, function and mechanisms in space time, absolute, relative and TO.

LÁMET - Law of Á, the Primary Mass Energy Dimension, in Actual Totality, The special and typical continuum form of TO.

LAMOL -. Law of More or Less, Its solution is, and it solves, the changing form and function of in and out interactions of TO.

LANERA - Law of never absolutely relative nor absolutely absolute, some of each must be in TO or TREQ.

LANERAP - Law of absolute and relative are only poles of the same continua, ABSTO-RELTO in TO.

LANEXAD - Law of Necessary Existence or Actuality, requires increasingly higher levels of essence and Direction. 25

LANSUP - Law of The Absolute Necessity of a Singular Unified Perspective and Presentation, the sum of representations. 39

LAOT - Law of the Absolute Necessity of Totality by TO, and its accompanying theory of truth, by the valued Relative, LART.

LAPAC - Law of proportionate action, universal and particular guided by REM and TO.

LAPDEL - Law of Dynamic Polarity of El, exclusion by TO is a highly typical and selective state driven by its polarities.

LAPRAUT - Law of Peripheral Reduction of ABSTO-RELTO, AU and TRU is typical and an important feature form of TO.

LAPRED - Law of Point of Reference, Relation, general-specific, convergent, definite.

LAPREM - Law of Á forms, dimensions, gradients and their Representation Measured by REM, Proof and test of TO.

LAPROD - Law of Proof and Practice Over excess Detail, loss and distraction of universal knowledge by valueless detail.

LAPIVOT - Law of the Pivotal Proof of TO, state of focus only limited by the finite power available, Archimedes l.

LAPRIC - Law of Actual Pure Reality, signifies how they are formed and exist together to fulfill TO, e.g., TES.

LAPRAUT - Law of Peripheral Reduction Physical Universe by TO. Disproved value of extremes of space time not by TO, e.g.,." God Particle."

LARPROT - Law of Á or SO Reduction Proof of TO and TRU, how to look at, and TO is formed and functions, OITIO.

LAPROIT - Law of Physical and Relative Universes, Reduction, out of and in interactions of TO.

LÂPRUT - Law of Absolutivity Proof of TO, being of RELTO, the special variety of TO, converges on absolutivity.12 14

LAPSCIT - Law of Potential and Actual Incremental States and Forces of Creation in Past and Future TO, 25

LAPTUB - Law of proportionate truth, unity and balance of TO, only one final truth, through TO, and TO with TOT-TOF.

LARÂR - Law of Reversible Absolute and Relative, this is a result of the fact that they are a continuum, only in TO-TOT.

LARCAD - Law of the absolute relative continua and duality, is the primary proof of TO, with a direct to indirect basis.

LARCAF - Law of relative and absolute correction and fusion by, and in, TO, 1.

LARD - Law of the Absolute Relative, what is above or about, yet inside relative, expressed as a primary law, PLAR.

LARDAC - Law of Absolute Relative Dependency, their essential interactivity and fusion by TO in TOT.

LARDOB - Law of Reversibility, Development often Replicates Being, collective, altering, progressive processes make TO

LARECIP - Law of Reciprocal Proof. signifies proofs of TO that reinforce, e.g., Á, SU-MOT-ALT-SPALT-SPAT.

LAREV - Law of Formal and Elemental Reversibility, PRUT by reversibility is of various and many of its features and aspects.

LAREX - Law of Absolute Relative Exclusiveness, TO is highly exclusive practically and relatively in the extended present.

LARF - Law of Relative Fixation, direct and indirect, centripetal and convergent set state and proof of TO. 19

LARFECT - Law of Relativity, Absolutivity, and forces that enable creation and evolution of actuality, that is TO. 24

LARIEX - Law of RELTO Alteration and Adaptation Assures EXclusive and Inclusive EXistence.

LARIT - Law of Alteration, Adaptation and Reversibility of all in TO, to be perfect requires changes typical for TO.

LARMIEX. - Law of Relative Massive Inclusion and Exclusion of future acceleration of TO, much acceptance in time.

LARMUT- Law of combined stages and Powers of Actuality and Reality Modes uniting TO.

LARNMAT - Law of Relative and Absolute Totality in the NeuroMuscular Form and Function in Actual Totality.

LAROS - Law of Relativity and Reality of the objective-subjective in TO, products of higher special TOV.

LAR - The Absolute and Relative Laws, collections of the greatest laws of absolute-relative continua in TO.

LARP - Law of Absolute and Relative Polarity, existence by means and extremes, opposite to a balanced spectrum in TO.

LARPET - Law of Absolutivity and Relativity converge, and come to life, by actual Pure Essences in TO, Primary Law of IO.

LARPT - Law of reversal and proof of TO, within a dimension, repeated return to TO, proves stability and continuity of TO. 16

LARS - Law of Reduction to Formal Simplicity, more or less need of humans to achieve optimal order and control.

LARUTO - Law of the two step identity absolute and relative totality, and their union in actual totality.

LARVIT- Law of mathematics by absolute and relative totality of varieties, TOV, and logic, of TO.

LASAP - Law of Sufficient Absolute Proof, Observer doesn't need to know every detail to fully comprehend or show TO.

LASCUT - Law of Accelerative Self Correction of all with Convergence on Unity in TO, greatest power by best human life.

LASCUTE - Law of Accelerative Self Correction in Unity, Equal to and Equilibrium with, the physical universe and absolute.

LASDET - Law of specified distribution of parts in any totality, or TO.

LASDIT - Law of the Spectrum of Same and Different in TO, its role in the absolute and relative, and their fusion in TO.

LASFA - Law of Specificity Focuses Actuality to determine and become TO, derives from the necessary, and APE of existence.

LASFACT - Law of Specificity Focuses Actuality and Converges on Reality to determine and become TO, by role of Á in ME.

LASOR - Law of Objectivity and Subjectivity Combined, or Law of Reality, S-R-O.

LÂSOT - Law of the non existence of absolutely independent form, function, mass, energy, observer, observed, DAOST of TO.

LÄSOT - Law of the primacy of observer-observed relativity, or the, Á, Subject Object, Mass Energy Specificity of TO.

LASPADT - Law of Symmetry, Proportions, form and deformation in Actual Totality, typical structure of TO.

LASPCOL - The biological law of self preservation and continuity of life. 12, 25

LASPRUT - Law of Summative Perspectives Prove TO-TOV, When we bring all differences they converge on the whole.

LASPU - Law of Speciation Unity, comprehension of the massive equilibrium of the whole to produce survival and continuity.

LASREPT - Law of Special Relativity Key Proof of, and basis for, TO, and disproof of any totality, or "theory of everything."

LASTIL - Law of the Suspended State of Structural Limits. Role of form in developing mass energy, emergence of life forms. 23

LATAN - Law of Language Notations of TO, absolutely necessary for new concepts it major new fields and sciences.

LATANIP - Law of the Absolute Necessity of Integrative modes, mechanisms, forces and processes of TO.

LATANOS - Law of the Absolute Necessity of Order and Standardization.

LATCEP - Law of TO Convergence in the Extended Present, a concentration of existence is shown mass and volume in time.

LATDOW - Law of Actual Totality from Disproof of other worlds, the three worlds of common human life.

LATEXP - Law of TO is All, or all by extension in time, and approaching perfection in quality.

LATID - Law of Totalities, Indefinite to Definite, worlds, universes, gods, Common inclination of all to become definite, in TO.

LATORM - Law of TO from the observer observed meaning of relativity, Knowledge of, proves existence of TO over others.

LATOZ - Law of TO always maintains its zero time focus in here and now, at any time past or future is some TO.

LATPUB - Law of proportionate truth in the unity and balance of TO, is especially evident in relative totality.

LATRASP - Law of Relative Speciation,, consists of many kinds of variations, gradations and levels, e.g., TOV-RELTO.

LÁTRO - Law of Logic's and Á Resolution of All, highest conceptual reality by the sum of the mass energies in TO.

LATSAV - Law of TO-TOT states and variations with time, understand all, foreseeable future and its variations. 27

LATSAK - Law of Time, time change, zero time depends on success of higher powers of knowledge and achievement.

LAZOT - Law of the zero time focus, or here and now of TO. It is critical this state of TO be learned and remembered.

LE - Law of Equality, the equation of precise relation between factors or parts, in TO, basis of algebra, and role of 'same'.

LEDEN - Law of Equilibrium of Dependence in Necessary Existence.

LENDS - Law of Enablement and Enhancement dynamics of speciation, TO exists only to the degree all are included.

LENEP - Law of Existent and Non Existent Parts, is a most fundamental law of parts. This applies to all entities.

LEQUAR - Law of Proportional and Abbreviated Reaction Equilibrium, Equal Abbreviated Reaction.

LERTO - Law of Quality Gradients, Lack, Loss or Less Reduction by TO.

LES - Law of Speciation, being all inclusive, not of the animal species alone, but all human life encompasses in TO.

LESEM - Law of Specific Energies and Masses, the state and dynamic relationship of what is predominant in mass and energy.

LET - Law of TO eternal, central power of TO, toward continuity, permanence, conditioned correction for short warping.

LETORD - Law of Equivalence of TO, its Varieties and Dimensional form with their Representation or model, SPALT.

LETOZ - Law of Existence and accommodation of time, time change and time zero within TO.

LEXAC - Law of the existence of an absolute center by TO, and with TOT continually adapts all to its absolute center.

LEXAPET - Law of Exfoliation, Socrates sought to exfoliate the essences of the great ideals of human life, so TO.

LEXAPET - Law of Exfoliation, by eliciting the actual pure essences of all into TO greater proofs can be found.

LEXIS - Law of Existence-s, non, partial, whole, objective, subjective, lower-higher and variable.

LEXS - Law of Exponential Sufficiency, is expansion toward completion of TO, central meaning of COGNOS advance of S-O.

LICPART - Law of the Internal Convergence of the person on, and proof of, absolutivity, by RELTO in actual totality.

LIDAR - Law of Individual Identity and Responsibility, it is the most specific form of LOPAN, and a person's duty.LIDE - Law of Integral and Differential Equality and Equilibrium, typical of the internal homeostasis of TO.

LÏDEX - Law of Indefinite that is The Definite Existence, is the beginning, development, and certainty and proof of TO.

LINCORS - Law of Incorporation of Speciation, mostly specie as the whole, consists of a typical unique body, TO, by RELTO.

LIOUT - Law or doctrine of the Logos, Imperative Order from convergent communicative powers of the Unity of Totality, TO.

LIPO - Law of Limits of ISMS, Oppositions, Polarities and Operations, its knowledge overcomes limits and enables TO.

Lipread - Law of the integrative processes in present reality of totality proper, a most crucial law, power and proof of TO.

LIRM - Law of Individual, best human life and TO Immortality, Largely from TO, also from the individual to TÍS-TES.

LO - Law of Order and Organization of Knowledge by the unified whole of TO, thoroughly developed with VAPO.

LOCET - Law of Logical Conceptual Elements of Language in TO, TO from the sum of logic, concepts, words, meanings.

LOCVAP - Laws of Constant and Variable Proportions, the constant is with the presence and stability of TO.

LOFT - Law of proof that there is only one final truth, practically a product of TO, and TO with TOT-TOF.

LOGAT - Law of, Gains to and Losses from of Actual Totality and its constituents, proves coherence and inclusiveness in TO.

LOGEM - Law of Logic the Essence of Mathematics, logic provides the basis on which math is formed and operates.

LOKPACT - Law of knowledge in life based on priorities, plans, preparations, actions, advances, and completion of the best TO.

LOLROT - Laws of Limits of Representation of TO, specific mass energy provides a good idea of the Form and function of TO.

LOLS - All things, and that of and in TO, have Limits, limits that are a determining feature of TO and its form.

LONECS - Law of New Concepts, and proofs, the new and unknown actual totality requires names for all of greatest value.

LONES - Law of Non Essentiality, more or less unnecessary for TO, all superfluous and unneeded do no exist in TO.

LONSCAD - Law of states of the natural order of combined dimensional form, mlt, ALT, ACEXIN and REM.

LOPAN - Law of Partial Necessity, value, quality and essential bond that exist between each part, as oneself, and TO.

LOPBIRT - Law of primacy of being by relative totality, produced by the whole dominance of human existence.

LOPDÄKT - Law of Perspective Precedence, Dominant role perspective knowledge of all has in forming TO. Yet limits, LOLS.

LOPU - Law of Unity of Opposites, the convergence of extremes that unifies, is given, and is produced by the continua of TO.

LORAMT - Law of Relative Absolute Massive greatness, power and acceleration by RELTO with ABSTO in actual totality.

LORVA - Law of Relative Value, basis for proportionality and relative magnitude in TO.

LOSEX - Laws of Successful Existence, the cure for all that is wrong, defective and diseased is primarily simple,

> 1. To exist by, and in, the one right world, TO. To know what the right world is, develop and know TO, actual totality.

> 2. Each person, and all people, to do no more nor less, than that needed to best fulfill their, and others roles in TO.

LOSID - Law of Solution in Deficit, a kind of disproof related to other disproofs by factors less than TO.

LOSTO - Law of Subjective TO Objective Dependence, all is more or less dependent in TO, especially, S-O and ME.

LOV - Law of Valuation, and Law of Total Value of TO, all has value by proportion of, and LOPAN in, TO.

LUDATO - Law of ultimate different things are actual totality, to complete the same and different there is balance of detail.

LUKOP - Law of Universal Knowledge from One, or Unity is Prime, the primary kind of knowledge all must have.

LUNEX - Law of Non Existence to Existence by Union, by the union of non, or less than existent, existence is produced. 24

LUSDATO - Ultimate Sum and Average of Same and Different in TO, a great help in distinguishing content of, and showing TO.

LUSITO - Law of ultimately same thing is the unity of actual totality, The opposite of LUSITO is LUDATO.

PLAR - Primary Law of the Absolute Relative, what is above or about, yet inside relative (LARD). 6, 15

PLO - Primary Law of Objectivity, Preeminence of contrasts, differences, heterogeneity with Á, adds distinction to detail ad TO.

SETLAW - Law of Systematic Unity in exclusive totality of TO, this is the Set State all is, by fixed limits, among TO.

SIL - Without Law, Sterile State and condition of all in TO, is and is not, as in contraries and support of TO.

STOP - Law of the Sterile State of Principles, Concepts, Laws, Language and Logic without TO.